Energy Resources and Energy Corporations

Energy Resources and Energy Corporations

DUANE CHAPMAN

Cornell University Press

ITHACA AND LONDON

Copyright © 1983 by Cornell University Press

All rights reserved. Except for brief quotations in a review, this book, or parts thereof, must not be reproduced in any form without permission in writing from the publisher. For information address Cornell University Press, 124 Roberts Place, Ithaca, New York 14850.

First published 1983 by Cornell University Press.
Published in the United Kingdom by Cornell University Press, Ltd.,
Ely House, 37 Dover Street, London W1X 4HQ.

International Standard Book Number 0-8014-1305-2
Library of Congress Catalog Card Number 82-74022
Printed in the United States of America
*Librarians: Library of Congress cataloging information appears
on the last page of the book.*

The paper in this book is acid-free and meets the guidelines for permanence and durability of the Committee on Production Guidelines for Book Longevity of the Council on Library Resources.

To my family, everywhere

Contents

Contents

PART V. PROBLEMS IN PUBLIC POLICY

Contents

Preface

This book is intended as an undergraduate text, and also for corporate analysts, citizen activists, and congressional staffers as well as for professors concerned with energy resources and public policy. The material is presently used in "Introduction to Energy Resources," a sophomore-level course at Cornell University, which usually includes some graduate students and many seniors and juniors. The book can also be used as supplementary reading in undergraduate and graduate courses concerned with energy or industrial organization.

This volume differs from other texts on energy economics in two ways. First, it is empirical rather than theoretical. The chapters on economic theory make extensive use of illustrative material and data on the petroleum industry and electric utilities. Appendixes show partnerships in oil well production, and there is an extensive discussion of the efficiency advantages of such joint operation for petroleum companies. Similarly, social cost is considered not only as a concept, but also empirically in terms of an extensive comparison of the actual known external costs of coal and nuclear power. A glossary and table of units define 200 common terms and concepts.

The second major difference from other texts is the author's perspective on the proper roles for the private and public sectors. For the private sector, I am of the general opinion that, in conventional economic terms, U.S. energy corporations performed well throughout the Growth Era—the period from roughly the 1860s to about 1973. During that time both utilities and petroleum companies were successful in providing increasing production levels; prices declined and the industry-government structure that developed was efficient in technological terms. The problems that arose in this period were those of worker health and safety, public health, and environmental protection. A major difficulty for the future was the evolution of an economic structure dependent on growth in both the production and consumption of energy.

[11]

Preface

In the years since the end of the Growth Era, the economic response of industry, government, and the public has not been adequate. Generally, economists believe that market forces will be an efficient mechanism for allowing efficient decisions to be made by business and consumers. With this belief I concur in the main, but I have concluded that in the energy sphere the role of market economics, although important, is limited, and that the combined effect of unresolved problems in several areas—declining domestic oil and gas production, health and environmental protection, transportation, and conservation—will necessitate a significant public role.

Part I of the book explains the economic concepts of competition, monopoly, growth, and maximum social welfare. Numerical illustrations adapted to electric utilities show how prices, production and consumption, profit, and social cost can be determined for each theory. Part II begins with description and analysis of the physical basis of energy, substitutability between different forms of finite and renewable energy, efficiency, demand elasticity, demand forecasting, and conservation. Basic patterns of consumption in households, commerce, and industry are examined, and the cost and efficiency of different forms of heating (including solar and wood) and insulation are compared.

Part III examines the basic economic dimensions of the oil and gas industry. Chapter 5, on the world petroleum market, shows U.S. and world estimates of proved and potential oil reserves, and describes the involvement of Russia, Britain, and the United States in Iranian government and oil, and the recent development of nationalized oil companies and agencies as the dominant global organizations for selling oil. Chapters 6 and 7 provide a picture of the organization of the major oil companies. Each stage of the oil business is introduced, and the economic incentives for cooperative activities are explained. The relationship of modern oil companies to the original Rockefeller Standard Oil Trust is examined, and the position of the British government as the controlling interest in the largest share of U.S. proved reserves is analyzed.

Chapter 8 considers natural gas and its technological and geologic affinities to petroleum, which render separate corporate production unfeasible. The economic history of natural gas price regulation is analyzed, and it becomes apparent why regulation could work effectively during the Growth Era but not since. The 1980s is a time of transition for both gas and oil. Chapter 9 examines the past and present structure of the industry in relation to economic theories of behavior and social welfare.

Part IV discusses electric utilities and the environmental costs of power generation, focusing on coal and nuclear power, which between them have fueled all growth in power generation since 1973. These two energy sources are analyzed in terms of reserves, industry structure and affiliation with the petroleum industry, comparative generating costs, and the significance of tax incentives. The economics of electric utilities is the subject of Chapter 12, with emphasis on the intricacies of cost-of-service regulation and its connections with revenue, tax normalization, fuel adjustment clauses, demand forecasting, and rate schedules. Chapter 13 then reviews the existing data on the social costs of coal and nuclear power generation. Extensive quantitative material is examined in terms of the dose-response relationships for air pollution and radiation, benefit-cost analysis of pollution control, risk analysis and back-end problems in nuclear power, and the major impact that environmental protection has on the costs of coal and nuclear power generation.

Questions of public policy are the central focus of Part V. New technologies are the subject of a detailed microeconomic analysis in Chapter 14, with particular attention given to gasohol, coal gasification, modes of transportation, solar energy, and super insulation. In Chapter 15, which is clearly differentiated from the rest of the book, I present my recommendations for future U.S. energy policy. The proposals include a plan detailing a reorganization of national energy consumption, more energy-efficient residential space heating, increased public and rail transportation, discontinuation of nuclear power, and increased solar and renewable energy in regions where economically feasible. Suggestions are also presented for means of providing revenue to meet the public costs of such a plan; they include comprehensive tax proposals, government participation in the ownership of major oil companies, and a social tariff based upon differences in wage and pollution control between the United States and countries producing goods to be imported into the United States.

Chapter 15, in its advocacy of certain public policy positions that may seem radical today, stands apart from the rest of the book. But it is the data and the logic essential to the first fourteen chapters that led me to the fifteenth. I suspect that the logic of events will place today's Chapter 15 in the middle of tomorrow's conservative consensus on energy policy.

It is a real personal pleasure to thank the many, many capable and conscientious people whose work at Cornell University, in the U.S. Congress, and elsewhere provides the basis for this book. They are, of course, not responsible for the opinions and interpretations here; many would offer strong disagreement. But I think nearly all would agree that a major change in the organization of energy use in America is necessary and inevitable. Kathleen Cole, Theresa Flaim, Lucrezia Herman, Timothy Mount, and Michael Slott, all of Cornell University at some point in the last six years, have each made major contributions to the preparation of this text, directly or through their own research. Olan Forker, Floyd Haskell, Kay Scheuer, Daniel Snodderly, and Bernard F. Stanton have all given significant support at some point in the preparation of the manuscript or the research on which it is based. The figures have been drafted by Joseph Baldwin. I am grateful to them all.

DUANE CHAPMAN

Ithaca, New York

Energy Units

Basic definitions and symbols
    Btu   British thermal unit; heat one pound of water by one Fahrenheit degree.
    calorie heat one gram of water by one centigrade degree.
    kWh   kilowatt hour; the basic measure of electric energy. A 100-watt bulb uses 1,000 kWh in 10 hours.

Abbreviations of magnitude
    k = 1,000
    M = 1,000 (Roman) or 1,000,000 (Greek)
    $\bar{\text{M}}$ = million; however, MM also means million
    $\bar{\text{M}}$Btu = million Btu
    MW = megawatt; million watts
    Q = quadrillion Btu; i.e., 10^{15} Btu
    cf = cubic foot; also ft^3
    Mcf and kcf = thousand cubic feet
    Tcf = trillion cubic feet

Metric/American equivalencies
    gram (g) = .035 ounces
    kilocalorie (kcal) = 3.97 Btu
    kilogram (kg) = 2.21 lb
    kilojoule (kJ) = .948 Btu
    metric ton = 1,000 kg; 1.102 American tons

Energy Units

Relationships

Energy type	*Normal unit*	*Average Btu content*
Electricity	kWh, kilowatt hour	3,412 Btu/kWh
Natural gas	cf, cubic foot	1,026 Btu/cf
Coal	lb, pound, or ton (ton = 2,000 lb)	22.14 M̄Btu/ton, 11,070 Btu/lb
Petroleum	bl, barrel, or g, gallon (bl = 42 g)	5.5 M̄Btu/bl (avg.) 131 kBtu/g (avg.)
crude oil		5.8 M̄Btu/bl
gasoline		5.3 M̄Btu/bl
residual fuel oil		6.3 M̄Btu/bl
residential heating oil		5.8 M̄Btu/bl
kerosene		5.7 M̄Btu/bl
Wood	cord (128 cf)	22.5 M̄Btu/cord
Glucose sugar	g, gram	gram = 4.1 kcal, or 16 Btu
Uranium-235	kg, kilogram	gram = 75 billion Btu

[16]

I

ECONOMIC THEORY: COMPETITION, MONOPOLY, AND GROWTH

1

Competition and Monopoly: The Conventional View

Four concepts in economic theory are examined in the first two chapters. They are termed "perfect competition," "profit monopoly," "growth monopoly," and "social welfare." The first three are theories about economic behavior, and the last is in the realm of economic philosophy: it claims to show how the public welfare can best be served.

Perfect Competition

Competition in economics has a rather precise definition. It means that no single buyer or seller, company or consumer, can control the prices at which products are bought and sold. It implies that there are so many buyers and sellers that no one of them can influence industry prices. These two assumptions imply a third: entry into the market is unrestricted, and any person or firm may become a producer, or buyer, or conceivably both.

Economic competition is not the same as *commercial rivalry*. As F. M. Scherer notes, business firms compete in advertising in order to create *product differentiation* for their brands in terms of consumer acceptance.[1] A product that is standard and uniform, such as gasoline, may nonetheless be viewed by consumers as having distinct properties. My favorite discussion on this point was held by Stanley Learned, president of Phillips Petroleum Company, and Rand Dixon, chairman of the Federal Trade Commission:

> Dixon, FTC: Do you exchange any gasoline [trade with other majors] in the East?
> Learned, Phillips Petroleum: We have some we exchange, yes.

[1]F. M. Scherer, *Industrial Market Structure and Economic Performance,* 2d ed. (Boston: Houghton Mifflin, 1980), p. 10. The concept of product differentiation originated with Edward Chamberlin in 1933; see note 4 below.

Dixon, FTC: That makes it hard for you to call yours Phillips, doesn't it?

Learned, Phillips Petroleum: As long as they meet our specifications.

Dixon, FTC: You put a little pinch of something in it that makes it a little different?

Learned, Phillips Petroleum: Yes; we have an additive which allows us to advertise. I don't know whether it does anything for the gasoline.[2]

If the president of Phillips Petroleum does not know whether additivies matter, how can the rest of us?

Gasoline additives serve primarily as a device for encouraging consumer loyalty. This is not wholly without economic reason, for brand loyalty in marketing gives stability to production and refining, and this in turn may lower costs. Since buyers perceive a difference between brands, companies can raise or lower prices by small amounts without having much influence on total consumption or market share.

Gasoline is clearly a physically uniform product whose differentiation permits commercial rivalry rather than economic competition.

Electricity is at the other extreme. The large majority of customers perceive no difference between Pacific Gas and Electric's electricity and New York State Electric & Gas's electricity. It might be argued that the existence of external social cost from air pollution or nuclear power affects the customer's view of the quality of the electricity. However, most buyers see electricity itself as a homogeneous product.

Energy-using appliances and automobiles are differentiated on physical bases as well as through advertising. Even those of similar type may vary considerably in durability, initial cost, fuel cost, and physical characteristics.

The electric utility industry can serve as a hypothetical example for illustrating economic theory. Figure 1-1 uses electricity to represent an energy product with a *competitive industry*. The *demand curve P* shows that lower prices for electricity will lead customers to buy greater amounts. At 15 mills per kWh, the industry will sell 3 trillion kWh. At 30 mills per kWh, it will sell 2 trillion kWh, and so on.

The *marginal cost* curve *MC* shows that costs of production are constant at 10 mills/kWh for the first ½ trillion kWh. (Marginal cost is the additional cost of producing another unit of electricity. This and other italicized terms are defined in the Glossary at the end of the book.) This constant 10 mills might reflect in a general way the availability of inexpensive hydropower. However, in the industry output range between ½ and 2½ trillion kWh, the marginal cost is a constant 40 mills/kWh, and this might represent existing fossil fuel and nuclear plants and more costly hydropower.

For the first ½ trillion kWh, *average cost AC* is also 10 mills per kWh. When the inexpensive hydropower is fully utilized, more costly steam genera-

---

[2]U.S. Senate, Committee on the Judiciary, *Petroleum Industry Competition Act of 1976*, Report to accompany S.2387, June 28, 1976, p. 15.

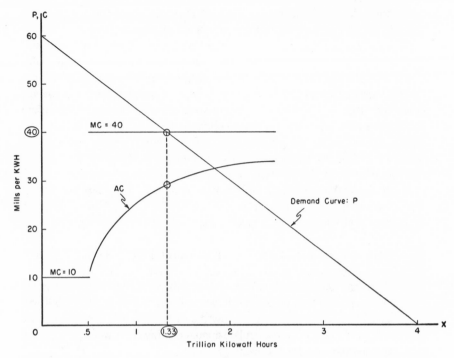

Figure 1-1. Competitive market: Marginal cost = price at 40 mills per kWh and output = 1.33 trillion kWh

tion (in Figure 1-1) causes marginal cost to increase, and average cost increases, but less rapidly. Graphically, average cost always follows marginal cost. (A mill, incidentally, is one-tenth of a cent. Figure 1-1 uses mills rather than cents for ease of conversion. If 1 trillion kWh is multiplied by 10 mills per kWh, the resulting number is $10 billion.)

Since marginal cost reflects the cost of additional production, the competitive industry would always increase production to the level where price and marginal cost were equal. Suppose this didn't happen; suppose production was at 1 trillion kWh. The price would be 45 mills (or 4.5¢) per kWh. Since the price realized from additional sales is 5 mills greater than the cost of more output, some producers would increase production until the total rose to the X = 1.33 level.

In the same way, suppose production was above the competitive level, say at 2 trillion kWh. Now marginal cost at 40 mills is above price at 32.5 mills. Obviously, with costs exceeding price, some part of this production is not paying for itself. The industry would reduce production until it sliced off this loss, and of course X = 1.33 trillion kWh is the production level at which further reduction in production would be unnecessary and unprofitable.

In economic theory, the production level where marginal cost and price are equal is the competitive equilibrium and is socially desirable.

Obviously, there are certain difficulties with this concept. We have studiously ignored black lung disease, respiratory ailments caused by air pollution, radioactive hazards, monopoly, government regulation, consumer ignorance, subsidies, and other economic goals besides money. However, economic theory holds that this competitive solution is efficient and therefore serves the public welfare. It is *Pareto optimal:* no one can be made better off without someone else being made worse off. As a consequence, production is done in the most efficient manner possible, consumers maximize their utility, and this is more or less the best of all possible worlds.

Economists who study these things have concluded that there are several logical requirements which must be met in order for a competitive industry to be optimally efficient. These conditions are:

(1) Every industry must be competitive. It may not be sensible to enforce competition upon coal if oil and electricity are monopolies.
(2) Economic knowledge must exist everywhere. Consumers must know how long a car will last, what the annual fuel and maintenance bill will be, and so forth. They must know the difference between a Btu and a kWh, and know how to calculate annual heating costs when they choose between gas and electric heating.
(3) Everyone—business and consumer alike—must be guided by the pursuit of maximum financial gain. For economic society, it doesn't matter whether a dollar is spent on my air fare to Europe or for the heating bill of a retired couple. Fairness is not part of the economic calculus.
(4) Consumers must be insatiable, and always desire ever higher levels of consumption.
(5) Economies of scale must not be pervasive; if so, monopoly becomes efficient.
(6) There must be no external social cost such as damage to public or employee health or general environmental degradation. If there is, then competitive industry develops a level of production and consumption which is above the desirable level.

If these conditions seem unlikely to be met in the real world, then there is no particular reason to suppose that policies based upon the presumption of a competitive industry are sensible.

Profit Monopoly

The theoretical view most commonly placed in opposition to the competitive position is the concept of profit monopoly. This theory shares with the competitive theory the assumption that profitability is the guiding light of economic behavior. It usually supposes that competition is desirable and possible. But it postulates the absence of competition, and the existence, for one reason or another, of conscious industry control over production and price levels.

Basically, monopoly power means the active ability to differentiate price and marginal revenue. A monopolistic industry can increase its revenue while it raises prices to customers. This revenue and profit increase occurs even if customers reduce the amount of purchases.

Marginal revenue is defined as the additional revenue going to a firm in response to producing and selling another unit of its product. This is the MR curve in Figure 1-2.

Table 1-1 might make this more evident. The first six columns show how price, revenue, and marginal revenue change with production levels. Table 1-1 is based upon the same demand and cost curves used in Figures 1-1 and 1-2. When production and sales grow from .4 trillion kWh to .5, price falls from 54 mills/kWh to 52.5 mills/kWh. Although there is an additional 100 billion kWh earning revenue, price has fallen, and thus the net effect on revenue is that it increases from $21.6 billion to $26.25 billion. So marginal revenue is positive, and is equal to the change in revenue ($4.65 billion)

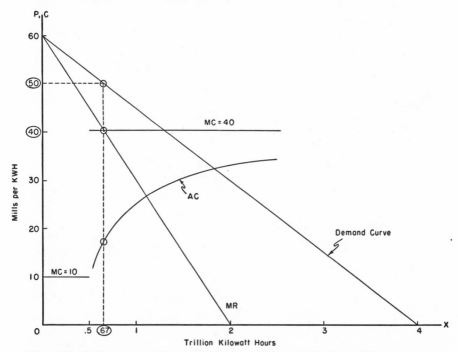

Figure 1-2. Profit monopoly: Marginal revenue = marginal cost of 40 mills per kWh, output·is ⅔ trillion kWh, and price is 50 mills per kWh

Table 1-1. Profit and marginal revenue

Production (trillion kWh) X	Price (mills/kWh) P	Revenue (billion $) R	Change in revenue ΔR	Change in production ΔX	Marginal revenue ΔR/ΔX	Marginal cost ΔC/ΔX
.4	54.0	21.60	–	–	–	10
.5	52.5	26.25	4.65	.1	46.5	10
.6	51.0	30.60	4.35	.1	43.5	40
.7	49.5	34.65	4.05	.1	**40.5**	**40**
.8	48.0	38.40	3.75	.1	37.5	40
.9	46.5	41.85	3.45	.1	34.5	40

divided by the change in output (.1 trillion kWh). Marginal revenue—the change in total revenue associated with changed production—is 46.5 mills/kWh. Since this exceeds marginal cost, profit can be increased by increasing production.

Recall that the marginal cost is always 40 mills/kWh for production between .5 and 2.5 trillion kWh. The closest marginal revenue gets to this is at .7, where marginal revenue is 40.5.

If production expands to .8 trillion kWh, marginal cost at 40 mills/kWh becomes higher than marginal revenue at 37.5. For a profit monopoly, then, the total output level with maximum profit will be near .7. Price at this output level will be near 49 mills/kWh.

Table 1-2 shows this same situation in more detail. The last column is total profit: revenue less cost. This table shows that the output level will actually be two-thirds trillion kWh with a maximum profit level of $21.66 billion. Nor-

Table 1-2. Maximum profit

Production (trillion kWh) X	Price (mills/kWh) P	Revenue (billion $) R	Cost (billion $) C	Profit (billion $) R – C
.1	58.5	5.85	1.00	4.85
.4	54.0	21.60	4.00	17.60
.5	52.5	26.25	5.00	21.25
.6	51.0	30.60	9.00	21.60
.66⅔	50.0	33.33	11.67	**21.66**
.7	49.5	34.65	13.00	21.65
.8	48.0	38.40	17.00	21.40
.9	46.5	41.85	21.00	20.85
1.0	45.0	45.00	25.00	20.00
1.33⅓	40.0	53.33	38.33	15.00
1.5	37.5	56.25	45.00	11.25
1.87	32.0	59.80	59.80	0
2.0	30.0	60.00	65.00	−5.00
3.0	15.0	45.0	105.00	−60.00
4.0	0	0	145.00	−145.00

mal profit includes a return on investment as a legitimate cost. Since this legitimate cost of capital is included in the cost terms, the $21.66 billion is excess *monopoly profit*.

Table 1-2 also shows that above 2 trillion kWh, revenue begins to decline. Figure 1-2 shows the same thing: marginal revenue is negative beyond 2 trillion kWh.

The figures and tables show the basic differences in behavior between competitive and profit monopoly industries. The profit monopoly produces and sells only half as much as the competitive industry, and its selling price is higher: 50 mills/kWh compared to 40 in the competitive case. If the competitive industry is a useful guideline, then the monopoly is not producing as much electricity as is desirable.

This raises serious problems for critics of energy industries. If the petroleum companies, utilities, coal companies, and nuclear companies are monopolies, then it follows from this view that production has been artifically restrained to extract excessive profits. But—if this view is held—it also follows that we should now be able to use even more energy than we do now. In other words, if corporations hold production below the desirable competitive level, more should be being produced and we should be consuming more than we are now. Two writers expressed such an opinion in the *International Socialist Review:*[3]

> The shortages of gasoline and heating fuels stem from the world domination of the oil industry by a handful of giant firms that jointly have the ability to set prices, restrict supplies, and manipulate reserves. It is a standard practice of monopolized industries to hold down production in order to drive up unit prices, and the oil companies are old hands at this practice.

However, economic statistics are unambiguous in describing American energy industries in the pre-1973 era as industries which were continually lowering prices in real terms and increasing production and sales at exponential rates. Figure 1-3 shows energy prices since the early part of this century—note their decline to 1972.

For petroleum, the picture of declining *real prices* held from 1860 to 1972. In 1967 dollars, crude oil sold at $54/barrel in 1860, at $8/barrel in 1900, and had fallen to $5/barrel in the early 1970s.

Energy production and sales grew at exponential rates in the 1947–1973 era, going from 33 to 75 quadrillion Btu.

There seems to be no factual basis for concluding that energy corporations acted as profit monopolies during the Growth Era. Perhaps this statement should be qualified—the data on production, consumption, and prices offer

[3]Steve Beck and Cliff Conner, "Energy Crisis: Bonanza for the Oil Giants," *International Socialist Review,* 35:1 (January 1974), 5.

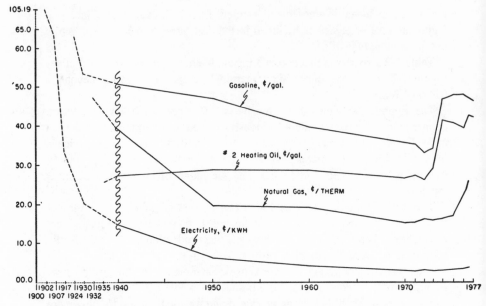

Figure 1-3. Residential energy prices, May 1976 dollars.
Sources: Data calculated from American Petroleum Institute, *Petroleum Facts and Figures* (1971) and *Basic Petroleum Data Book* (1977); U.S. DoE, *Monthly Energy Review;* U.S. Department of Commerce, *Historical Statistics of the United States* and *Survey of Current Business;* U.S. President, *Economic Report of the President* (1981); Edison Electric Institute, *Statistical Yearbook of the Electric Utility Industry;* American Gas Association, *Gas Facts: A Statistical Record.*

no empirical basis for concluding that energy corporations were profit monopolies. Moreover, even if the profit monopoly analyses were correct, the corollary that more energy should be being produced and consumed does not seem to fit with the growing recognition of the need for energy conservation.

Monopolistic Competition

The perfect competition and profit monopoly concepts have been described in the preceding sections as being mutually exclusive. However, another view, developed originally by Edward Chamberlin, postulates a blending of these two concepts in the theory of *monopolistic competition*. The essential argument here is that under certain conditions a noncompetitive industry with few firms will not have monopoly profit. Each firm may determine its output of differentiated products according to the monopoly rule of marginal revenue being set equal to marginal cost. The existence of multiple firms and the possibility of additional entry leads, in this theory, to an equilibrium in which each firm's production level causes its average cost to equal price.[4]

Stable monopolistic competition as usually described requires that marginal cost be rising, and that average cost be fully convex (i.e., U-shaped). If economies of scale are such that average cost is always declining, monopolistic competition collapses and becomes simple monopoly. If firm average and marginal costs both always increase above some low production level, monopolistic competition must give way to simple competition.

A significant aspect of the Chamberlin theory is his conclusion that monopoly may be associated with average cost pricing, and excess profit may be absent even in the presence of certain monopoly characteristics. These two points lead to a third major view of corporate behavior, to be examined in the next chapter.

[4]The special conditions noted above mean that if an equilibrium exists for monopolistic competition, each competitor's U-shaped average cost curve must be tangential to its own demand curve at the industry price level. Edward Chamberlin's *The Theory of Monopolistic Competition* was published in 1933 (eighth edition—Cambridge: Harvard University Press, 1962). General discussion can be found in such texts as Scherer, pp. 14, 15, and Hal Varian, *Microeconomic Analysis* (New York: Norton, 1978), pp. 65–68.

2

Growth, Monopoly, and Social Welfare

Growth and Monopoly

The perfect competition and profit monopoly concepts have dominated American economic theory. Nevertheless, there has been continuing interest in a third view of corporate behavior which holds that growth and size are the major interests of large corporations. W. J. Baumol of Princeton held this opinion:

> I have been surprised at how consistently the firms with which I have had dealings at least appear to have held to some sort of sales and growth objectives . . . and . . . I believe that to him [the businessman] sales have become an end in and of themselves.[1]

A second perspective is that of J. K. Galbraith:

> Above a certain profit threshold the members of the technostructure are better rewarded by growth itself. The accepted economics is a remarkable barrier to understanding the most basic tendency of modern economic society. That is for constituent firms to become vast and to keep on growing.[2]

This suggests the third theory, that corporations in noncompetitive industries may pursue maximum size and sales as long as they do not lose money.

[1]William J. Baumol, *Business Behavior, Value and Growth* (New York: Harcourt, Brace & World, 1967), pp. 51–52 and 46.

[2]John Kenneth Galbraith, *Economics and the Public Purpose* (Boston: Houghton Mifflin, 1973), p. 83.

In Figure 2-1, the growth-maximizing industry produces 1.87 trillion kWh. This is the point where the average cost curve (AC) is equal to price. This is the maximum sales level that the industry may have without financial loss. If it attempted to push sales beyond the 1.87 level, it would operate at a loss and eventually face the problem of choosing between bankruptcy or reducing production to 1.87 trillion kWh. If it produced at a lower level—say 1.5 trillion kWh—the industry would see that, since price was greater than average cost, it could increase production without losing money.

An important point to make is that sales maximization requires monopolistic conditions. The region between the competitive equilibrium (sales at 1.33, price at 40 mills) and sales maximization (sales at 1.87, price at 32 mills) has marginal cost in excess of price. If monopoly did not exist, a new firm could enter the industry selling at a lower price, say 30 mills. The monopoly could not keep its production at 1.87 and sell at 30 mills because its revenues would not cover its costs. It would be forced to meet its competitor's price, and reduce production.

Since sales maximization requires monopolistic conditions for the industry to behave differently from a competitive industry, we shall term this concept the *growth monopoly*. In comparison to perfect competition and profit monop-

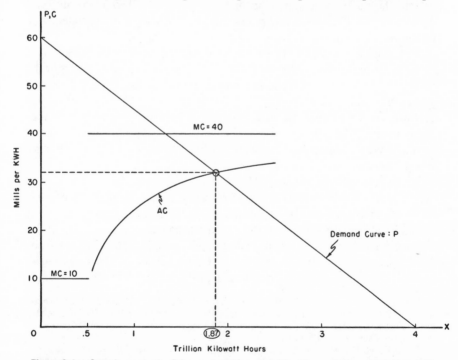

Figure 2-1. Growth monopoly: Price = average cost at 32 mills per kWh and output is 1.87 trillion kWh

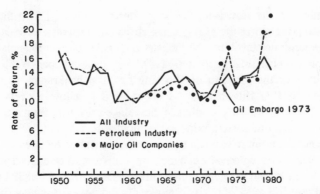

Figure 2-2. Profitability: net income to stockholder equity, percentage return all industry, petroleum industry, and major oil companies 1950–80.
Sources: See footnote 3.

oly, the growth monopoly has the lowest prices, no monopoly profit, and the highest revenues and production. This is clear if one compares Figures 1-1 and 1-2 and Table 1-2 and with Figure 2-1.

The growth monopoly concept seems to answer certain questions that are beyond the power of competition or profit monopoly theories to answer. Growth monopoly predicts low prices and high consumption, characteristics of the pre-1973 Growth Era which were noted in Chapter 1.

With respect to profitability, petroleum companies were generally average or below average from 1950 to 1972. This evidence supports the growth monopoly concept, and does so in a predictable pattern in Figure 2-2 for the years 1950–72.

However, the profitability statistics for 1973–80 do not conform to the earlier pattern. The industry experienced above-average profit rates in four of the eight years. Major oil companies (studied in Chapter 6) generally performed worse in bad years and better in good years.[3]

For electric and gas utilities, regulatory agencies generally work toward a normal after-tax rate of return of 13–15 percent per year. Average cost pricing rather than marginal cost pricing characterizes the basic approach.

Empirical support for the growth maximization theory comes from the work of Silvio Flaim and Timothy Mount, who examined the historical data for the petroleum industry for 1960–73.[4] Figure 2-3 shows that growth monopoly theory closely approximates actual sales before the OPEC embargo

[3]U.S. Senate, Committee on Finance, "Oil Company Profitability," Committee Print, February, 12, 1974, is the source of the 1950–72 data. Profitability data for 1973–80 are from the annual *Fortune Double 500 Directories* (Trenton, N.J.) and from the Annual Reports of the major oil companies analyzed in Chapter 6.

[4]Silvio Joseph Flaim and T. D. Mount, *Federal Income Taxation of the United States Petroleum Industry and the Depletion of Domestic Reserves*, Solar Energy Research Institute, October 1978, pp. 64–68.

and its aftermath. In fact, the growth monopoly theory predicted total sales of 56.3 billion barrels for the 14 years, and actual sales were 56.2 billion barrels. A competitive industry would have sold a lesser 48.6 billion barrels, and a profit monopoly a far lower 28.1 billion. Naturally, prices would be correspondingly different. In 1973, the actual price was $14.63 per barrel. The growth monopoly hypotheses predicted $14.16, perfect competition $19.08, and profit monopoly $22.40.

However, Figure 2-3 also shows that the accuracy of the growth monopoly theory ceased in the 1973–77 period. Does this suggest that after 1973 the petroleum industry began to pursue different objectives?

Perhaps we could end this discussion by noting that the growth monopoly theory seems to work better than the others, that is, it comes closest to reflecting/describing our real world. We could also note that it supports industry contentions of bringing the lowest possible prices to consumers, and supports liberal and socialist critics to the extent that it is evidence for the existence of monopoly power. Yet industry supporters will not accept such weak evidence as presented to date on the question of competition, and liberal and socialist critics are unlikely to be comfortable with the implications of normal profits and low (too low!) prices.

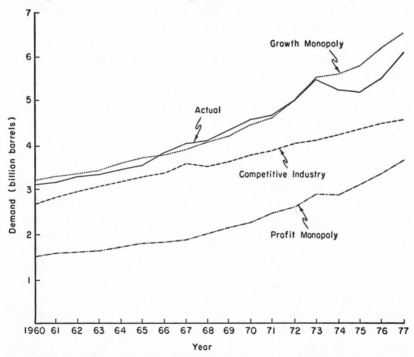

Figure 2-3. Predictions of the quantity of oil demanded, using alternative objective criteria. *Source:* Flaim and Mount

Several chapters hence I will again take up this subject of competition and monopoly, which becomes more meaningful after the industry is carefully analyzed. But it is first necessary to understand how economics judges questions of social welfare and to determine whether there is some logical guideline by which behavior can be judged.

Social Welfare: The Economic Optimum

The preceding discussion of competition, profit monopoly, and growth monopoly has been about theories of behavior. They are empirical, in the sense that they attempt to explain actual industry behavior. Growth and profit monopoly are not "normative" concepts; they do not by themselves define a desirable course for public policy. The perfect competition concept, however, is ambivalently viewed both as actual behavior and as nonexistent but desirable behavior.

The concepts of social value and social cost form the basis for the economist's view of the social optimum. When efficiency is looked at from a national perspective, *social cost* is defined as the cost to society of producing and consuming a good. Therefore, it includes not only the market costs of production already discussed but *external social costs* and *subsidies* as well. These nonmarket concepts are particularly relevant for energy because the magnitude of external impact and subsidies is probably greater for energy than for any other sector of the economy.

External social costs are unintentional by-products of economic activity with negative impact on human welfare. Damage to health, property, and the environment is the major dimension of external social cost. There is no acceptable method of measuring such damage in monetary terms, and little research has been done on the subject. However, Table 2-1 shows selected aspects of external cost.

Although highway safety has improved with the 55-mph speed limit since 1973, deaths still total 50,000 per year. The insurance industry provides one measure of the economic cost of highway accidents, estimating the loss in 1979 at $56 billion. Since automobiles alone require 20–25 percent of the country's energy for their manufacture and use, they are simultaneously a major component of energy consumption, a significant source of external social cost, and a leading cause of death in the United States.

The air pollution problem has lessened since 1970, when the Clean Air Act became national policy. This improvement is a major success for industry effort and federal regulation, and is studied intensively in Chapter 13. Most areas of the United States show lower levels of particulates, sulfates, hydrocarbons, and carbon monoxide. On the debit side, nitrogen dioxide levels have not improved, and most urban areas have not reached existing standards for pollution control.

Table 2-1. Some monetary estimates of energy-related external social costs

Category	Amount
Traffic accidents, 1979 (lost wages, funeral and medical expenses, legal fees, property damage)	$56 billion
Air pollution damage, 1978, extrapolated from: a. Council on Environmental Quality b. Business Roundtable	$85 billion $30 billion
Black lung disease, projected to 1990	$ 9 billion
West Valley nuclear waste cleanup, 1978	$ 1 billion

Sources: See footnote 5.

The most serious effect of air pollution is its impact on respiratory systems. The effects are particularly acute for persons with asthma and respiratory diseases, the elderly, children, and perhaps smokers. In preparing for the 1982 review of Clean Air Act standards, the U.S. Council on Environmental Quality (a federal environmental agency) and the Business Roundtable (a corporate group) each evaluated the apparent benefits of pollution control. Each study supposes approximately a one-fifth reduction in air pollution below 1970 levels. This is the basis for the extrapolation of apparent total damages in Table 2-1.[5]

Black lung disease is a serious occupational hazard for underground coal miners. Lung impairment may be found in more than 90 percent of miners who had worked underground for 30 years before the 1969 Coal Mine Health and Safety Act.

Other external damage from coal mining is difficult to value. Mining—including surface mining—remains the most hazardous occupation in the country in terms of on-the-job deaths and serious injuries. The petroleum industry is second in terms of injury and death rates per man-hour worked. There are no available estimates of land damage from surface or deep coal mining. (Health and safety problems in coal mining are summarized in Chapter 10.)

Civilian nuclear power has to date experienced a better safety record than other energy industries. Control of radioactive hazards in federal facilities to

[5]Tables 13-2 and 13-3 and the discussion in Chapter 13 review both studies in detail. Sources for Table 2-1: U.S. Department of Commerce, Bureau of the Census, *Statistical Abstract of the United States: 1980;* Lewis J. Perl and Frederick C. Dunbar, *Cost-Effectiveness and Cost-Benefit Analysis of Air Quality Regulations,* prepared by National Economic Research Associates for The Business Roundtable, November 1980; A. Myrick Freeman III, *The Benefits of Air and Water Pollution Control,* prepared for the Council on Environmental Quality, December 1979; United Mine Workers of America, "An Analysis of the Condition of the Black Lung Disability Fund," 1981; U.S. Council on Environmental Quality, *Environmental Quality,* Annual Report of the Council, 1979 and 1980; U.S. Department of Energy, *Western New York Nuclear Service Study, Companion Report* (late 1978).

date has been so successful that—notwithstanding the probability of greater cancer incidence from exposure—overall employee on-the-job health may be better at these facilities.

However, there is growing evidence that nuclear waste disposal is becoming the basic question to answer about the future continuation of nuclear power. The West Valley facility is one example. Located 30 miles south of Buffalo, New York, it is a storage location for several types of radioactive waste. The most hazardous material is in a tank that is slowly corroding, and some solution must be found. It remains to be seen whether the West Valley site can be decontaminated sufficiently so that it may ever (literally) be used for some other purpose. The necessary technologies do not yet exist. The waste problem is equally difficult for operating power reactors. Their spent fuel simply accumulates at reactor sites, awaiting a federal solution.

As a consequence of these problems, nine states representing 87 million citizens have joined together to offer critical commentary on federal nuclear waste policy.[6] California has prohibited new nuclear plants unless some safe waste disposal method is found.

(The West Valley facility has been operated by a petroleum company, Getty Oil. As will be noted in later chapters, the petroleum industry is significantly integrated into the nuclear fuel business. The general subjects of reactor safety and waste fuel and their economic significance are important, and are examined in Chapters 11 and 13.)

This very brief summary of certain aspects of external social cost is hardly exhaustive. No discussion is attempted here of the carcinogenic impact of petrochemicals, of the water and land impacts of power plant siting, of liquefied petroleum gas transport and storage, of uranium mining, of such proposed fuel technologies as shale oil fracturing or coal slurry transport, and many other problems.

Nevertheless, it is clear that external social cost is a major and continuing element of public policy.

Direct subsidies to energy production are sizable. Table 2-2 summarizes a recent Battelle Laboratory study. The $217 billion total has many components. The largest recipient of subsidies has been the petroleum industry, which has, through various deductions and credits, been made exempt from considerable tax liability. The electric utility industry is second in the magnitude of subsidies. Its subsidies have been through tax exemption for both private and public utilities, low-interest and tax-exempt bonds and loans, and direct federal investment in transmission facilities.

If we consider the available information, it is clear that the social cost of energy is considerably higher than its market cost alone. Figure 2-4 illustrates

[6]The attorneys general of Hawaii, Illinois, Missouri, New Mexico, New York, Ohio, and Texas; the California Energy Commission; and the Maine Department of Environmental Protection offer their evaluation in *Comments on the Draft Report to the President by the Interagency Review Group on Nuclear Waste Management*, December 1, 1978.

Table 2-2. Subsidies to energy production over the 1947–77 period, in 1977 dollars

Category	Amount
Electricity:	$57 billion
Tax subsidies:	
private: $15 billion	
public: $16 billion	
Bonds and loans: $21 billion	
Federal investment: $6 billion	
Nuclear power: Reactor development, regulation, uranium enrichment	$18 billion
Coal: Depletion allowance, transportation	$10 billion
Petroleum:	$101 billion
Depletion and other tax allowances: $50 billion	
Expenditures, pricing, services: $51 billion	
Natural gas: Tax subsidies	$17 billion
Hydropower: Federal expenditures and loans	$15 billion
Total	$217 billion

Source: Battelle Pacific Northwest Laboratory, *An Analysis of Federal Incentives Used to Stimulate Energy Production,* prepared for U.S. Department of Energy, December 1978.

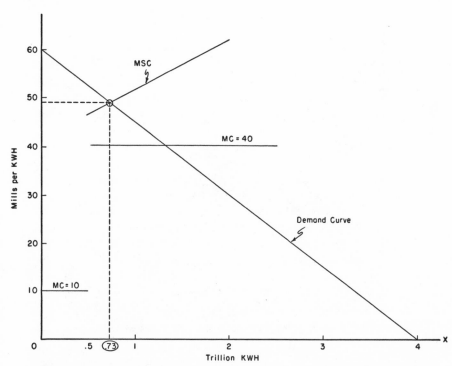

Figure 2-4. The social optimum: Price = marginal social cost at 49 mills per kWh and output is .73 trillion kWh

marginal social cost (*MSC*). It includes the subsidy received by the industry, the market costs of production, and the external social cost of energy use. Economists frequently assume that the demand curve is equivalent to marginal social value. Therefore, the intersection of marginal social cost and marginal social value define the optimum level of sales at .73 trillion kWh.

The assertion that the demand curve was equivalent to marginal social value follows from the economists' view that the value of consumption is reflected by the area under a demand curve. Suppose we have three persons who might buy a car. Mr. Ackley will pay $15,000 if he needs to. Ms. Brown will pay less, $10,000. Professor Cardiff is the least willing: he feels he can only afford $5,000 at the very most. Figure 2-5 indicates this three-person demand curve. If the price is $15,000, only one person will buy: Ackley. The value to him is block *A*, $15,000. If the price is $10,000, Brown will buy

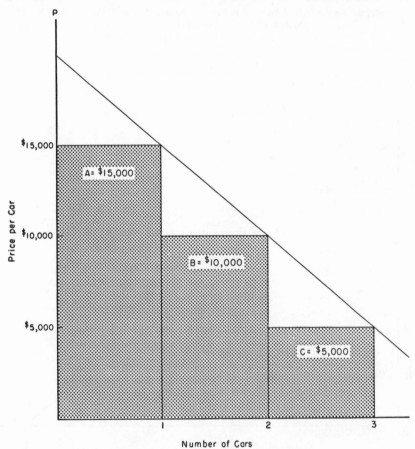

Figure 2-5. Consumer value is the total value to individual consumers

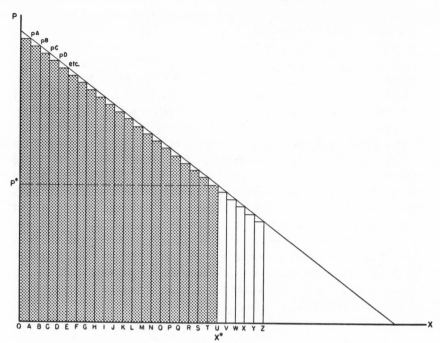

Figure 2-6. Social value = consumer value: Willingness to pay

also, and the value to her is block *B*. Finally, if the price is $5,000, Cardiff will buy, adding block *C*. At the $5,000 price, the value to all three persons is *A* + *B* + *C*: $30,000. The value to all three consumers is considerably higher than the price because most consumers—if they had to—would be willing to pay more.

Notice that the shaded area *A* + *B* + *C* almost fills the area under the demand curve. With more customers, the value to consumers becomes very close to being the area under the demand curve. In Figure 2-6, twenty-six potential consumers are shown, *A–Z*. Just as in Figure 2-5, *A* is willing to pay the most, then *B*, then *C*, and so on. Buying a car is worth P^A to *A*, P^B to *B*, P^C to *C*, and so on. At an actual price of P^*, all consumers *A–U* will buy, leaving *V–Z* out of the market.

Just as in Figure 2-5, the total value to consumers is the value to each person buying: *A* + *B* + *C* + *D* and so on through *Q* + *R* + *S* + *T* + *U*. The important thing to see is that the sum of these shaded blocks *A–U* is very close to the area under the demand curve. When the demand curve is based upon millions of consumers, the shaded area is almost exactly equal to the area under the demand curve.

This kind of reasoning has led economists to conclude that *social value* is equal to the sum of the value to individual consumers. Therefore the value to society of using an economic activity is equal to this same demand curve area. The theory leads to interesting results. How do perfect competition, profit monopoly, and growth monopoly compare in terms of social welfare? One method is to define *net social value* as the difference between social value and social cost for each level of production. Net social value is the total gain to society from production and consumption by all consumers, and is reduced by external cost. When net social value is the highest possible, the *social optimum* is attained, as in illustrated in Figure 2-7.

Net social value could also be depicted as the difference between total social value (the area under the demand curve) and total social cost. Notice that the net social value curve changes abruptly at $X = .5$ kWh. One reason is because of the assumption that marginal market cost increases from 10 mills per kWh to 40 mills per kWh. A second reason is that external social cost begins at $X = .5$. Since total social cost includes marginal market cost, external social cost, and subsidy, marginal social cost increases rapidly from $X = .5$ (see Figure 2-4). Consequently, net social value grew little beyond $X = .5$ and is at a maximum at $X = .73$. Thereafter, as external social cost and total social cost grow rapidly, net social value declines and even becomes negative above $X = 2$.

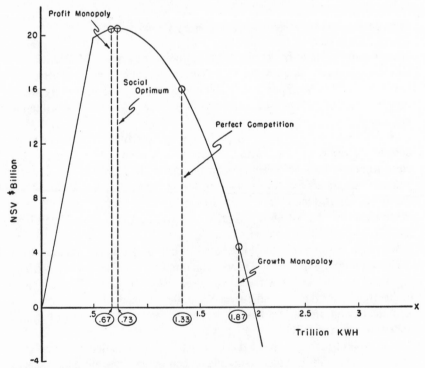

Figure 2-7. Net social value and the social optimum

Like Figure 2-4, Figure 2-7 indicates the social optimum with maximum net social value when $X = .73$, and net social value is at $20.5 billion. The competitive industry sales (see Figure 1-1) are at $X = 1.33$, and net social value is $16.0 billion. For growth monopoly, sales are pushed to 1.87, and net social value is $4.3 billion (see Figure 2-1).

Profit monopoly (Figure 1-2) has output at .67 trillion kWh and net social value at $20.4 billion. The profit monopoly solution is very close to the social optimum! This possibility has been noted by William Nordhaus, who was a member of President Carter's Council of Economic Advisers, and Robert Solow, former president of the American Economic Association. Nordhaus said: "But as Robert Solow has noted, monopolists are the conservationists' best friends: higher prices lead to lower consumption, a stretching out of finite resources, and possibly even lower prices in the future."[7]

The effects on income distribution are quite different. In the profit monopoly (Figure 1-2 and Table 1-2), excess monopoly profit is $21.66 billion in excess of costs. The price of 50 mills/kWh is almost three times the average cost of 17.5 mills. This excess profit may be assumed to flow to corporate management and owners, and no compensation is received by the victims of external social cost.

The social optimum (Figure 2-4) has a comparable price of 49 mills per kWh. But in this instance, we should assume that the interaction of nonmarket social cost and political dynamics has led to a public policy that does not permit the existence of monopoly profit. We may assume that this public policy compensates the victims of external social cost, or else requires the industry to control pollution, or else collects the monopoly profit as a public revenue. It is equally likely that public policy directed at this problem may utilize some components of each of these three policy approaches.

Summary

The four theories of actual and desirable economic behavior have uncomfortable implications. The growth monopoly theory calls for the lowest consumer prices and the highest levels of production and consumption. Yet, because of subsidies paid by the public, and external social cost, growth monopoly has the least net social value of any case, and is the furthest from the social optimum. In economic theory, the competitive industry is most efficient in social terms. But the existence of considerable subsidies and external social cost logically argue that a competitive industry—if it existed— would produce and sell excessive amounts of electricity. The profit monopoly case had the production and sales level closest to the social optimum, but with

[7]William D. Nordhaus, "Resources as a Constraint to Growth," *American Economic Review*, 64:2 (May 1974), 25.

a presumed consequence of greater income inequality and an absence of compensation for victims of external social cost.

The empirical evidence on the performance of energy industries appears to give a slight edge to the growth monopoly theory as that which most closely describes reality—at least to 1973. The basic points supporting this conclusion are the Flaim-Mount analysis, the decline of real energy prices over 114 years to 1973, the historical exponential growth of energy use, and the normal or below-normal rates of reported profits for oil companies and utilities.

However, there is a major obstacle to acceptance of the growth monopoly theory. The data on concentration in national energy markets generally show that the national petroleum industry is one of the least concentrated of major industries. In national markets, leading oil companies have less apparent control over production than do the leading companies in any other large manufacturing activity.

The next several chapters will develop an understanding of energy consumption and the technology and economics of petroleum. While undertaking this discussion, I shall be particularly interested in the interaction of economies of scale and cooperative activities in production, transportation, refining, and management. Then we shall again take up the question of economic behavior.

Appendix 2-A. Mathematical Illustration

The figures in Chapters 1 and 2 that show a competitive industry, a profit monopoly, a growth monopoly, and the social optimum can also have a numerical version. The demand function is

(1) $P = 60 - 15X$

Since revenue is price times quantity,

(2) $R = P * X = 60X - 15X^2$

Marginal revenue—the rate of change of revenue with respect to output—is

(3) $MR = \dfrac{dR}{dX} = 60 - 30X$

Marginal cost of production is 10 mills per kWh for the first .5 trillion kWh, then 40 mills per kWh:

(4) $MC = 10, \quad 0 < X \leq .5$
        $MC = 40, \quad\quad .5 < X$

Consequently total cost, the integral of marginal cost, is

(5) $\quad TC = \int_0^x 10dX = 10X, \quad 0 < X \le .5$

$\quad\quad TC = \int_0^x 40dX - 15 = 40X - 15, \quad .5 \le X$

The constant term -15 makes the two total cost equations equal at $X = .5$. Average cost is total cost per unit of output:

(6) $\quad AC = \dfrac{TC}{X} = 10, \quad 0 < X \le .5$

$\quad\quad AC = \dfrac{TC}{X} = 40 - \dfrac{15}{X}, \quad .5 \le X$

Marginal external social cost is nonexistent for the first .5 trillion kWh and then grows at a rate equal to $10X$:

(7) $\quad MESC = 0, \quad 0 < X \le .5$
$\quad\quad MESC = 10X - 5, \quad .5 \le X$

The marginal subsidy is a constant 6.67 mills per kWh.

(8) $\quad MSUB = 6.67, \quad 0 < X$

Marginal social cost is the sum of marginal cost of production, marginal external social cost, and subsidy.

(9) $\quad MSC = MC + MESC + MSUB$
$\quad\quad MSC = 16.67, \quad 0 < X \le .5$
$\quad\quad MSC = 41.67 + 10X, \quad .5 \le X$

Total social cost is the integral of marginal social cost.

(10) $\quad SC = \int_0^x 16.67dX = 16.67X, \quad 0 < X \le .5$

$\quad\quad SC = \int_0^x (41.67 + 10X)dX - 13.75 = 41.67X + 5X^2 - 13.75,$
$\quad\quad\quad\quad\quad\quad\quad\quad\quad\quad\quad\quad\quad\quad\quad\quad\quad\quad\quad\quad .5 \le X$

The constant term of -13.75 brings the two social cost equations into equality at $X = .5$.

Social value is the area under the demand curve, the integral of the demand function in Eq. (1):

$$(11) \quad SV = \int_0^x (60 - 15X)dX = 60X - 7.5X^2, \quad 0 < X$$

Net social value is the difference between social value (Eq. (11)) and social cost (Eq. (10)):

$$(12) \quad NSV = SV - SC$$
$$NSV = 43.33X - 7.5X^2, \quad 0 < X \le .5$$
$$NSV = 18.33X - 12.5X^2 + 13.75, \quad .5 \le X$$

The social optimum is the output level where net social value is highest. This can be found by differentiating Eq. (12) to find the maximum NSV, and this is equivalent to finding the point where price equals marginal social cost.

$$(13) \quad P = MSC, \quad \text{or } 60 - 15X = 41.67 + 10X \text{ at } X = .73$$

The numerical values for the other terms for the social optimum solution area in column one of Table 2-3.

For growth monopoly, the goal is maximum revenue subject to the requirement that there be no loss:

$$(14) \quad \text{maximum for } R = 60X - 15X^2,$$
$$\text{but } R \ge TC \text{ means } 60X - 15X^2 \ge 40X - 15$$

The solution is beyond the $X = .5$ level, and is equivalent to average cost pricing:

$$(15) \quad P = AC, \quad \text{or } 60 - 15X = 40 - \frac{15}{X} \text{ at } X = 1.87$$

See Table 2-A-1 for the other numerical values. This table also shows the values for the monopoly profit and competitive markets. For monopoly profit, the goal is the maximum difference between revenues and total cost, and this maximum is beyond $X = .5$ and is defined by the equivalence of marginal revenue and marginal cost.

$$(16) \quad MR = MC, \quad \text{or } 60 - 30X = 40 \text{ at } X = .66\tfrac{2}{3}$$

The competitive solution requires price to equal marginal cost:

$$(17) \quad P = MC, \text{ or } 60 - 15X = 40 \text{ at } X = 1.33\tfrac{1}{3}$$

[42]

The competitive market solution with total revenues exceeding total costs was not discussed in the text. The assumption here is that regulatory agencies require marginal cost to equal marginal price, and any excess profit is taxed by federal and state income taxes.

Table 2-A-1. Numerical illustrations of social optimum, growth monopoly, profit monopoly, and a competitive market

	Social optimum	Growth monopoly	Profit monopoly	Competitive market
Price, mills/kWh	**49.0**	**32.0**	50	**40**
Sales, trillion kWh	.73	1.87	.66⅔	1.33⅓
Revenue, $ billion	35.8	59.8	33.3	53.2
Marginal revenue, mills/kWh	38.0	3.9	**40**	20
Marginal cost, mills/kWh	40	40	**40**	**40**
Total cost, $ billion	14.2	59.8	11.7	38.2
Average cost, mills/kWh	19.5	**32.0**	17.5	28.7
Marginal external cost, mills/kWh	2.3	13.7	1.7	8.3
Marginal subsidy, mills/kWh	6.7	6.7	6.7	6.7
Marginal social cost, mills/kWh	**49.0**	60.4	48.3	55.0
Social cost, $ billion	19.5	81.7	16.3	50.7
Social value, $ billion	40.0	86.0	36.7	66.7
Net social value, $ billion	20.5	4.3	20.4	16.0
Excess profit, $ billion	21.6	0	21.7*	0
Total subsidy, $ billion	4.9	12.5	4.4	8.9

*Actual value is revenue ($33⅓) less total cost ($11⅔), equal to $21⅔, rounded to 21.7. Excess profit at the social optimum is slightly less.

II

ENERGY USE

3

Aggregate Consumption, Substitutability, and Efficiency

> A policy that focuses disproportionately on reducing consumption is essentially a policy of despair and carries substantial risk of creating even greater problems in the future. A policy of despair is not necessary in a country with the resources and technology of the U.S.
> —William Tavoulareas, president of Mobil Oil

The preceding section on economic theory described concepts with contradictory predictions about levels of energy use, costs, and profit rates for energy corporations. These aspects of industry behavior will not be discussed again until Chapter 9. The subjects in this section include aggregate energy consumption, substitutability between energy forms, system efficiency in energy utilization, demand elasticity, forecasts of future levels of use, conservation, and future transportation patterns. Chapters 3 and 4 develop a quantitative analysis that leads to a conclusion unlike Tavourlareas's.[1]

Chapter 3 relies considerably upon the works of other authors. Several are used extensively, particularly Earl Cook's *Man, Energy, Society,* and Ehrlich, Ehrlich, and Holdren's *Ecoscience.* Kathleen Cole's thesis provided most of the data for the discussion of comparative costs.[2]

Introduction

Civil War America was a place and time that we today would find foreign. The difference is especially evident in energy use. In 1860, firewood was

[1]William Tavoulareas and Carl Kaysen, *A Debate on a Time to Choose* (Cambridge, Mass.: Ballinger, 1977), p. 8.

[2]Earl Cook, *Man, Energy, Society* (San Francisco: W. H. Freeman, 1976); Paul Ehrlich, A. H. Ehrlich, and J. P. Holdren, *Ecoscience: Population, Resources, Environment* (San Francisco: W. H. Freeman, 1977); Kathleen Lynn Cole, "Tax Subsidies and Comparative Costs for Utilities and Residential Heating in New York," M. S. thesis, Cornell University, Department of Agricultural Economics, 1981.

Energy Use

Table 3-1. Sources of energy consumed in mechanical work, 1860

	Horsepower hours, billion	Btu trillion	Percent
Wind	2.1	5.3	15
Water	1.3	3.3	10
Wood	0.7	1.8	5
Work animals	7.6	19.3	56
Coal	1.8	4.6	13
Total	13.5	34.4	99

Sources: Sam H. Schurr and Bruce C. Netschert, Resources for the Future, *Energy in the American Economy, 1850–1975* (Baltimore: The Johns Hopkins University Press, 1960), pp. 54, 55; J. Frederic Dewhurst and others, *America's Needs and Resources: A New Survey* (New York: Twentieth Century Fund, 1955), pp. 1114, 1115.

more common than coal: firewood provided five times the energy supplied by coal. Natural gas was not used, and electricity generation was unknown. Energy production for mechanical work in this era is shown in Table 3-1. Note the importance of work animals. The energy used in mechanical work was insignificant compared to that used in heating. Overall, fuel wood provided 2.6 quadrillion Btu of energy in 1860, most of which was for heating. This is 5 times the total energy derived from coal in the same year.

Oil was discovered and produced in Pennsylvania in 1859 in small quantities, and was used for lamplighting and medicines. Transportation was by foot, horse, train, or sail boat. The population numbered 32 million, and the mountains and plains of the interior were not yet states. The territories of Alaska, Hawaii, and Puerto Rico had not yet been acquired.

1980 is rather different. America uses one in every four barrels of oil produced throughout the globe and nearly one in every three kilowatt hours of

Table 3-2. World energy production and U.S. energy consumption, 1980

Type of energy	World production	U.S. consumption	U.S.% of World 1980	(1978)
Crude oil, billion barrels	23.1	6.2	27%	(30%)
Electricity, trillions kWh (1979)	8.0	2.4	30%	–
Natural gas (dry), trillion cu ft	52.8	20.1	38%	(39%)
Nuclear power, billion kWh (non-communist countries)	585	251	43%	(52%)
Coal, billion tons	4.16	0.70	17%	(17%)
Total energy, quadrillion Btu	286.7	76.2	27%	(28%)
Population billion	4.3	0.23	5%	(5%)

Sources: World Bank, *World Development Report 1981* (New York: Oxford University Press, 1981); U.S. Department of Commerce, Bureau of the Census, *Statistical Abstract of the United States: 1980;* U.S. Department of Energy, Energy Information Administration, *International Energy Annual, 1980,* September 1981; United Nations, *1979 Yearbook of World Energy Statistics.*

Table 3-3. U.S. primary energy consumption, 1979

Type of energy	Amount consumed	Consumption in Q's	Percentage of total	Per capita consumption
Coal	681 million tons	15.1	19%	6,200 pounds
Natural gas	19.5 trillion cubic feet	19.9	25%	88,600 cu ft
Petroleum products	6.7 billion barrels	37.1	47%	1,300 gallons
Electricity				
(total utilities)	(2.2 trillion kWh)	(7.7)		10,200 kWh
hydropower	280 billion kWh	3.2	4%	
nuclear power	255 billion kWh	2.8	4%	
Total		78.2	100%	355 M̄Btu

Population was 220.6 million. Rounding makes slight differences in sums and totals, and .1 quadrillion Btu (Q) from other forms is not shown.
Source: U.S. DoE, *Monthly Energy Review.*

electricity. We use more than one-third of global natural gas production, and almost one-half of the nuclear power generated outside the communist countries. In aggregate, for all forms of energy, the United States consumes one-fourth of total world production. This is accomplished by a country with only 5 percent of the world's population.

While it appears that American consumption is relatively high, the figures for 1980 show a decline from the peak reached in 1978 or 1979, America's historical maximum for energy consumption. In Table 3-2, these statistics are shown for 1980, and 1978 percentages are given in parentheses.

Table 3-3 gives an introduction to aggregate energy use in the United States. (Woodfuel and active solar heating are not included here. Detailed sectoral consumption statistics are in Table 15-1.) It is significant that oil and natural gas supply more than two-thirds of America's energy. In 1973, the first year of active OPEC price control, American natural gas and petroleum consumption was 77 percent of total U.S. energy use. The relative importance of oil and gas has been continually declining since 1973.

The period of energy growth in the United States was above all the era of growth in oil and natural gas, from 1859 to 1973. It is a diverting coincidence that the early history of oil and natural gas is associated with rural western New York, close to the location where this book was written. One writer 150 years ago reported that oil had first been seen by Europeans in America in the 1600s near Seneca Lake, a few miles west of Ithaca, New York.[3]

By the mid-1800s, Seneca Oil had become commercial and was sold primarily as a cure for rheumatism, blindness, skin ailments, and constipation, and as an insecticide. Indians had earlier used this crude petroleum for the same medicinal purposes. The going rate was 75¢ to $1.00 per gallon in Pittsburgh in 1843, about $14 in current dollars. Under other names (British Oil, American Oil, Rock Oil, and so on), it was sold for the same uses.

[3]S. P. Hildreth, *The American Journal of Science and the Arts*, July 1833, p. 64.

But it had also been recognized that this oil, which collected in seeps or springs was useful as an illuminator (when burned) and as a lubricant. For both purposes, it was equivalent in quality to animal fats—whale or bovine— but the quantity was limited and the price high. So, in 1858, the Seneca Oil Company was formed in northwest Pennsylvania to drill for oil. In 1859, manager Edwin Drake's well succeeded. That year, 3,000 barrels of oil were produced in the United States, most of it in Pennsylvania.

Twenty years earlier, in 1839, John D. Rockefeller was born near Richford, New York, a small rural village seventeen miles southwest of Ithaca. The Rockefeller family was very poor, its poverty made more difficult by William Rockefeller's frequent absences, the family's many moves, and William's eventual departure from his family. John D. was working at age 16 in Cleveland. According to popular Cleveland legend, in 1860 he represented a Cleveland group that sought firsthand information on the financial potential of the new Pennsylvania fields. Rockefeller visited the oil field. Upon his return, he advised the group to stay out of the business, and they followed his advice. Rockefeller then set up a partnership and refinery of his own.[4] By 1880, when Edwin Drake died, having lost his oil earnings in speculation and lived in harsh poverty until he received a special pension from the Pennsylvania legislature, the Standard Oil Company for some years had been the world's largest.

Every part of America has its own local history of energy development. Each region can single out its own roles. Fredonia, New York, might emphasize its first commercial use of natural gas for lighting in 1821; New Englanders might prefer to note their region's role in supplying whale oil for lighting. Palo Alto, California, is the site where electric lights were first used; Detroit was the first home of mass-produced automobiles. Texas and the Gulf regions began producing inexpensive oil at Spindletop in 1901. Coal was mined in colonial Virginia at the beginning of the eighteenth century. Moving beyond a provincial focus on American industrial history, we have already noted the American Indian use of petroleum; we should also note that Arizona's Hopi mined and burned coal in A.D. 1000. The recorded consumption of fossil energy extends backward in time to the Roman use of Sicilian oil and the Semitic use of asphalt in Mesopotamia 5,000 years ago. Energy use is intertwined with human history, across the breadth of the country and the world.

What is energy? How do we use it, and how has it come to be used in the particular ways that it is? Who makes it and owns it? Are energy industries

---

[4]William Hoffman, *David* (New York: Dell, 1972), p. 267; John T. Flynn, *God's Gold: The Story of Rockefeller and His Times* (Chautauqua, N.Y.: Chautauqua Press, 1932), pp. 93, 98; Allan Nevins, *John D. Rockefeller: The Heroic Age of American Enterprise* (New York: Scribner, 1940), pp. 172, 176. Flynn and Hoffman report this story as fact. Nevins describes it in less detail, and notes Rockefeller's lack of confirmation.

monopolistic or competitive? Is government policy independent? What present significance does the Standard Oil Company have? How do domestic and foreign policies interact with energy companies? Finally, the ultimate questions: Who should own energy resources, and how should those remaining be managed? In time, we shall take up each of these questions, more or less in order.

Energy as a Concept

Energy is an idea, a concept developed by the human mind to group together a large number of actions. It includes things known to humanity since its earliest times: such things as muscle power in man or animals, or wind power in a sail, or the burning of animal fat as a source of light, or the burning of wood for heating a home or cooking or warming water. Energy is embodied in steel, plaster, and nylon. Perhaps less obviously, it is embodied in wooden houses and furniture and in cotton or woolen clothes.

In modern usage, energy means the uses of common sources of power: nuclear power, hydropower, electricity, oil, natural gas, coal, wood, and solar energy. The modern and historical forms of energy share the earliest meaning of the ancient Greek word *energia,* which meant activity in work. So we usually mean by energy the ability to become or the process of becoming something else. Coal (with oxygen) becomes heat, carbon dioxide, and air pollution. Heat can be captured by water, becoming steam. Steam passes through the blades of a turbine, turning it. The turbine, now possessing mechanical energy, turns within the generator, transforming mechanical energy into electrical energy. The electrical energy, forced through a resistant heating system in the home, becomes heat again. Or heat in the home may be released directly by burning wood, transforming the *hydrocarbons* within the wood into products of oxidation and heat. The energy within the wood was originally collected as a result of the conversion of inorganic chemicals and water into cellulose. The coal burned in a generating plant is itself a product of photosynthesis and solar radiation, apparently originating hundred of millions of years ago in early plant life forms.

Wood is termed a *renewable resource,* and coal a *finite resource;* yet we see that coal and wood energy are both embodiments of solar energy. And if we wish to develop this concept of energy, we may note that every form of it is an embodiment of stellar energy, deriving from solar and galactic processes.

The renewable energy sources are all driven by solar radiation. Hydropower (and windpower, such as it is) arises from the annual climatic variations associated with the oceanic absorption of solar energy, evaporation and precipitation over land areas, and the rotation of the earth.

[51]

The *fossil fuels* (coal, oil, natural gas) are all ancient remnants of photosynthesis from the earth's early life. Therefore, from the perspective of our own lifetimes, their supply is finite and limited.

Nuclear heat depends upon the process by which radioactive material (embodying energy in a very dense form) decays into more ordinary forms. Nuclear fuel is itself prepared by the application of large amounts of energy to uranium or similar materials in complicated chemical processes; by this means, the radioactive material is concentrated so that it may later decay. But, the original source of uranium and other nuclear fuels is the most elemental form of stellar energy—the original creation of the sun, the earth, the solar system, and its elements. Hence nuclear fuel is finite.

Yet energy takes one other form. The preceding illustrations all portray energy as the process of becoming something else. One other kind of energy is the creation of physical structure to resist decay. Obvious examples are furniture, clothes, horses, cars, and so on. Yet, note that this materialization of energy is based upon the same processes or substances already mentioned. Current photosynthesis makes wood, and prepares food for humans and animals. Organic processes create cotton, wool, silk, and fur. Fossil fuels provide a source of hydrocarbon structures, which, with the application of energy, provide the physical bases for plastics, synthetic rubber, synthetic fabrics, and so on.

From a distant perspective, all our energy is solar energy in different forms. All energy sources, renewable and finite, have their origins in the basic elements and processes that began with the creation of the sun and solar system from the original stellar nebula.

From a closer perspective, our energy is clearly of two types. One is the renewable kind that arises from the flow of natural cycles. The major forms of renewable energy are (obviously) human and animal energy; plant and animal processes that create fibers for clothing, materials such as wood, thatch, and hides for shelter; industrial processes utilizing hydropower; and direct solar radiation. Finally, solar energy and photosynthesis create the food chains for plant, animal, and human life.

The fossil fuels and nuclear power are readily recognizable as finite energy resources. Less well known members of the finite group are geothermal heat and certain synthetic fuels. Geothermal heat rises from the earth's inner core and is probably not generated continuously. The coal-based synthetic oil and gas fuels are all derivatives of a fossil fuel and thus are also finite.

We can perceive another aspect of the interconnectedness of energy forms by comparing the oldest and the newest. Sunlight is not only the oldest energy form used by humans for heat and light (and of course for plant and animal production), it is also the most accessible and—when used in its received form—requires no machines or equipment. The source of sunlight is the continuing decay of the sun through the fusion of hydrogen nuclei. The basic result is the creation of helium and the transformation of a small amount of mass into solar energy. The newest and most complex energy form—nuclear power—is a cousin of solar energy, deriving its energy from elemental nuclear processes. And, as we pointed out above, uranium was formed more or less at the same time as the sun, billions of years ago, in the original process of creation.

We may see another aspect of interconnectedness by comparing the energy processes in humans, animals, plants, and the fossil fuels. In humans and animals, carbohydrates and fats are major energy sources. They consist of various combinations of carbon, hydrogen, and oxygen. When oxidized, the carbohydrates or fats are transformed into water and simpler molecules which are primarily waste products. In plants, structure is provided by cellulose, a plant form of carbohydrate similar to the starch found in animals and humans. Both wood and cotton are primarily cellulose. In Table 3-4, the common nature of starch, corn sugar, and beef fat is evident in their composition of carbon, hydrogen, and oxygen. (It should be noted that this table is an economist's view of the chemical basis of energy, which has been purposely simplified.)

Crude oil is a mixture of hydrocarbon chemicals. The basic similarity among the different oils is the importance of carbon and hydrogen and the absence of oxygen. Table 3-4 shows the molecular bases for heptane and other components of oil. Comparing the chemical components of oil with those of cellulose and other carbohydrates suggests the process by which oil has been made. It is generally believed that oil is the decayed remains of primitive plant and animal life, perhaps primarily microorganisms. Thus, today's crude oil is the remnant of plant and animal life 500 million to one billion years ago.

Natural gas consists primarily of methane. Methane is one of the simplest

Table 3-4. Simple components of energy

Source	Type	Major components	Illustrations	
			Form	Frequency of components
Human, animal	carbohydrates and fats	carbon, hydrogen, oxygen	beef fat	$C_3H_5(C_{17}H_{35}CO_2)_3$
Plant	carbohydrates	carbon, hydrogen, oxygen	cellulose sugar	$C_6H_{10}O_5$ $C_{12}H_{22}O_{11}$
Oils	hydrocarbons	carbon, hydrogen	octane paraffin heptane butylene	C_8H_{18} $C_{12}H_{26}$ C_7H_{16} C_4H_8
Natural gas	hydrocarbon	carbon, hydrogen	methane	CH_4
Coal	carbon	carbon	anthracite	C

and lightest hydrocarbons, and thus vaporizes from liquid into gas at $-238°F$. This explains why natural gas is frequently found in association with oil: it is an evaporated gas of their common source. Table 3-4 also suggests that oil or natural gas might be made purposely from animal and plant carbohydrates. This is in fact the nature of the geological process, and apparently the process of creating fossil fuels of carbon and hydrogen from organic carbohydrates is one major source of the earth's current stock of oxygen.

Although the historical process has taken hundreds of millions of years, the same reactions can take place in a much shorter period. For example, the City of Los Angeles has sunk gas wells into its garbage dumps. Also, decomposition of organic waste can be controlled to produce gas. American dairy and poultry farms are beginning to experiment with this process. And in Brazil, ethyl alcohol for automobiles is produced from sugar cane and manoic. The intention here is to provide a renewable source of fuel usable in cars, trucks, and industrial boilers.

Coal is primarily carbon, the result of the decomposition of trees, ferns, and plant debris. The most common form is bituminous coal, which still retains significant fractions of hydrogen, oxygen, nitrogen, and sulfur in association with the carbon. Anthracite coal is more purely carbon; the "coalification" process has gone further. Peat and lignite are "young" forms of coal in which the forms of the original plant material are still evident, and the fractions of associated elements are larger.

It is evident that Table 3-4 has arranged our energy sources in two dimensions. One dimension is chemical complexity, in which the order goes from most complex—human and animal carbohydrates and fats—to least complex, coal. The second dimension involves time: renewable resources are listed first, and the older, finite resources follow.

To summarize at this point; first and most important, energy is a concept which can encompass a broad variety of living and mineral substances, and these things have definite patterns of physical relationships to each other.

Second, energy forms are interconnected in their various uses. We have alluded to this broad principle of substitutability, to the use of whale oil or kerosene in lamps, and so on. This is made more explicit in Table 3-5, which illustrates the comparable uses of the broad classes of energy sources. Certainly many energy sources and uses are not included there. No recognition is given to the role of energy in transforming metal ores into usable objects, and there is no mention of weaponry.

One aspect of the organization of energy uses in Table 3-5 is analogous to that of the energy forms in Table 3-4. The oldest sources and uses are renewable, and the modern sources are based upon fossil fuel use. Electricity is an exception, drawing upon an old renewable form (hydropower), the fossil fuels, and the newest finite form, nuclear power. However, Table 3-5 is sufficient to make evident the broad range of substitutability between different forms of energy. Just as Table 3-4 shows the physical relationships between

Table 3-5. Common energy sources and direct uses

Activity: method, process, or function	Renewable sources				Finite sources			
	Human	Animal	Plant	Physical	Oil	Natural Gas	Coal	Electricity including nuclear power
Transportation (method or mode)	walking	horse, ox	triffids gasohol	sail	car, plane, rail, truck, boat	–	rail	rail
Mechanical power	arm power	horse, ox	–	moving water, wind	internal combustion engine	———steam engine———		electrical motors
Heating (energy source or appliance type)	bundling	dung	wood, peat burning	sunlight	oil furnace	gas furnace	coal furnace	resistance heating, heat pump
Clothing and fabric fibers	–	wool, fur, hides, silk	cotton, bark	–	——nylon, orion, polyester——		–	–
Structural (material for housing)	–	mammoth bones, coral rock, hides	wood, thatch	marble, concrete, caves	asphalt	–	–	–
Furniture and utensils	–	bones, shells, skins	wood, grasses	stone	———plastics———			–
Lighting	–	tallow candles, whale oil	pitch	sunlight	kerosene	gas lamp	carbon lamp	incandescent and fluorescent bulbs

renewable organic and finite fossil energy, so Table 3-5 shows the similarity in uses.

Energy Measurement

The physical *substitutability* between the different kinds of energy is the basis for using the British thermal unit or the calorie as a basic unit of measure for all forms of energy. The British thermal unit (Btu) is the amount of heat necessary to raise the temperature of one pound of water one degree Fahrenheit. The calorie is also defined in reference to heating water, and is the energy necessary to raise the temperature of one gram of water one centigrade degree. The two are obviously related, and 1,000 calories are equivalent to 3.97 Btu.

The list of energy units following the Preface shows some of the equivalencies in different energy forms, and is helpful in indicating relative amounts of energy associated with various natural and human activities. The list gives basic definitions and magnitudes, and is a useful reference throughout the text. Note two inconsistencies in it that reflect common American usage. The letter "g" can represent either gallon (as in mpg) or gram (as in kg). The letter "M" can mean either thousand (as in Mcf) or million (as in MW). Here, to avoid this confusion, $\bar{M}$ will be used to denote million.

How much energy is a lot? we can judge our use from quite different extremes. As Table 3-2 indicated, American energy consumption is 75-80 Q, and world energy consumption, very roughly, is probably 275 Q. Solar energy reaching the earch is about 2.6 million Q in a year. So, our conventional energy consumption is a very small one-hundredth of 1 percent of the solar energy the earth receives. But from another perspective things look quite different. A person walking 3 miles will take an hour to do so, using perhaps 200 kilocalories (800 Btu), the energy in about 50 grams (1.76 ounces) of sugar. If the person drives a new fuel-efficient car that delivers 30 miles per gallon of gasoline, the 3-mile trip requires about 13,000 Btu and 13 ounces of gasoline.

These comparisons and equivalencies provide a more thorough understanding of the nature of substitutability in energy consumption. This leads to the engineering concept of efficiency, which has important economic implications.

Efficiency

A simple measure of *efficiency* is the proportion of energy supplied which goes to accomplish work. This means simply

$$\text{efficiency} = \frac{\text{energy utilized}}{\text{energy supplied}}$$

In a house, a natural gas furnace may have an efficiency of 80 percent. This means that if 146 M̄Btu of gas enter the house, 117 M̄Btu go to heat it. The remaining 20 percent is lost in the smoke going up the chimney.

The concept of efficiency can also be applied to processes that convert one kind of energy to another. An electric resistance element converts electricity to heat, and does so at nearly 100 percent efficiency. Each kWh of electricity produces about 3,412 Btu of heat in the house. However, converting fossil fuel energy to electricity is less efficient. It takes 10,500 Btu to generate one kWh of electricity. This means 10,500 Btu produce 3,412 Btu, so the efficiency is 32 percent.

These efficiencies have all been measured at a single stage of a process. System efficiency can be examined by linking the separate efficiencies of each stage. In Table 3-6, the upper section indicates that 90 percent of the energy in natural gas leaving a field will reach a home. The lost 10 percent is used in pumping the gas, or in leakage. In the home itself, 80 percent of the gas is utilized in heating. The cumulative system efficiency of the two stages is simply 90% * 80%, or 72%; that is, 72 percent of the natural gas from the gas field is delivered as heat in the house.

Compare this to the middle section of Table 3-6, where the natural gas is used to generate electricity and electricity heats the home. The overall system efficiency is a far lower 35 percent. In two of the stages, the gas-fired electric system is superior. It loses less natural gas in transportation, and is 100 percent efficient in home conversion. Clearly, electric heating loses the system efficiency ranking because of losses in the power generation stage.

Table 3-6 illustrates two basic rules of system efficiency in residential energy consumption. First, the direct burning of fossil fuel—natural gas, oil or perhaps coal—is always physically more efficient in heating than is the use of electric resistance heating operated on electricity generated from fossil fuel. This does not mean that a fossil fuel is always preferable—gas may be

Table 3-6. System efficiency for home heating

Stage of system	Stage efficiency	Cumulative system efficiency
Natural Gas		
1. natural gas production and transportation to home	90%	90%
2. gas burned in home furnace	80%	72%
Gas-generated electricity		
1. natural gas production and transportation to utility	95%	95%
2. power generation	40%	38%
3. transmission of electricity to home	92%	35%
4. electric resistance home heating	100%	35%
Gas-generated electric heat pump system		
1. natural gas production and transportation to utility	95%	95%
2. power generation	40%	38%
3. transmission of electricity to home	92%	35%
4. electric heat pump home heating	230%	80%

unavailable, coal prohibited, and oil too costly. In the strict sense of physical efficiency, however, electric resistance heating is less efficient.

The second basic rule is that natural gas appliances are always ahead of electrical appliances in terms of system efficiency. This is true for cooking, water heating, and clothes drying, as well as for space heating.

It would be incorrect, however, to conclude that an electric system will always be less efficient. For example, in the last section of Table 3-6, a heat pump replaces electric resistance heat. Now the overall system efficiency is 80 percent. The key is the 230 percent figure for stage 4 for the heat pump. The heat pump is similar to a refrigerator. It uses electrical energy to transfer heat from the outside into the house; it cools the outdoors when it heats the indoors. The 230 percent efficiency means that one Btu of electricity can transfer 2.3 Btu from the outside into the house. Obviously, a heat pump is a different kind of machine from a furnace, and it is more appropriate to use the term *coefficient of performance* than the term efficiency in discussing a heat pump.

The potential for the heat pump may be greatest in the urban temperate regions of the country, which have sizable cooling and heating requirements.

In industry, the stage efficiency of an electrical process may be so much greater than that of a combustion process that the system efficiency for the electric method is higher. For example, electric steel and glass furnaces have system efficiencies that are higher than those of their coal and natural gas counterparts.

Table 3-7 ranks various methods of home heating according to their end-use efficiency in typical houses in central New York in 1980. It is clear that end-use efficiency and total system efficiency differ significantly, as was seen in Table 3-6.

There are several problems with using Table 3-7 as a basis for decisions

Table 3-7. Annual heating fuel cost, 1980, Central New York, old house requiring 117 M̄Btu of heat energy

System type	Efficiency or performance factor		Price in $/M̄Btu (1980 $)	Annual fuel cost
	end use in home	total system		
Electric heat pump	1.6	0.56	$12.60	$ 921
Electric resistance	1.0	0.35	$12.60	$1474
Natural gas	0.8	0.72	$ 4.00	$ 585
Fuel oil	0.7	0.60	$ 6.84	$1143
Solar/natural gas	0.5 (solar)	?	$ 4.00 (gas back-up)	$ 293 (gas)
Wood	0.5	?	$ 4.44	$1039
Insulation, substantial	1.7	?	$ 4.90	?

about what type of heating in preferable. First, we need to determine which economic group is making the decision. A renter would prefer solar heating. But the performance rating of .5 means only 50 percent of the 117 M̄Btu is provided by solar energy, so a *back-up* system is needed to supply the other 50 percent. If this back-up were natural gas, the system would have a fuel bill of $293.

The renter's second choice would be an all-natural gas system, with a $585 fuel bill. The heat pump would be third, then wood burning, then fuel oil; electric resistance heating would be the last choice.

But suppose the builder or the builder and first owner are making the decision. They will prefer the system with the lowest installation cost. The preferred ranking is electric resistance heating, natural gas, wood, fuel oil, electric heat pump, and solar/natural gas.

Another problem with Table 3-7 is its assumed level of insulation and heating requirements. The house was assumed to be one built before the mid-1970s. If this old house could have a substantial improvement made in its insulation level, it would require 59 M̄Btu per year instead of 117 M̄Btu. If this more substantial amount of insulation had been used when the house was built, the cost would have been about $2,000, equivalent to $4.90/M̄Btu.[5] So more insulation would have been sensible if the house was to be heated electrically or with fuel oil. Once a house is built, however, additional insulation becomes costly, and it may not be economically justified. It is a paradox of the present federal incentives that the government rewards insulation expenditures after construction, when insulation is less economically efficient.

There is still another deficiency in Table 3-7. It does not integrate initial purchase cost and annual fuel cost to determine overall cost. Furthermore, it does not consider income tax deductions and credits or property tax exclusions that will affect overall cost, but it is this overall cost that may be of primary interest to a home-owning family.

The location for the data presented in Table 3-8 is Long Island, New York. This table differs from Table 3-7 in several ways. First, the Long Island house uses 84 M̄Btu of heat energy instead of the 117 M̄Btu assumed for central New York. Second, fuel cost is considered over a 20-year period rather than in 1980 only. Third, wood burning is not considered for Long Island, but solar heating is included. Fourth, the Long Island climate is more suitable for heat pumps than the central New York climate, and the coefficient of performance on Long Island is 2.3 rather than 1.6.

The mortage payments in Table 3-8 are always 14.2 percent of the initial cost. This assumes a 13 percent interest rate for 20 years.

---

[5]A 13% interest rate for 20 years defines a 14.2% mortgage factor. This 14.2% times the $2,000 insulation investment gives an annual cost of $284. The energy saving is 58 M̄Btu each year, so the cost is $4.90 per MBtu. Chapter 14 explains insulation, heat energy, and the method of economic analysis in greater detail.

Table 3-8. Total consumer cost for heating, future inflation, Long Island, New York, new house requiring 84 M̄Btu of heat energy

| Type of system | Purchase cost, 1980 | Annual costs, 1980–1999 | | | | Builder: Initial cost | Renter: Fuel cost | Owner-occupied: Total annual cost to household | Public: Total cost to economy |
		fuel cost	mortgage payment	tax effect	total consumer cost				
Natural gas	$ 2,500	$ 877	$ 356	$ −142	$1329	2	2	1*	1*
Fuel oil	3,200	1511	456	−182	2109	3	4	2	2
Heat pump	5,500	1391	783	−313	2357	4	3	4	3
Electric resistance	1,800	3199	256	−103	3651	1*	5	5	4
Solar/natural gas	16,500	439	2349	−1771	2318	5	1*	3	5

Note: Total consumer cost includes annual property and sales taxes and maintenance costs. These are $238 for the gas system, $324 for oil, $299 for electric resistance heating, $496 for the heat pump, $1,301 for the solar/gas system. For total public cost, add the tax effect to total consumer cost. The asterisk (*) emphasizes the least-cost system for each criterion.

The tax effect in Table 3-8 shows the effect of deductions, credits, and exclusions for taxes. For all systems, this includes the effect of interest expense deductions for New York and federal personal income taxes. The solar system tax effect also includes the federal sales tax credit and the New York property tax exemption.

The result in Table 3-8: Natural gas heating is the least costly for the owner-occupied home. Electric resistance heating is the most costly. The solar-gas system is in third place, near the electric heat pump.

Kathleen Cole has examined this problem of comparative costs from many perspectives, and certain conclusions seem rather definite. First, natural gas is always least costly. Second, an appropriate renewable system is frequently in second place. This is wood space heat in the northern rural United States, and solar water heating in much of the coastal and southern United States. For example, Cole finds solar/gas and solar/electric water heating to be less costly on Long Island than conventional electric water heating. Electric heat pumps are less costly than electric resistance space heating. Electric resistance space and water heating is usually the most costly option available in terms of overall cost to the home owner or renter.

The problem of social cost has arisen here: the tax subsidies in Table 3-8 have a significant effect on customer cost.

Perhaps the most significant conclusion is that present-day houses should have been built with much higher levels of insulation, and reducing the amount of energy required to heat a home may be more important than selecting the type of heating system. This problem is studied in depth in Chapter 14.

4

Demand and Conservation

Conservation is primarily philosophical in meaning; it involves an attitude about the human community's relationship to the natural environment. Americans have long given it two emphases, which have often been in conflict. One line of thought, developed in the early part of this century by John Muir and the Sierra Club, saw the natural environment as having special significance for humanity's adaptation to society. Natural environments, this view holds, often merit their own special rights and privileges, which should protect them from major development activities. The second view of conservation is represented by Gifford Pinchot. It holds that development of natural resources is in general a positive thing, but asserts a public interest in the matter that is in addition to (or in conflict with) private rights to development.

Sometimes the two viewpoints will coincide: to promote strip mine regulation or advocate continued federal ownership of federal forest land. At other times the two aspects of conservation are in sharp conflict. In fact, these two views first entered into serious opposition over an energy problem, the Hetch-Hetchy project. Simply stated, the one view valued the Hetch-Hetchy valley as the equal of the other yosemite, Yosemite Valley. The second view valued the site for its potential as an inexpensive source of water and power.[1]

From my perspective, it seems that the concept of energy conservation is a modern marriage of the earlier schools of thought. Energy conservation seems to presume that energy is necessary for economic welfare, but at the same time it carries an important element of "less is better." This latter tendency associates the use of finite energy resources (uranium, gas, oil, coal) with environmental impact, and perhaps as well to unarticulated values regarding social structure.

Econmists try to reconcile the two views of conservation while working with the economics of private development. The concepts of amenity rights,

[1]John Muir, "Let Everyone Help Save the Famous Hetch-Hetchy Valley," pamphlet, San Francisco, 1909.

options, and generational equity have developed to reflect the values of the first school. And the public regulation and development perspective has led to the working out of cost-benefit analysis and other methods that attempt to weigh public and private gains and losses. The basic aim has been to develop criteria for overall economic efficiency which consider relevant aspects of monetary and nonmonetary values. It should be noted that the analysis of the interaction of public and private values has long tradition in economics, being expressed in a comprehensive manner by S. V. Ciriacy-Wantrup in 1952,[2] and arising out of earlier work in the United States and Europe. It should not be presumed, however, that economics has succeeded in this endeavor to create a ledger for the debits and credits of wilderness, health, income, and assets. I mean only to say that economics has addressed the problem.

All this is introductory to the problem of defining *energy conservation* in an economic framework. In one meaning energy conservation is any act that reduces energy use, or reduces it below what was otherwise anticipated. If energy prices rise and less energy is used, energy conservation is taking place. But does this mean that as an energy price falls and demand rises, waste is taking place? If energy conservation is a positive attribute of social behavior, what is its negative, its opposite? Do we value equally growth or decline in use of clean and dirty fuels, or renewable and finite resources?

An economist is unlikely to provide interesting answers. Our approach is to attempt to place these things in the calculus of public and private gain and loss, and to see if there may be clear conditions under which specific actions may promote or retard economic efficiency. (Again, recall that we mean efficiency in the context of both, and perhaps conflicting, public and private values.)

Conservation Economics

In economic terms, at least six types of energy conservation can be described. Energy conservation could mean shifting demand functions with unchanged coefficients, greater energy price response, lesser income response, reduced demand in reaction to price increases, or a reduction in the time required for demand to adjust to new circumstances. Finally, in a broad context, we could consider conservation to be any lower level of energy use than that which had been expected. These six concepts are shown in Figure 4-1.

Each part of Figure 4-1 depends upon the prior existence of a *demand function*. This concept is described more fully in the next section of this chapter. Temporarily, assume simply that a demand function shows the rela-

---

[2]S. V. Ciriacy-Wantrup, *Resource Conservation: Economics and Policies* (Berkeley: University of California Press, 1952).

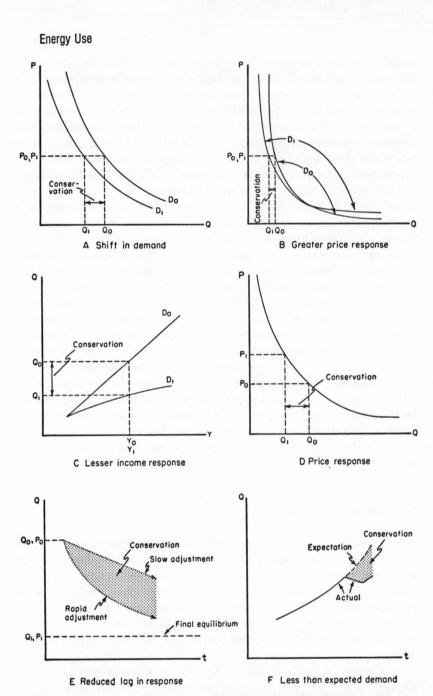

Energy Use

A Shift in demand

B Greater price response

C Lesser income response

D Price response

E Reduced lag in response

F Less than expected demand

Figure 4-1.   Types of energy conservation A–D show new energy Q_1, resulting from the various types of conservation. The amount of conservation is always the difference between Q_0 and Q_1. In E and F, conservation is the product of changing energy use over time and is equal to the shaded area.

tionship between the quantity of energy used, the price paid for it, and the income of buyers. This is similar to the simpler definition of demand curve used in Chapters 1 and 2.

Figure 4-1-A shows a shift in an energy demand function in which the parameters are unchanged. The curve simply shifts, and less energy is consumed at the same price level. This apparently occurred in 1974.[3] Figure 4-1-B shows demand function D_1 with a greater responsiveness to price than function D_0. In fact, there are indications that price responsiveness itself increases as prices rise.[4] Figure 4-1-C shows demand function D_1 with a lesser income response. In a manner analogous to the price response, the income response may decline as incomes rise. Figure 4-1-D simply shows a fixed demand function with less energy being purchased at the higher price. This latter phenomenon explains much of the geographic variation in energy use and part of the recent change in energy growth rates.

Figure 4-1-E shows differential change over time in response to different lengths of time necessary for adjustment in consumption. Starting in year 0 with price P_0 and energy use Q_0, the price increases to P_1. The new equilibrium level of use (other things being constant, which of course they are not) is Q_1. In general, because of the slow rate of replacement of machines, vehicles, and appliances, and the time necessary for consumer classes to determine their responses to changed economic conditions, the movement from Q_0 to Q_1 is not immediate but takes place over many years. If the time necessary for adjustment is short, energy use will be lower in succeeding years than it would with a longer adjustment period, although both paths eventually reach Q_1.[5] Public education programs may have a substantial effect upon this time lag in response, and this lag may itself decline as prices rise.

Figure 4-1-F simply shows conservation defined as departure from previous trends. In addition to the economic influences enumerated here, Figure 4-1-F may reflect temporary climatic or business cycle variation, or long-run so-

---

[3]Timothy D. Mount and Timothy Tyrrell, "Econometric Modelling," Appendix B in National Academy of Sciences (NAS), *Alternative Demand Futures to 2001*, Report of the Demand and Conservation Panel to the Committee on Nuclear and Alternative Energy Systems, 1979.

[4]Timothy D. Mount, Duane Chapman, and Timothy Tyrrell, "Electricity Demand in the United States: An Econometric Analysis" (Oak Ridge, Tenn.: Oak Ridge National Laboratory, June 1973), p. 10. Suppose $Q = \alpha P^\beta$. In Fig. 4-1-A, the shift in demand means α_1 is less than α_0, but the price response β is unchanged. Fig. 4-1-B has β_1 (the price response) with a greater magnitude than β_0 in the range of actual demand.

[5]In its simplest form, the geometric adjustment function is

$$Q_t = \lambda Q_{t-1} + \sum_{j=1}^{n} \beta_j X_{jt}$$

This is an equation in logarithms, with Q_t being energy demand in year t, λ the lag coefficient, X_j's the n economic influences, and β_j's the short run elasticities. A relationship where $\lambda = .9$ will take much longer to adjust than one where $\lambda = .6$.

[65]

cietal changes in birth rate, household formation, geographic patterns, and so on.

Energy Demand Functions

During the last decade, electricity has experienced greater growth in demand than have other energy forms. This suggests that electricity should be a major focus in understanding demand and conservation, and I will use it to illustrate the most significant characteristics of demand functions. The basic concept is the *elasticity*. This is the percentage change in demand associated with a 1 percent change in a particular economic factor.[6] For example, a −1.2 elasticity indicates that a 1 percent increase in electricity prices will cause a 1.2 percent reduction in electricity demand. Electricity demand is a function of income, population, natural gas and oil prices, and climate, so separate elasticities exist for each of these factors.

Table 4-1 summarizes the relationship of residential electricity demand to the more important economic variables. All the prices and incomes are the "real," inflation-adjusted values discussed in Chapters 1 and 2.

The absolute magnitude of the price elasticity is greater than any others in Table 4-1. Consequently, the future path of electricity prices will be the single most important economic factor in determining future levels of use. Figure 1-3 showed changes in energy prices over time, with their historical minima achieved in the early 1970s.

Figure 4-2 indicates recent developments since the end of the Growth Era in 1973. Note that in 1979 residential electricity price had been declining slightly for two years, and the deflated 1980 price equaled the 1974 price. This is certainly not the case for residential natural gas and heating oil, for these fuels have experienced major real cost increases. The lack of increase in real electricity price has been a major economic influence in maintaining upward pressure on electricity consumption. The increases in real heating oil and natural gas prices have, as the elasticities in Table 4-1 indicate, had a positive influence on electricity demand. Population and income levels also are major positive influences on demand. Population continues to grow but at a lesser rate than in the Growth Era. Deflated per capita income has shown little growth since 1973—about 1 percent annually to 1982. In fact, real weekly earnings for production workers have declined since 1973, and in 1982 had fallen to their 1961 level. Median family income has also been in decline.[7]

[6]The definition in the text is adequate for the discussion here, but it is not complete. An advanced student will recall that this elasticity estimate is in fact applicable only at a single point: the elasticity is actually the ratio of the proportional change in demand to the proportional change in the economic factor when the factor itself changes by a very small amount.

[7]U.S. Council of Economic Advisors, *1980 Annual Report* (January 1980), pp. 229, 245, and 232; U.S. Department of Commerce, Bureau of Economic Analysis, *Survey of Current Business,* August 1982; *Business Statistics,* 1973.

Table 4-1. Electricity demand by residential customers

Factor	Relationship to electricity demand growth	Illustrative elasticity	Likely pattern for the future
Electricity prices	Demand responds negatively to price increases, and demand changes more than price	−1.2	Stable or declining for several years, then increasing
Population	Demand is proportional to population	+1.0	Declining growth in population
Per capita income	Demand responds positively to income growth, but the change is less than for income	+0.6	Stable for last several years, future is uncertain
Natural gas and oil prices	Electricity demand is increased when natural gas and oil prices increase	+0.3	Natural gas will move toward the Btu price of heating oil. Heating oil prices will rise in periods of economic growth and stabilize in recessions
Renewable resources: solar, wood, etc.	Electricity demand is reduced when use of these resources rises	?	Solar water heating is competitive with electricity in the Sunbelt now, and wood heating is competitive with electric heat in the rural North
Insulation	Demand falls as insulation increases	?	Higher levels of insulation are competitive now, and its use will continue to increase

Source: The author's generalization of such empirical estimates as those made by Mount and Tyrrell, in "Econometric Modelling."

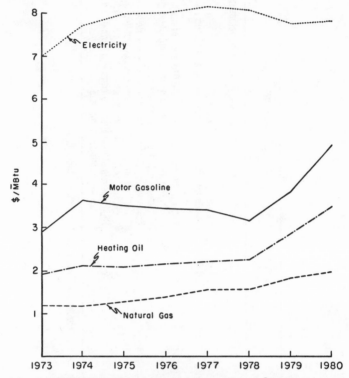

Figure 4-2. Real, deflated residential energy prices since 1973, $/MBtu (1972 dollars).
Source: U.S. Department of Energy, Energy Information Administration, *Monthly Energy Review.*
1973 heating oil price has been estimated according to the average ratio of heating oil price and
motor gasoline price for the 7 years (1974–80) for which data was available

The divergent paths followed by real per capita income, real median family
income, and real weekly wages for production workers present a paradox.
One explanation is that family size is declining. Another possibility is that
income equality in the United States has lessened since 1973. We have,
however, no basis to suppose that variations in income equality or family size
will influence electricity or energy demand.

Renewable energy use and serious efforts at insulation and conservation are
in their infancy. Table 4-1 cannot provide elasticity estimates for these factors. However, Tables 3-7 and 3-8 indicate that woodfuel is less costly than
electric resistance heating in the rural North, and solar water heating is (in
combination with natural gas) less costly than electric water heating.[8] These
factors will impose a negative influence on electricity demand.

[8]Actually, Chapter 3 discusses space heating only. The analogous figures from Cole's work
show $293 annual equivalent cost for gas water heating, $407 for solar/gas, $596 for solar/
electric, and $901 for electric water heating. As in Table 3-9, these are calculations for Long
Island, New York.

[68]

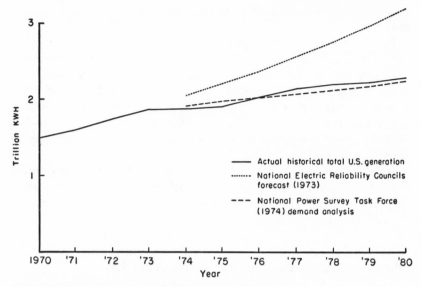

Figure 4-3. Total U.S. electricity generation, actual and forecasts.
Sources: Chapman et al., *Power Generation,* p. 17. and DoE, EIA, *Monthly Energy Review.*

Taking all these factors together, I expect electricity demand to continue to grow at a very slow rate. Figure 4-3 is one of my favorites. It shows two estimates of 1980 electricity demand prepared in 1974. The high curve shows the predictions made by electric utilities, which reached 3 trillion kWh by 1980. The lower curve is a specific planning prediction made by my colleague Timothy Mount and me: we recommended 2.2 trillion kWh as a basis for planning 1980 capacity requirements.[9] Actual generation is satisfactorily close to our recommendation. Generation in 1980 was 2.3 trillion kWh.

An examination of demand factors for other energy forms would be similar. I would expect the price, income, and population elasticities to be comparable for natural gas and heating oil.

Gasoline is distinct: the physical unavailability of transportation systems using other energy forms may make gasoline consumption somewhat less responsive to price and income changes. However, variations in energy use requirements for alternative transportation modes may make a significant difference in the future. This possibility is examined in a later section of this chapter.

The economic factors influencing industrial energy demand are basically the same. The most significant difference is that industry may utilize coal or oil in addition to the common residential fuels. In addition, industry is subject to substantial environmental regulation of air pollution emissions. Conse-

[9]Duane Chapman, Timothy Mount, John F. Finklea, et al., *Power Generation: Conservation, Health, and Fuel Supply,* U.S. Federal Power Commission National Power Survey, March 1975.

quently, it may be required to use natural gas or electricity even if coal is less costly.

In considering the effect of price and income on aggregate consumption, one must realize that the net price response is lower than the individual price responses. This is because an increase in the price of electricity will have a negative impact on the demand for electricity but a positive impact on the demand for natural gas and other energy forms. If we take −1.2 as representative of the price elasticities for each energy type, and +0.3 to +0.6 as the elasticity of one energy's prices with respect to another's demand, then the overall net price elasticity is probably between −0.5 and −1.0.

Before we go on to compare forecasts made for total energy consumption, it will be enlightening to examine U.S. energy requirements against those of a European country. This provides a physical basis for understanding differences in aggregate forecasts.

International Comparisons: Sweden and the United States

Several years ago, Lee Schipper and Allan Lichtenberg studied the patterns of energy use in the United States and Sweden.[10] Their work inspired considerable interest in the subject at Resources for the Future and the National Academy of Sciences.[11] One result of this concern was the finding that per capita GNP and per capita energy use have been closely related. A second point of interest had to do with top rankings: The United States is first neither in relative energy use, nor in per capita income. Canada has surpassed the United States in terms of energy consumption per capita, while Sweden has greater national income per capita and lesser energy use. But it is still the differences between Sweden and America that reveal the most significant points about the potential for energy conservation. Table 4-2 provides a picture of the comparative differences as well as some indication of changes over time.

The social data in Table 4-2 show important similarities and contrasts between the economies of the two countries. Literacy and life expectancy are equivalent, although the sex gap in life expectancy is larger here! GNP per capita was about $10,000 in both nations in 1978, although Sweden has more newspaper readership, more teachers per student, lower infant mortality, and

[10]Lee Schipper and Allan J. Lichtenberg, "Efficient Energy Use and Well-Being: The Swedish Example," *Science*, 3 December 1976, pp. 1001–1013.

[11]Sam H. Schurr, et al., Resources for the Future, *Energy in America's Future: The Choices before Us* (Baltimore: The Johns Hopkins University Press, 1979), pp. 101–104; Joel Darmstadter, Joy Dunkerley, and Jack Alterman, *How Industrial Societies Use Energy: A Comparative Analysis* (Baltimore: The Johns Hopkins University Press, 1977); National Academy of Sciences, *Energy in Transition 1985–2010*, Final Report of the Committee on Nuclear and Alternative Energy Systems (CONAES), 1980, pp. 107–110.

Table 4-2. Sweden and United States energy use and economic structure, 1970s

	U.S.	Sweden
Social data		
GNP per capita, 1971	$5,100	$4,400
GNP per capita, 1978	$9,590	$10,210
Population density per sq. mi., 1979	60	48
Calorie supply per capita, daily, 1977	3,600	3,200
Literacy	99%	99%
Teachers per 1,000 students	34	60
Population per physician	583	616
Infant mortality per 1,000 births	15	8
Life expectancy, men	69 years	72 years
Life expectancy, women	77 years	78 years
Population per hospital bed	152	71
Television sets per capita	.6	.4
Newspapers per capita	.3	.6
National budget, % of national income	20%	30%
Basic industry, 1971, kg/p.c.		
Steel production	620	680
Steel production, 1978	579	519
Cement	342	430
Paper	224	540
Aluminum	17	9
Industrial energy use, 1971, kWh/kg		
Steel	7	5
Cement	2	2
Paper	10	7
Aluminum	18	18
Residential heating, 1971		
Average degree-days	5,500	9,200
Heat energy, kWh/sq. m/dd	.05	.03
Rooms per person	1.5	1.5
Average area, sq. m	115	110
Transportation, 1971, miles per person		
Local travel (less than 30 miles)		
auto	4,850	1,825
bus	112	460
train	64	85
Intercity travel (more than 30 miles)		
auto	4,200	3,225
bus	122	25
train	21	356
air, including international	733	246
Total passenger travel inc. other	10,100	6,221
Rail travel, miles per person, 1977	48	419
Persons per car	2.3	3.4
Persons per car, 1976	2.0	2.9
Miles per gallon	14	24
Average car weight, tons	1.9	1.2
Energy prices, 1970		
Gasoline, retail	35¢/gal	61¢/gal
Gasoline tax in retail price	12¢/gal	42¢/gal
Gasoline, before tax	23¢/gal	19¢/gal

(*continued*)

[71]

Table 4-2. Continued

	U.S.	Sweden
Gasoline, retail, 1979	91¢/gal	$1.97/gal
Gasoline tax in retail price, 1979	14¢/gal	95¢/gal
Gasoline, before tax, 1979	77¢/gal	$1.02/gal
Electricity	2.8¢/kWh	2.1¢/kWh
Distillate fuel oil, 1979	74¢/gal	78¢/gal
Residential natural gas	87¢/M̃Btu	$5.50/M̃Btu
Coal	$13/ton	$18/ton
Total Energy, 1971		
Net consumption per capita	342 M̃Btu	204 M̃Btu
Net consumption per capita, 1978	358 M̃Btu	187 M̃Btu
Proportion as electricity	8%	12%
Proportion in		
transportation	24%	13%
commercial	14%	18%
residential	18%	24%
industry	36%	49%
feed stocks	6%	4%
trade	2%	−8%

Sources: World Bank, *World Development Report 1980* (Washington, August 1980); U.S. DoE, EIA, *Monthly Energy Review* and *International Petroleum Annual; The World Almanac and Book of Facts 1980* (New York: Newspaper Enterprise Association, 1979); *National Petroleum News,* Fact Book Issue, 1980; Schipper and Lichtenberg.

more hospital beds than does America. The United States leads Sweden in television sets per person (as well as other appliances), per capita caloric intake, and population density.

In terms of basic industry, Sweden exceeds the United States in per capita cement and paper production and lags behind in aluminum production. In 1971, Sweden produced more steel per capita than did America, but this was reversed in 1978.

Essentially, we have a picture of two modern industrial countries which have, on the whole, rather parallel lines of economic organization. But, as the last part of the table indicates, energy consumption per capita is much lower in Sweden than in the United States. In fact, energy consumption declined in Sweden in the 1970s, while it rose slightly in the United States. Since Swedish per capita GNP exceeded American GNP in this period, it is clear that lower living standards have not caused Sweden's lower energy use.

The middle sections of Table 4-2 summarize the major sources of differences in energy use between the two countries.

For travel, Swedes rely more on buses and trains and less on automobiles and airplanes than do Americans. In total, they travel less, even though population density is lower so that Swedes presumably need to travel greater distances than Americans to visit friends and family. Swedes own fewer cars per person, drive lighter automobiles, and travel farther on a given volume of gasoline than do Americans. Although the data on frieght transportation energy are not shown in Table 4-2, they are analogous to the passenger data.[12] The implication for the United States is that travel and transportation habits need to change; the next section of this chapter offers a partial solution to this problem.

Single family dwellings in each country are comparable in size, but those in Sweden require less heating energy. Swedish superiority in this area is not due to the country's milder climate! According to Schipper and Lichtenberg, Swedish insulation requirements were about double the U.S. requirements at the time of their study.

The Schipper-Lichtenberg analyses of industry reported that the two countries' energy use per unit output was equivalent in some sectors, such as aluminum and cement, but that in other industries, such as steel and paper, Sweden required less energy per unit of output than did the United States.

Three aspects of energy consumption differ qualitatively between the two countries. Sweden uses more electricity in proportion to fossil fuels than does the United States. Furthermore, Sweden makes major use of two technologies that are rare in most of America. The first is cogeneration, the combination of electricity generation and heat supply from a single plant. At a typical thermal generation plant in Sweden, steam or hot water heat for buildings will use 24 percent of the input energy, electricity generation will use 29 percent, and the remaining 47 percent is left as waste heat. In America, on the other hand, a typical plant will transform its fuel energy into 33 percent electricity and 67 percent waste heat.

The second is district heating, a system in which a central plant produces steam for several nearby buildings. While this occurs in some locations in the United States, such as Cornell University in New York State, the incidence of its use is not significant. Sweden, however, uses district heating to generate 19 percent of its residential heating.

In the previous section of this chapter, the concept of price elasticity was examined in the context of electricity and energy demand. Table 4-2's section on energy prices indicates important differences in that area between the two countries. Of the various energy prices reported, only electricity was less costly in Sweden than in the United States. Also, in Sweden, electricity comprised a greater portion of total energy consumption.

It is particularly interesting to observe that the pre-tax gasoline price of 19¢ per gallon in 1970 in Sweden was less than that in the United States. But gasoline taxes were so high that Sweden's retail price in 1970 (61¢ per gallon) was considerably higher than that in the United States. In 1979, Sweden collected a 95¢-per-gallon tax, and the retail price was more than twice the American retail price. Retail prices for natural gas and coal are also much

[12]Schipper and Lichtenberg, Table 5.

[73]

higher in Sweden. The differences in retail prices between these two countries were, in my judgment, the major economic influence responsible for the greater demand for and consumption of energy in the United States.

This discussion would not be complete without our noting that energy supply plays an important role in energy consumption in Sweden and other European countries. Western Europe—even with its North Sea oil—produces only one-sixth of the oil it consumes. Natural gas is similarly limited, as is coal. Given these realities of economic geography, Western Europe's energy consumption differs from the United States's in several important ways. First, nuclear power has been more significant there. Second, the major European oil companies have been almost entirely organized around the enterprise of owning and buying oil in other countries, a fact that will be an important part of the picture of the world petroleum market developed in the next chapter. Third, energy conservation has always been an integral part of European policy. Given our greater energy resources, it is surprising to observe that Europe has adapted to OPEC oil pricing and control at least as well as America has.

Transportation Energy

Suppose we consider work done by Swedish and U.S. cars when each used a gallon of gasoline. The average American car, according to the table, weighed 1.9 tons and moved 14 miles on a single gallon. This is 27 ton-miles per gallon. The Swedish car was not superior in this comparison, because its work was a comparable 29 ton-miles per gallon. The difference in energy

Table 4-3. Energy requirements for transportation

	Historical: 1966–1970	Potential: 2000
Passenger travel: passenger-miles per gallon		
Intercity		
bus	83	125
train	65	125
auto	35	60
plane	18	30
Urban		
bus, diesel	40	50
bus, electric	n.a.	100
rail	50–60	70
auto	18	27
Freight transportation, intercity, ton-miles per gallon		
rail	200	240
truck	50	75
plane	10	12

Source: Richard A. Rice, "Toward More Transportation with Less Energy," Technology Review, 84:2 (November-December 1981), 66–78.

Table 4-4. Transportation patterns, intercity travel

	1960	1973	1979
Air	4.4%	10.6%	14.5%
Auto	90.4	86.7	83.1
Total air and auto	94.8%	97.3%	97.6%
Bus	2.5	2.0	1.7
Train	2.8	0.7	0.7
Total bus and train	5.3%	2.7%	2.4%
Total passenger miles, trillions	0.781	1.345	1.551

Source: Motor Vehicle Manufacturers Association, *Facts and Figures '80* (Detroit: MVMA, 1980).

required for personal transportation in the two countries arises from the greater use of automobiles and the greater average weight of cars in the United States.

A major problem for the United States is our dependence upon automobile and air travel. Roughly one-sixth of total world oil production is used by American vehicles and planes. Richard Rice reviewed both historical and potential fuel requirements for various kinds of transportation. Passenger travel is summarized in Table 4-3, as is freight transport. Bus and train travel used half the fuel per passenger-mile that car travel required, in urban transportation as well as intercity travel. Air travel has required twice as much fuel per passenger mile as auto travel. The rankings are similar for freight transportation. When Rice turned his attention to the future, he visualized technological improvements in every form of transportation. Again, he found rail superior to autos, trucks, and planes. Buses and trains are comparable in terms of future energy requirement per ton-mile.

Given this picture of fuel usage, the pattern of transportation in 1973 and 1979 is rather surprising. The modes requiring less energy actually decreased slightly in relative popularity (see Table 4-4).

While my opinion is clear that a future shift to rail and bus transportation is necessary, most authors think otherwise. Our Demand and Conservation Panel in the National Academy of Sciences Committee on Nuclear and Alternative Energy Systems (CONAES) concluded that the potential for mass transportation is limited. The final CONAES Academy report shared this opinion.[13] So did Stobaugh and Yergin in the first edition of *Energy Future.* They concluded then: "Pooling and mass transit do deserve considerable commitment, but both lack the convenience desired by the American public. This is a fact of American life."[14]

[13]The CONAES report is cited in footnote 11. Our panel report is National Academy of Sciences, *Alternative Demand Futures to 2001,* Report of the Demand and Conservation Panel to CONAES (Washington, 1979).

[14]Robert Stobaugh and Daniel Yergin, eds., *Energy Future: Report of the Energy Project at the Harvard Business School* (New York: Random House; 1979), p. 148.

Stobaugh and Yergin emphasized fuel economy in automobiles, recommending, for example, better air pressure in tires and better engine tuning.

Overall, the transportation problem is one of the few major subjects on which I am in clear disagreement with informed technical opinion. I believe that Stobaugh and Yergin and CONAES are incorrect, and that basic changes will have to be made in transportation. I recognize, of course, that widespread automobile and airplane travel are major components of America's high living standard. If there is a shift to bus and rail travel, mobility will be reduced. My concern is that the absence of a realistic assessment of the transportation problem may lead to a situation in the 1990s in which we still lack effective bus and rail services, and many of us will be unable to afford auto and air travel. This basic problem is revisited in Chapters 14 and 15 on public policy.

Forecasting Future Energy Demand

It is of considerable importance to identify the boundaries that will define future energy demand, and to make this kind of forecast with confidence.

Figure 4-4 shows that simple extrapolation was formerly a reliable guide. In fact, for most of the Growth Era, the rates of growth themselves were increasing. Energy consumption was not merely increasing; it was accelerating. The broken line in the middle of Figure 4-4 indicates what present levels of energy consumption would be if the 4.2 percent growth rate of 1967–73 had continued. If this had happened, our present consumption would be about 105 Q. Instead, it is about 75–80 Q. Figure 4-4 illustrates the "less-than-expected-demand" kind of conservation discussed above.

The clear majority of industry and government analysts continue to expect significant but lower rates of growth. The solid bar defines the range of average forecasts given by Exxon, the U.S. Department of Energy, and Data Resources, Incorporated, an important consultant to government and industry.[15]

Our Demand and Conservation Panel with the National Academy of Sciences-CONAES study adopted a different philosophy. The panel identified several future price patterns and conservation policies, and worked to determine the future levels of energy consumption that would be consistent with these policies.

In Figure 4-4, one NAS Demand Panel value is 73 Q in the year 2000, about equal to the 1973 level. In fact, this 73 Q figure is lower than present energy consumption. This level of energy demand was defined as Scenario A.

---

[15]U.S. Department of Energy, Energy Information Administration, *Annual Report to Congress*, 1979, Vol. 3, p. 160.

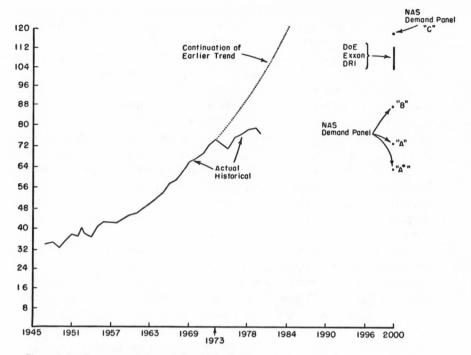

Figure 4-4. Future energy demand (quadrillion Btu)

It assumed very large increases in real energy prices: for example, that electricity prices more than tripled in real terms, oil prices increased fourfold, and natural gas prices increased to 11.5 times their 1975 levels—again, in real dollars. These high energy costs are associated with major improvements in energy efficiency. Automobile mileage rises from 13 mpg in 1973 to 37 mpg, heating requirements per square foot decline by 37 percent, and industry improvement in energy per unit of production ranges from 27 percent for iron and steel to 42 percent for construction. Similar improvements are hypothesized for appliance usage, freight transportation, and other sectors.

We also assumed that real GNP would double, and that the U.S. population would rise to 279 million by the year 2010. As a result of increased energy efficiency, this scenario indicates that higher population and income can coexist with a stable level of energy consumption.

Scenario B assumes the same economic and population growth. However, the increase in energy prices is much smaller: coal, electricity, and petroleum prices double in real terms, and natural gas price increases sixfold. The 2000 consumption level is 87 Q.

The panel also examined a C scenario of 117 Q near the DoE-Exxon-DRI range, and assumed this higher level of demand was possible only if energy prices did not increase. Another scenario, D, has energy demand rising to 134 Q in 2000, but this requires energy prices to decline in real terms. Such a development could be possible given a presently unforeseen breakthrough in synthetic fuel technology, or considerable subsidies for synthetic fuel use. Scenario D does not appear in Figure 4-4 because it is off the graph.

Finally, the panel speculated about the effect of aggressive conservation policies interacting with high energy prices. The result here was a speculative reduction in energy use to 80 percent of present consumption while economic growth continues. This lower level of use (approximately 62 Q in Scenario A*) may be associated with significant changes in economic structure. Such changes might include increased use of solar and renewable energy, extensive building codes and automobile mileage requirements, efficiency standards for appliances, a reorganization of the geography of work and living patterns, decreased automobile travel, and greater use of labor-intensive goods and services.

At this point, an independent observer has no clear logical basis for selecting one of these forecasts as most probable. However, several relevant conclusions are possible. First, it is very improbable that the smooth, predictable growth of the pre-1973 period will be repeated. Second, real energy prices will increase, and, in the near future, the rates of increase will be highest for petroleum and natural gas, and lower for coal and electricity. Third, public policy with respect to conservation will interact with energy price levels. In other words, future use will not simply be a reaction to economic variables. Future use will depend upon conscious policy decisions. Finally, the trajectory of aggregate energy consumption since the Growth Era is within the range defined by the Demand Panel's A and B scenarios. Neither the Exxon-DoE-DRI forecast nor the panel's maximum conservation scenario (A*) is supported by the past seven years' experience.

I think that a continuation of present trends will lead to a future level of use in the range of 70–90 Q. A further rise to the Exxon-DoE-DRI 100 Q-plus level probably requires major increases in subsidies to energy corporations. A decline to the Demand Panel's A* level of 62 Q would probably require a considerable increase in energy prices and more aggressiveness in implementing conservation policies.

Conclusion

Chapter 3 introduced the physical dimensions of energy use, substitution, and efficiency. This chapter defines the economics of demand and conservation. Much of what is seen as conservation is, in fact, change in demand that is caused by economic factors, particularly price changes. Elasticities are quantitative measures of these responses and must be differentiated according to their short-run or long-run impact.

In the early 1970s, many observers continued to predict an extension of the era of accelerated growth in demand. The use of demand functions, however, relates empirical elasticity estimates to expected economic factors, and this approach to demand gives more accurate results.

A detailed examination of differences in energy use in Sweden and the United States indicates several specific areas where American energy use may decline in the future. Higher Swedish prices for gasoline, oil, electricity, and coal contributed to decreased energy requirements. Although Sweden is more dependent on imported energy than the United States, it appears that Sweden's per capita income has risen above the U.S. level.

Transportation and petroleum consumption continue to comprise a major problem. One-sixth of the total world oil supply is being used by American vehicles and airplanes. In Europe, the absence of significant fossil fuel resources has led to a greater emphasis on nuclear power as well as on conservation.

I believe that energy requirements dictate a shift to rail and bus transportation, but most observers do not agree.

Many analysts expect U.S. energy consumption to reach the 100 Q level by the end of the century. In particular, recent forecasts by Exxon and the Department of Energy are at this level. I think this unlikely unless subsidies to energy corporations are increased above present levels. A decline in energy use, as examined in the National Academy of Sciences CONAES Panel analysis, would be possible only if conservation were given much greater emphasis.

In the long run, America may be required to turn to renewable resources, conservation, public transportation, and coal. The economic dimensions of these subjects will be examined in later chapters. Nevertheless, it is clear that world petroleum economics will continue to dominate American and global energy production and consumption in the near future. This is the subject of the next chapter.

III

PETROLEUM, NATURAL GAS, AND OIL COMPANIES

5

The World Petroleum Market

I owe my throne to God, my people, my army—and to you!
—Shah Mohammed Pahlavi to Kermit Roosevelt, U.S. CIA

Iran owes me—us, the Americans and the British who sent me—absolutely
nothing . . . there is no debt, no obligation.
—Kermit Roosevelt to Shah Pahlavi and General Fazlollah Zahedi

Earlier chapters have shown that the Growth Era from 1859 to 1973 was characterized by exponential growth in energy consumption and declining real energy prices. This is as true for petroleum in particular as it is for energy in general.

European governments and corporations have always been concerned about foreign oil supply, because they lack America's original endowment. For the United States, foreign oil did not become important until the 1950s. For 89 years—from 1859 to 1948—the United States was a net oil exporter. But as it became apparent that low-cost oil would be produced outside the United States, American oil companies became interested in major production efforts in foreign countries. For several years, the industry obtained imported crude oil at costs equal to or less than U.S. costs. As a consequence, petroleum product prices continued to decline in the 1950s, 1960s, and early 1970s.

However, this access to low-cost foreign crude was dependent upon the willing participation or acquiescence of exporting countries in the pricing and production plans of the major international corporations.

Although Mexico had challenged corporate ownership by its 1938 nationalization, an effective international boycott eliminated Mexico from a major role in world oil markets until the 1970s. Iran became the second exporting country to nationalize its oil resources, doing so in 1951. The histories of the United States and Iran have been linked since. Examining the economic background of this Iran-American relationship is instructive. America, it becomes evident, has been only a recent participant in Iranian affairs.

The leading industrial regions of the world have all played strong roles in Iranian history. Perhaps least known in the United States is Russia's past

[83]

participation in Iranian political processes. The common border between the two nations has provided Russia with a geographic incentive for involvement that has spanned its czarist and Communist governments. In the last century, Russia consolidated its occupation of the Caspian Sea region in the Caucasus and of Georgia, Tiflis, and Baku. In this century, the 1907 Anglo-Russian Convention divided Iran into two spheres of influence; Great Britain occupied the south while Russia occupied the northern provinces on its borders.

The discovery of oil in western Iran in the British sphere led to the creation of the Anglo-Persian Oil Company. The British government, in turn, became directly involved, seeking to secure a stable source of foreign oil for its navy and its domestic economy. The Anglo-Persian Oil Company was the predecessor of the British Petroleum Company; the British government owns, directly, 39 percent of the company.

The Bolshevik Revolution in Russia led to a brief Soviet Republic of Gilan in northern Iran in 1920. This government collapsed with the withdrawal of Russian troops. In 1941, however, Russia returned to Iran's northern provinces in a military occupation that coincided with a similar action by Britain in the south. After the end of World War II, Russia again increased its troop strength in the north. An "Autonomous Republic of Azerbaijan" was created, and its prime minister was Jafar Pishevari, a Communist.

Iran's prime minister, Ahmed Qavam, proposed a joint Russian-Iranian oil company. It appears that Russia overestimated the support for its policies in Iran. The USSR withdrew its armed forces in the spring of 1946, and the Azerbaijan Autonomous Republic ceased to exist by the end of the year. The proposed Russian-Iranian oil company was never formed.

The role played by Great Britain in Iranian history is comparable, but larger. The concession Britain obtained in the southwest gave the British the powers of government: customs, airports, roads, communications, police. These functions were in addition to the control of Iranian oil.

British control of Iranian oil became an important element in Iranian politics. In the early 1950s, the desire for nationalization became a political force. On March 3, 1951, then–Prime Minister Ali Razmarah told the Majlis (Iran's parliament) that nationalization was impractical and illegal. He was assassinated on March 7. The Majlis confirmed nationalization on March 15, and Mohammed Mossadegh became prime minister in April.

Nationalization was followed by international boycott. All the European and American oil companies refused to purchase Iranian oil from its new owners. British Royal Air Force planes forced a Panamanian ship to surrender its Iranian oil in Aden. The boycott policy was favored by both Labour and Conservative governments in Britain, and Democratic and Republican presidents in the United States. Russia did not attempt to end the boycott.

In 1953, the United States and Great Britain undertook a joint effort to eliminate Mossadegh. This effort was coordinated by a U.S. Central Intelligence Agency group headed by Kermit Roosevelt.[1] The new prime minister, Fazlollah Zahedi, hid in the basement of an American CIA agent, Fred Zimmerman, during the change in government. Notwithstanding Kermit Roosevelt's statement quoted above, the new government and the Shah gave American companies a major role in Iranian oil production, as shown in Table 5-3 in a following section. American CIA interest in Iran was shown more explicitly in 1973, when Richard Helms resigned the directorship of the CIA to become America's ambassador to Iran until 1977. Kermit Roosevelt later became director of the Gulf Oil Corporation's Washington office for government relations, a post he held from 1958 to 1964. Roosevelt was a Gulf vice-president from 1960 to 1964.

The United State's involvement in Iran from 1953 to 1979 does not appear significantly different from the earlier activities of Britain or Russia. It would seem illogical to suppose that American private corporations are uniquely responsible for this kind of action. The historical record shows similar activities by a national oil company (British Petroleum), a socialist government (the Labour party in Britain), and a Communist government (Russia), and support for American oil companies by Republican Presidents Eisenhower, Nixon, and Ford, and Democratic Presidents Truman, Kennedy, Johnson, and Carter.

In retrospect it can be seen that Iran's attempt at retribution, the seizure of American hostages, gained it nothing. Instead, Iranian actions encouraged both the Russian occupation of Afghanistan and the Iraqi invasion of Iran. Similarly, it resulted in greater American military presence in the Persian Gulf region.

Iranian history is a dramatic representation of a common phenomenon. Russia, Britain, and America have each been active participants in a competition of global interest.

A similar situation obtained in Angola during that country's recent civil war. The American government and the Central Intelligence Agency supported the UNITA and FNLA groups publicly. John Stockwell, the agent in charge of operations there, reported a total expenditure of $32 million. At the same time, Gulf Oil maintained good relations with the Marxist MPLA—also supported by Cuba and Russia—in order to continue its concession in Cabinda. In March 1976, Gulf paid $125 million to the MPLA, and continues to hold a major share in Angolan oil. Communist China, on the other hand, apparently supported the UNITA and FNLA groups against the Marxist MPLA. (The abbreviations mean National Union for the Total Independence

[1]Kermit Roosevelt, *Countercoup: The Struggle for Control of Iran* (New York: McGraw-Hill, 1979); quotations at the beginning of the chapter are on pp. 199 and 201. Much of the discussion of political economics comes from this source, and from Yahya Armajani, *Iran* (Englewood Cliffs, N.J.: Prentice-Hall, 1972) and Joseph M. Upton, *The History of Modern Iran: An Interpretation* (Cambridge: Harvard University Press, 1961).

[85]

of Angola [UNITA], National Front for the Liberation of Angola [FNLA], and Popular Movement for the Liberation of Angola [MPLA].) South Africa, whose military participation was more direct, also supported UNITA and FNLA.[2] One line-up thus featured support from Communist China, South Africa, and the U.S. CIA, while the other team was backed by Cuba, Russia, and Gulf Oil.

The difficulty of understanding the significance of such influences is indicated by this analogy made more than 20 years ago by Joseph Upton:

> One would have a clear conception of the situation in Persia at various times since 1900, if one could imagine say Massachusetts of a hundred years ago as an independent nation with the following features: Belgians administer the customs service on the eastern seaboard and along the northern frontier; Swedish officers command the state police as well as the municipal police of Boston; Russian officers staff and command the National Guard, supplied with Russian equipment at public expense; Hungarians with broad executive powers administer the Treasury and collect taxes with a Hungarian chief in each county seat; the Dutch own and operate the only telegraph line in the country with Indian employees; French professors lecture in French at Harvard; the Dutch own and manage the only bank (the New England Trust Co.), staffed largely with Puerto Ricans, with a branch in each county seat; a Dutch company owns and manages with its own personnel the only large industrial operation (textiles), employing a large number of Indians; and the Dutch and Russian governments maintain large legations in Boston with consulates in the larger cities, such as Worcester, Springfield, etc., all guarded by substantial detachments of uniformed Dutch or Russian troops who escort legation or consular officials whenever they leave their compounds. These are only some of the types of foreign activity present at some time in Iran, by no means all of them. While some citizens of Massachusetts as, for example, those with enough initiative to learn Dutch, Hungarian, etc., and so to obtain employment, or those who profited because of influence or knowledge useful to the foreigners, might not find the situation unpalatable, by and large the people of Massachusetts would certainly resent the state of affairs and try to change it, despite the likelihood that some of the activities might actually be highly beneficial to them.[3]

One modest benefit of the growing importance of autonomous control by producing governments is the lesser need to follow such bizarre groupings. Whether such matters are truly relegated to history remains to be seen.

The object of this dimension of global politics is, of course, oil.

World Oil Production and Reserves

Table 5-1 shows that Saudi Arabia, Kuwait, and Iran together hold nearly one-half of the world's *proven reserves* of 671 billion barrels. The Middle

[2]The information here is from the book by the agent in charge, John Stockwell, *In Search of Enemies* (New York: Norton, 1978).
[3]Upton, p. 32.

Table 5-1. Major countries and regions with proven crude oil reserves estimated as of Jan. 1, 1982, billion barrels (Total world estimate: 670.7)

Country or Region	Amount	Percent of world total	Cumulative percent
Saudi Arabia	164.6	25%	25%
Kuwait	64.5	10%	35%
Iran	57.0	8%	43%
Mexico	57.0	8%	51%
Abu Dhabi	30.6	5%	56%
U.S.	29.8	4%	60%
Iraq	29.7	4%	64%
Libya	22.6	3%	67%
USSR	63.0	9%	76%
China	19.9	3%	80%*
Asia-Pacific	19.2	3%	
Western Europe	24.6	4%	
Middle East/Arabic	398.3	59%	
Africa	20.7	3%	
U.S./Canada	37.1	6%	
Latin America	85.0	13%	
Communist countries	85.8	13%	

Source: Oil and Gas Journal, December 28, 1981.
*Rounding error causes differences in sums.

East and Arabic countries together hold 59 percent of proven world reserves. No other region has as much as 15 percent. It is surprising to learn that the regional distribution indicated in Table 5-1 has changed little in 30 years. The only significant changes from 1950 show a Middle East/Arabic increase, and a North American decline in holdings from 30 percent of proven world reserves to 6 percent now.[4]

World consumption reached a maximum in 1979 at a level of 23 billion barrels per year, and has declined slowly since then. Proven reserves of 671 billion barrels are equivalent to 30 years of consumption at current global rates. This statistic is less informative than it appears to be, because of the relationship of proven reserves and potential reserves. Before we look at these different concepts of oil reserves, however, Figure 5-1 should be examined. First, it shows a consumption decline from 1979 to 1981, the first such occurrence since the 1974–75 decline. It may be that the world has just passed—or is about to pass—the year of historical maximum global oil production. This possibility will be investigated in the following section on King Hubbert's theory of production cycles. A second aspect of Figure 5-1 is the proportion of global production consumed in the United States. In 1925, America used 78 percent of world production—all of which was produced in

[4]The 1950 statistic is from Norman A. White, *Financing the International Petroleum Industry* (London: Graham & Trotman, 1978), p. 14.

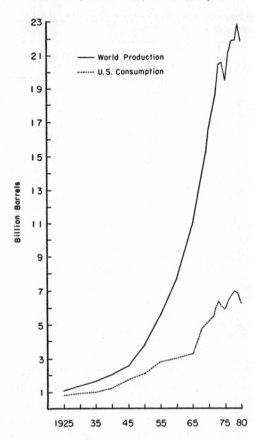

Figure 5-1. Crude oil, world production and U.S. consumption. *Sources:* Calculated from data in American Petroleum Institute, *Basic Petroleum Data Book* and *Petroleum Facts and Figures* (1971), *International Petroleum Annual*, *Oil and Gas Journal*, *Business Statistics, Survey of Current Business.*

the United States (see Figure 5-4 in the next section). In 1973, the United States used 31 percent of world production, and in 1980 22 percent.

It is probable that at least as much oil is remaining to be discovered as is now known to exist in proven reserves. Proven reserves are quite distinct from *potential resources,* and both definitions require careful study:

Proven reserves are amounts which are known to be recoverable under existing economic conditions and technology. They are located in specific known reservoirs that have been tested and have demonstrated the ability to produce.

Potential resources are estimates not specific to individual areas, and based upon general geological knowledge about the possible distribution of the resource.

[88]

Table 5-2. Possible potential reserves and original oil in place, Jan. 1, 1982, billion barrels

	United States	World
Cumulative production through 1981	130	470
Proven reserves, Jan. 1, 1982	30	671
Additional potential reserves	83	938
Possible total original recoverable oil	243	2,079

Sources: See footnote 5.

Proven reserves are essentially an inventory. The companies or governments which hold the reserves know that this oil can be produced, and refining, transportation, and marketing can be planned accordingly. They can be described as "on the shelf," in the sense that they are available for use. Potential resource estimates are quite a different matter. They are generalizations and extrapolations from past experience to future possibilities. Professional geologists actually divide this concept of potential resources into several subcategories such as probable, possible, speculative, and undiscovered reserves. Here, the two categories of "proven" and "potential" will be sufficient.

Table 5-2 shows current estimates of U.S. and world potential oil reserves. In 1981 U.S. production of crude oil was 3.1 billion barrels, and the consumption of petroleum products was 5.8 billion barrels. Global production and consumption was 20.5 billion barrels. On this basis, if proven and potential reserves are combined, the world would have 78 years supply at its current rate of production and the United States would have 36 years. However, the potential estimates need additional qualification. First, their speculative nature means they are much less knowable. The U.S. statistic of 83 billion barrels of potential reserves awaiting discovery is actually a mean estimate. The Geological Survey considers a 1 in 20 probability that the value is below 64 billion barrels. But—they also consider a 1 in 20 probability that the value is above 105 billion barrels.

A second qualification is the nature of oil recovery. The *recovery rate* defines the proportion of the original oil in place which will be extracted. In the past, it has averaged 32 percent in the United States.[5] This means that America's cumulative production of 130 billion barrels (Table 5-2) was extracted from an original volume of oil on the order of 400 billion barrels.

[5]U.S. General Accounting Office, *Analysis of Current Trends in U.S. Petroleum and Natural Gas Production,* December 7, 1979, p. 24. Sources for Table 5-2: *Oil and Gas Journal,* December 28, 1981; mean value of potential U.S. oil resources from U.S. Department of the Interior, Geological Survey, *Estimates of Undiscovered Recoverable Resources of Conventionally Producible Oil and Gas in the United States, A Summary,* Open File Report 81-192, 1981; current world data for 1981 estimated from earlier National Academy of Sciences, *Mineral Resources and the Environment,* prepared by the Committee on Mineral Resources and the Environment (COMRATE), Commission on Natural Resources, National Research Council (1975).

However, *enhanced recovery* increases the recovery rate. Primary recovery depends upon water or natural gas to force oil to flow through producing wells. Enhanced recovery means that the oil field operators use additional techniques, such as steam, chemical, or water injection to increase the oil flowing to wells. The degree to which enhanced recovery will increase ultimate recovery is unclear, but an aggregate average value of a 5 percent increase in the recovery rate might be a useful illustration.

If Table 5-2 is reasonably accurate, about one-fifth of human society's original crude oil endowment has been consumed. For the United States, the comparable statistic shows 53 percent of the original oil in place has been consumed. On a per capita basis, it is clear that America was generously endowed, with an original oil-in-place figure of 1,000 barrels for each person now living in the United States. The comparable figure for the rest of the world is 450 barrels per capita.

Hubbert's Theory

King Hubbert's analyses of production cycles are dramatic now, and were even more so when authored in the 1960s.[6] Hubbert reasoned that the history of production for a large region would be analogous to the pattern over time for a specific oil reservoir. His basic assumptions were that production begins and ends at zero, that the rate of increase in production is at first exponential and then the rate of increase declines. Production reaches its peak, after which it declines slowly, then declines at an exponential rate, and finally terminates after a period of limited production. Hubbert's geological knowledge led him to apply these logical characteristics to the actual historical data for national production.

Figure 5-2 is a direct reproduction of Hubbert's 1969 analysis. Note that, for the coterminous United States, he anticipated the peak year to be between 1965 and 1970, and the peak production to be 3 billion barrels. He foresaw an ultimate recovery of 190 billion barrels for the United States; this included 25 billion barrels in Alaska.

In fact, U.S. crude oil production peaked in 1970 at a rate of 3.5 billion barrels. The coterminous 48 states show the same pattern, the peak being 3.4 billion barrels. Figure 5-3 shows U.S. crude oil production since 1955. The

---

[6]M. King Hubbert, *Energy Resources,* A Report to the Committee on National Resources of the National Academy of Sciences and National Research Council (Washington, 1962); and "Energy Resources," in *Resources and Man,* Committee on Resources and Man, National Academy of Sciences–National Research Council (San Francisco: W. H. Freeman, 1969), pp. 157–242.

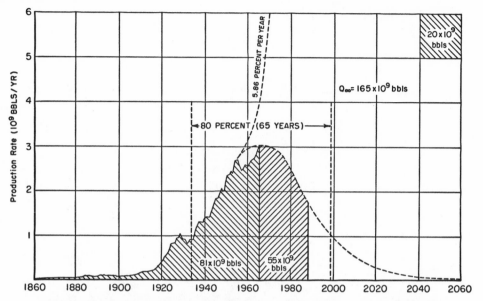

Figure 5-2. Hubbert's analysis: Complete cycle of crude oil production in the United States and adjacent Continental Shelves, exclusive of Alaska.
This figure is a reproduction of Figure 8.17 from M. King Hubbert, "Energy Resources," in National Academy of Sciences, Committee on Resources and Man, *Resources and Man* (San Francisco: W. H. Freeman, 1969), and appears with the permission of both the author and the National Academy Press, Washington, D.C.

pattern hypothesized by Hubbert in Figure 5-2 is readily identifiable in Figure 5-3.

This Hubbert pattern is applicable to the production profile for each of the major producing states. Texas, Louisiana, Oklahoma, and California all peaked between 1967 and 1972. Notwithstanding the growth in offshore oil production, Texas production in 1980 was 75 percent of its peak 1972 production of 1.3 billion barrels.

The decline in oil production in major producing states interacted with continuing acceleration in American oil consumption to create the major events that marked the end of the Growth Era. Figure 5-4 combines production and consumption data. It illustrates America's century as a net exporter of petroleum. In 1965, the United States still imported less than 20 percent of its total consumption, but imports were accelerating even more rapidly than consumption.

By the early 1970s conditions existed which made possible the events that revolutionized world petroleum markets.

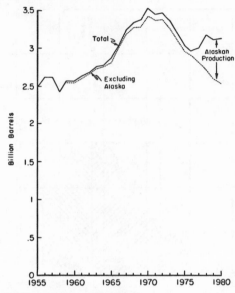

Figure 5-3. U.S. crude oil production, 1955–1980.
Sources: American Petroleum Institute, *Basic Petroleum Data Book;* U.S. DoE, *Petroleum Statement Annual* and *Monthly Energy Review.*

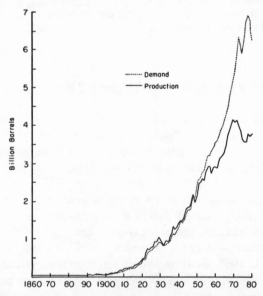

Figure 5-4. U.S. petroleum, total production and total domestic demand, 1860–1980.
Sources: American Petroleum Institute, *Petroleum Facts and Figures;* U.S. Department of Commerce, *Business Statistics* and *Survey of Current Business;* Sam H. Schurr and Bruce C. Netschert, *Energy in the American Economy, 1850–1975:* (Baltimore: Johns Hopkins University Press, 1960).

World Petroleum Marketing

Before 1973, American and European ownership of foreign oil was easily defined. The Arabian-American Oil Company described its concession in Saudi Arabia in these terms: "The Company is granted the exclusive right to prospect for, manufacture, transport, deal with, carry away and export oil and oil products, and the right to use all means and facilities it deems necessary or advisable to carry out the purposes of the enterprise."[7] This description, written in 1960, also indicated that Aramco expected to exercise such control until 1999. In Table 5-3, Aramco's parent partners are shown to be Exxon, Mobil, Texaco, and Standard of California.

Table 5-3, however, is dated. In 1950, the eight companies listed held 82 percent of crude oil production outside their own North American–European territories. Nineteen years later, the figure had fallen to 70 percent.[8] In the 1980s, producing governments and their state oil companies participate in at least two-thirds of world crude oil production.[9] It is clear that the control American and European oil companies once held over oil reserves and production in the Middle East, Africa, Asia, and South America is now part of history.

Other major companies also participated in these areas on a more limited basis. Amerada Hess, Atlantic-Richfield, Continental, Getty, Marathon, Occidental, Phillips, Standard of Indiana, and Standard of Ohio were among the lesser majors involved in these concessions.

If we examine current reserve data from Table 5-1 in the context of concessions from the earlier era, it can be seen that these eight companies—in just the 10 countries in Table 5-3—would hold 80 percent of 1982 proven oil reserves outside North America and Communist countries. The objective was clear: control of global oil for use in America and Europe. For the last years of the Growth Era, the effort was successful.

Table 5-3 also indicates the nature of working partnerships between major oil companies. Every company was a partner at least once with every other company. Certain patterns are evident. Standard of California and Texaco are formal partners in Caltex, and had identical percentages in four of the five shared concessions. CFP, the French company, nearly always shared concessions with British Petroleum, and usually with Shell as well. Mobil and Exxon were always involved with each other in their seven concessions.

---

[7]Arabian American Oil Company, *Aramco Handbook* (The Netherlands, 1960), p. 136.

[8]These statistics were calculated from data in M. A. Adelman, *The World Petroleum Market* (Baltimore: The Johns Hopkins University Press, 1973), pp. 80–81; and from American Petroleum Institute, *Petroleum Facts & Figures* (Washington, 1971) pp. 548–557. White (p. 17) had estimated that the largest seven (the Table 5-3 companies except CFP) controlled 90% in 1952.

[9]White, p. 18.

Table 5-3. World oil concessions before 1973

Country	Company									Total
	Exxon	Shell	BP	Mobil	Texaco	Socal	Gulf	CFP	Others	
Saudi Arabia	30%			10%	30%	30%				100%
Kuwait			50%				50%			100%
Iran	7	14%	40	7	7	7	7	6%	5%	100%
Iraq	12	24	24	12				24	5	101%
Abu Dhabi onshore	12	24	24	12				24	5	101%
Abu Dhabi offshore			67					33		100%
Libya	23	4	5	4	6	6			51	100%
Venezuela	42	26		3	5	2	11		11	100%
Nigeria		31	31				37			99%
Indonesia	3	16		3	33	33			12	100%
Algeria		9						27	64	100%

Sources: U.S. Senate, Committee on Foreign Relations, Multinational Corporations, 1973, Part 5, p. 289; Adelman, pp. 80, 81.

Table 5-4. North Sea leases

	Exxon	Mobil	Texaco	Socal	Gulf	Stand. Oil (Ind.)	Atlantic Richfield	Shell	Conoco	Tenneco	Sun	Occidental	Phillips	BP	Unocal	Amerada Hess	Marathon	Ashland	Cities Service	Getty	CFP	Others
Exxon	12	2	—	—	1	5	1	74	2	—	—	—	—	1	—	2	—	—	—	—	—	10
Mobil	2	20	14	24	2	4	2	—	—	—	—	—	1	3	—	17	—	—	—	1	—	34
Texaco	—	14	15	7	73	2	1	—	—	—	—	—	—	—	—	—	—	—	—	—	—	84
Socal	—	24	7	6	6	1	1	2	—	—	—	—	—	—	—	—	—	—	—	—	—	16
Gulf	1	2	73	6	27	2	—	—	—	—	—	—	3	9	—	—	—	—	—	1	—	45
Stand. Oil (Ind.)	5	4	2	1	2	—	—	—	—	—	—	—	—	—	—	56	—	—	—	—	—	146
Atlantic Richfield	1	2	1	1	—	—	14	2	1	—	—	—	2	1	2	1	—	2	1	—	—	26
Shell	74	—	—	2	—	—	2	39	2	—	—	—	—	—	—	2	—	—	—	1	—	18
Conoco	2	—	—	—	—	—	1	2	—	—	—	—	—	2	2	1	1	—	1	4	—	57
Tenneco	—	—	—	—	—	—	—	—	—	—	15	—	4	—	—	—	—	—	3	1	3	21
Sun	—	—	—	—	—	—	—	—	—	15	—	—	1	—	—	—	—	—	—	1	—	19
Occidental	—	—	—	—	—	—	—	—	—	—	—	—	—	—	—	—	6	3	—	—	—	9
Phillips	—	1	—	—	3	—	2	—	—	4	1	—	13	—	—	—	—	—	—	—	13	77
BP	1	3	—	—	9	—	1	—	2	—	—	—	—	39	—	—	—	—	—	—	—	29
Unocal	—	—	—	—	—	—	2	—	2	—	—	—	—	—	—	4	—	—	—	—	—	14
Amerada Hess	2	17	—	—	—	56	1	2	1	—	—	—	—	—	4	—	—	—	—	—	—	57
Marathon	—	—	—	—	—	—	—	—	1	—	—	6	—	—	—	—	—	2	—	2	—	8
Ashland	—	—	—	—	—	—	2	—	—	—	—	3	—	—	—	—	2	—	—	—	—	5
Cities Service	—	—	—	—	—	—	1	—	1	3	—	—	—	—	—	—	—	—	—	—	—	11
Getty	—	1	—	—	1	—	—	1	4	1	1	—	—	—	—	—	2	—	—	—	—	11
CFP	—	—	—	—	—	—	—	—	—	3	—	—	13	—	—	—	2	—	—	—	—	80
Others	10	34	84	16	45	146	26	18	57	21	19	9	77	29	14	57	8	5	11	11	80	257

Sources: For sectors belonging to the United Kingdom, Netherlands, W. Germany, and Denmark: Casenove and Co., *The North Sea*, November 1974. For the Norwegian sector: Norway Ministry of Petroleum and Energy, *Concerning the Activity on the Norwegian Continental Shelf*, Storting Report No. 53, 1979–80. Note: The numbers given represent *leases* for individual blocks held by the combinations listed. Since companies are often part of consortia involving several major and/or minor companies, individual blocks themselves may be counted more than once. The number given for Others-Others represents the total number of *blocks* held by one or more unlisted company(y)(ies); it does not include leases held in Major-Minor-Minor combinations. Table prepared by Lucrezia Herman.

The next chapter describes the economic motivations for partnerships in exploration and production: lower costs and lower prices to consumers. However, it is appropriate to a discussion of global activities to illustrate the partnership pattern in another area: the North Sea. In Table 5-4, North Sea leases are summarized by ownership. The 20 largest U.S. companies are shown, and so is CFP. U.S. subsidiaries are grouped with domestic or foreign parents: Sohio with BP, Skelly with Getty, etc. Every major company is a partner with at least one other major. Typically, a major oil company will be partners with five to ten other major companies in North Sea leases.

Given these data on international oil marketing, it is logical that the seven major companies and Standard of Indiana are responsible for more than half of U.S. crude oil imports.[10] Similarly, the countries that have been discussed in this chapter continue to be the major sources of crude oil imported in the United States.

World Costs and Prices

Actual production cost information in OPEC areas is not publicly available. Several years ago, however, M. A. Adelman examined historic and prospective costs, and arrived at the cost estimates in Table 5-5. While Venezuela and African regions are more costly than Persian Gulf areas, it is evident that prices now collected by OPEC countries for their oil are on the order of 30-times the oil's cost.

Price data for Saudi Arabian oil world trade are shown in Figure 5-5. This Saudi oil is termed *benchmark crude* because of its good quality and high production. In 1981 dollars, the average official price for Saudi Arabian oil was $3.56 per barrel in 1970, something below the cost of American crude oil. This rose slowly into early 1973 ($4.55 per barrel), and was increased abruptly to $17.58 per barrel by July 1974. It is not generally known, but real crude oil prices declined for the next four years to $15.34 in 1978. The following years have seen another major increase following the Iranian disturbances, and another decline in the early 1980s.

Comparing Figure 5-5 with Table 5-5 shows, simply, monopoly power at work. This is of course the purpose of OPEC, the Organization of Petroleum Exporting Countries. At its formation in 1960, however, OPEC did not hold the power it does today. The original members were Saudi Arabia, Iran, Iraq, Kuwait, and Venezuela. Membership has expanded to 13 nations with the addition of Algeria, Libya, Qatar, United Arab Emirates (which includes Abu Dhabi), Indonesia, Nigeria, Ecuador, and Gabon. It is not often recalled, but in 1960 oil consumption was expanding while prices declined. OPEC

[10]In January-September 1979, 54%. Data collected by Kathleen Segerson in 1980.

Table 5-5. Adelman estimates of average supply cost for 1970–85, $/barrel, 1981 dollars

Saudi Arabia	$0.44
Kuwait	0.44
Iran	0.44
Iraq	0.44
Libya	1.18
Venezuela	1.40
Nigeria	1.01
Algeria	1.25

Source: Adelman, p. 76, adjusted by GNP inflation index. See also Fadhil J. Al-Chalabi, *OPEC and the International Oil Industry: A Changing Structure* (Oxford: Oxford University Press, 1980), p. 112; and Leonard Mosley, *Power Play: Oil in the Middle East* (New York: Random House, 1973), p. 391.

intended to reverse half of this: it intended to halt price decline, and raise prices if possible. For 13 years OPEC made slow progress. The 1967 Arab-Israeli war had no long-term effect on OPEC strength, but the 1973 Arab-Israeli conflict galvanized the Arab members of OPEC into becoming a strong force for global price and production controls. As Figure 5-5 indicates, OPEC was successful in 1973–74, lost ground to inflation from 1974 to 1978, made major price increases from 1978 to 1981, and began losing ground again in 1981.

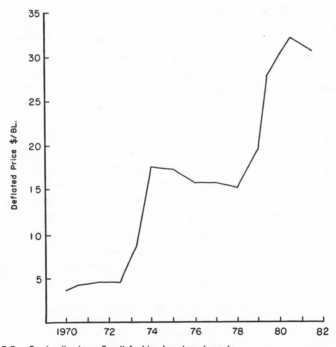

Figure 5-5. Crude oil prices, Saudi Arabian benchmark crude

The early 1980s represent the downside of the global oil-economy cycle. Declining consumption in the United States and Europe creates excess supplies at then-existing prices. Both spot and contract prices as well as production decline.

However, OPEC is weakened by two other factors. First, Mexico and the United Kingdom have not been affiliated with it. Neither was a major producer in 1973, but each now contributes about 5 percent of world production. These two countries are the only noncommunist nations where production has grown significantly. In terms of proved reserves, the United Kingdom has about 15 billion barrels as its North Sea share. Mexico, as indicated in Table 5-1, is a clear world leader in oil resources. Since OPEC's share of noncommunist production has fallen from 68 percent in 1973 to 54 percent in 1981, the presence or absence of Mexico and the United Kingdom will be important for the organization's future.

OPEC's second problem is its own divisions. The Iran-Iraq war has reduced production by 75 percent in those two countries. Such are the problems of successful global monopoly in producing at 50¢ to $1.50 and selling at $30 to $35.

During OPEC's early ineffective period (1960–73), Middle East oil prices were about 40 percent less than American crude prices. American producers promoted an Oil Import Control Program which reduced the importation of inexpensive crude oil.[11] The life of the program—1959 to 1973—almost exactly coincided with the early ineffective period of OPEC. Seen in hindsight, the majors' efforts to obtain OPEC oil at low cost while reducing its use seem contradictory. One explanation: the very large international companies considered foreign sources highly significant; hence their drive to lower Middle East prices. The lesser American majors had greater interest in their U.S. production; hence their interest in limiting U.S. imports from the Middle East.

One anecdote is irresistible: the "Brownsville turnaround." Overland imports from Mexico were exempt from import control, a fact that created an incentive to ship Middle East oil to Mexico for relay to the United States. Apparently, however, major Mexican ports were so far from the United States that such a transfer was very costly. So, Middle East oil would be unloaded from tankers in Brownsville, Texas, and put into tank trucks. These trucks would then drive south into Mexico, turn around, and drive north back into Brownsville so the oil could be qualified as Mexican overland![12]

Although these policies, viewed narrowly, seem to be contradictory, it should not be forgotten that during this period the industry obtained for the

[11]Described in Walter S. Measday, "The Petroleum Industry," in *The Structure of American Industry*, ed. Walter Adams, 5th ed. (New York: Macmillan, 1977), pp. 145–147; and John M. Blair, *The Control of Oil* (New York: Pantheon, 1976), pp. 171–186.
[12]Blair, p. 176.

United States a reliable domestic and foreign production system, and this system supplied accelerating amounts of oil at consumer prices and company costs which were lower than those ever seen before or since.[13]

William Nordhaus has offered a complex but penetrating insight into the current nature of global oil prices.[14] First, he finds spot market prices to be linked to capacity utilization. (The *spot market* refers to transactions in oil sales that do not involve long-term contracts.) When oil production capacity utilization is below 97 percent, Nordhaus finds spot prices are below official prices. However, when utilization rates rise above 97 percent and approach 100 percent, he finds that spot prices approach double the official prices. In the second stage, OPEC responds to large differences between spot and official prices by raising official prices. This creates the "ratchet" effect. Prices either go up, or stay level, but never decline. (Of course, a level price in dollars per barrel may mean that inflation creates a real price decline.) This part of Nordhaus's theory (that OPEC nominal market prices do not decline) will be given a strong test if U.S. petroleum consumption decreases and Mexican oil production increases. Finally, in terms of macroeconomics, Nordhaus defines several relationships that link oil demand and price, output, employment, wages, and infla on. One implication of this analysis may be that Western countries feel forced to create recession and unemployment to counteract inflation and oil price increases. This possibility raises several problems for national economic policy, the subject (briefly) of Chapters 14 and 15.

Exxon, BP, Shell: Worldwide and U.S.

Exxon, BP, and Shell epitomize the global nature of world oil. Their revenues in 1980 approximated $225 billion, and their operations spanned every continent. Table 5-6 summarizes their basic statistics for 1979. Note that Standard Oil of Ohio is a BP subsidiary, and consequently BP is the largest owner of domestic proved oil reserves. Since the British government holds 39 percent of BP, the United States now finds that the largest single block of domestic American oil reserves is controlled by the British government. Taken together, the three owned 34 percent of the 27.05 billion barrels of proved reserves existing at the end of 1979. Royalty oil[15] adds about another 5 percent to their controlled reserves. These three companies sold 27 percent of all the petroleum products sold throughout the noncommunist

[13]See Figure 1–3.
[14]William D. Nordhaus, "Oil and Economic Performance in Industrial Countries," *Brookings Papers on Economic Activity 2*, W. C. Brainard and G. L. Perry, eds. 1980, pp. 341–388.
[15] Royalty oil is discussed in Chapter 6 following.

Table 5-6. Exxon, Shell, and BP in 1979

	Exxon	Royal Dutch/Shell	British Petroleum
Countries of origin	United States	United Kingdom, Netherlands	United Kingdom
Ownership information	Chase Manhattan 1.7% Mfg. Hanover 1.3 J.P. Morgan 1.1 TIAA/CREF 1.0 Fayez Sarofim 0.9	U.K. owners 40% Netherlands owners 22 Swiss owners 16 U.S. owners 11	British government 39%
Revenue	$85 billion	$53 billion	$41 billion
Net income	$4 billion	$7 billion	$5 billion
Profit rate on investment	19%	33%	41%
Crude oil acquisition	1.62 bill. bl.	1.67 bill. bl.	1.19 bill. bl.
Product sales	1.94 bill. bl.	1.90 bill. bl.	0.96 bill. bl.
Major U.S. subsidiary	—	Shell USA (69%)	Standard Oil/Ohio (53%)
Reserves in U.S.	3.00 bill. bl.	2.32 bill. bl.	3.89 bill. bl.
Percent of U.S. total reserves	11%	9%	14%

Sources: Company Annual Reports for 1979; U.S. Senate, Committee on Governmental Affairs, *Structure of Corporate Concentration*, Committee Print, December 1980.

world. In the United States, either Shell or Exxon is typically the country's leading gasoline retailer.[16]

The three largest stock owners in Exxon are American banks. The fourth is a nonprofit public employees' retirement fund: TIAA/CREF. The fifth, Fayez-Sarofim, is a Houston investment company. Each of Exxon's owners also owns stock in other major oil companies. Fayez-Sarofim is the most common investor being a leading owner in each of America's five other largest oil companies.

Chapter 7 gives an overview of ownership, debt, and management relations and affiliations among American oil companies. Briefly put, the economic incentives that lead to such affiliations are profitable for the providers of capital, and lead to lower costs of production for the industry.

I would anticipate a similar picture of the private owners in Royal Dutch/ Shell and British Petroleum if such ownership data were available.

Conclusion

Summarization is difficult; each observer of world oil markets sees events according to individual perspective. I offer my view here with the realization that others are equally valid.

(1) Global politics oriented toward acquisition of oil resources has been a continuing factor of the world economy in this century. The British government owned 51 percent of British Petroleum in 1914, and encouraged BP's foreign development activities for most of the following 60 years.

American oil corporations became particularly active in foreign areas as the United States changed from an exporter to a major importer of oil. Formal recognition of the dual public-private role of the American oil companies was shown by the development of the foreign tax credit. This provision allows the companies to deduct semi-royalty excise taxes directly from their U.S. corporate income taxes. The foreign tax credit was implemented in the early 1950s to allow American oil companies to reduce their U.S. tax payments while expanding their payments to the Saudi Arabian government. It is widely believed that this tax provision was developed so that the United States could support Middle East Arab governments through oil companies, while publicly supporting the Israeli government.[17]

(2) The ten countries with greatest proven crude oil reserves have 80 percent of the 670 billion barrel total. The Middle East and Arabic countries control 59 percent. According to geological estimates of possible additional— as yet undiscovered—oil, almost one trillion barrels may be found, and are

[16]*National Petroleum News*, Fact Book Issue, 1982, p. 110.
[17]U.S. Senate, Committee on Foreign Relations, Subcommittee on Multinational Corporations, Hearings, *Multinational Corporations and United States Foreign Policy*, 1974, Part 4; Anthony Sampson, *The Seven Sisters: The Great Oil Companies and the World They Made* (New York: Viking, 1975), pp. 110–112.

now viewed as potential resources. To date, the world may have used about one-fifth of its ultimately recoverable resources. The United States may have used more than one-half of its similar original endowment. The amount recoverable may be increased by enhanced recovery.

(3) King Hubbert used geological reasoning to project peaks and declines in U.S. and world oil production. The U.S. historic maximum was reached in 1970, and the original major producing states all reached their peaks in the years between 1967 and 1972. Alaskan production will continue to increase for a few years, but will then decline. Hubbert's generalized function fits American data. Worldwide, 1979 production of 21.8 billion barrels will have been, or is near, the maximum the world will experience.

(4) In 1950 the eight major international oil companies owned 80 to 90 percent of noncommunist crude oil production outside their own North American–European territories. No major exporting country controlled its own oil resources. The Iranian nationalization was ineffective, and American oil companies joined British Petroleum as major participants in Iran in the 1950s. In the Middle East and in other oil-exporting regions, American companies began to produce significant amounts of oil. Concessions were generally held by cooperating partnerships of American and European groups. The government-owned companies in Europe were frequently partners with American companies. American and European governments assisted their corporations in forming these partnerships and in obtaining control of foreign crude oil sources. However, in the 1980s, producing governments and their state oil companies participate in at least two-thirds of world production in the major export regions.

(5) Crude oil and petroleum product prices reached their 114-year historical minimum in 1973. This long decline in costs and prices was ended by the OPEC price increases in 1973–74. The period of low-cost imported oil in the United States coincided with the peak years of production in the major low-cost areas in the United States. The Growth Era ended in 1973 in this sense: the United States can no longer increase production from low-cost domestic sources, and American oil companies do not today control production and pricing decisions in OPEC countries.

(6) OPEC was formed in 1960 but was largely ineffective until 1973. In the 1980s, OPEC exerts considerable monopoly pricing power. Oil selling at $32 per barrel probably costs 50¢ to $1.50 per barrel to produce.

(7) American oil companies do not dominate pricing and production decisions affecting U.S. petroleum products. OPEC members set prices on oil imported into the United States. As for domestic oil, the largest controlling interest in proven reserves in the United States is the British government, because of its 39 percent control of British Petroleum and BP's 53 percent ownership of Standard Oil of Ohio. Together, BP and Sohio own nearly half of the Prudhoe Bay oil, and their proven reserves in the United States exceed that of Exxon or any other American company.

6

The Petroleum Industry in the United States

We mean to secure the entire refining business of the world.

Such is the attitude that the independent oil producers believed motivated John D. Rockefeller and associates in the 1870s.[1] The monopoly position gained by Rockefeller's Standard group is generally known. By 1889, Standard had 84 percent of the market for refined products in the United States.[2]

Yet a major paradox arises from the fact that Standard secured its monopoly position by achieving major economies of scale, as a consequence of which oil prices declined and sales grew at exponential rates throughout the Standard Trust era. Figures 6-1 and 6-2 show price and production data for the years 1859–1911, the dates marking the birth of the oil industry and the formal dissolution of the Trust. It is evident that production grew at exponential rates, about 10 percent per year, while prices declined.

The railroads promoted Standard's control, finding that their shipping costs were greatly reduced with Rockefeller oil. The independent oil producers did not have predictable shipping needs, did not always have barrels, and had hundreds of shipping points. Rockefeller had a small number of shipping points with regular and large, predictable quantities. Solid oil trains could be sent from Cleveland to New York. The railroads estimated that they could reduce their investment costs in rail cars by two-thirds in dealing with Rocke-

---

Discussion and data in this chapter have in part appeared previously in Duane Chapman, Theresa Flaim, Kathy Cole, Jan Locken, and Silvio Flaim, *The Structure of the U.S. Petroleum Industry: A Summary of Survey Data,* Committee Print, U.S. Senate, Committee on Interior and Insular Affairs, Special Subcommittee on Integrated Oil Operations, 1976; and in Duane Chapman, "The Problem of Growth and Monopoly," U.S. House of Representatives, Committee on Interior and Insular Affairs, Subcommittee on Energy and the Environment, Hearings, *Public Energy Competition Act,* September 1977, pp. 306–358. Theresa Flaim's Ph.D. thesis ("The Structure of the U.S. Petroleum Industry: Concentration, Vertical Integration, and Joint Activities," Cornell University, 1977) covers statistical data on the period prior to 1975 in a comprehensive and thorough manner.

[1]Attributed to members of the Standard Oil group by Ida M. Tarbell, *The History of the Standard Oil Company,* 2 vols. (New York: McClure, Phillips, 1904), 1:159.

[2]Tarbell, 2:396, showing the 1900 Report of the Industrial Commission.

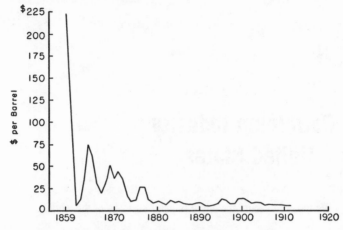

Figure 6-1. Crude oil prices during the Standard Trust, January 1981 dollars

feller, and so they gave the Standard group a 70¢ discount off a nominal rate of $2 a barrel. Rockefeller shipped 4,000 barrels a day, every day.

In fact, Rockefeller business was so profitable that the railroads began **taxing** other shippers through the device of the *drawback*. Between 20¢ and 35¢ per barrel was added to the rate of independent shippers, and this amount was passed along to the Rockefeller group. By the late 1870s, independent shippers had to pay Rockefeller a fee in order to have their oil shipped on U.S. railroads.[3]

The historic facts hardly sit well with either the competitive or profit monopoly theories discussed in Chapters 1 and 2. Here we find in the early history of the petroleum industry that competition was excessively costly, and that economies of scale interacted with the Rockefeller monopoly to bring about lower costs and prices. While monopoly power is evident (recall the 84 percent control in 1889), there is no evidence of restricted production or excessive prices.

The modern petroleum industry began in 1911 with the forced dissolution of the Standard Trust. Thirty-three companies were formed where one had existed before. This, plus the incentives of the economies of large size and vertical integration, may have made the present form of the U.S. petroleum industry inevitable. While these economies could have been captured by a single company (the Standard Trust, for example) or by a nationalized industry, the combination of forced separateness and economies of scale and vertical integration has created a unique pattern of cooperative activities in the oil industry.

[3]Affidavit of James H. Devereaux, president of the New York, Pennsylvania, and Ohio Railroad, November 13, 1880: Tarbell, 2: App. 3.

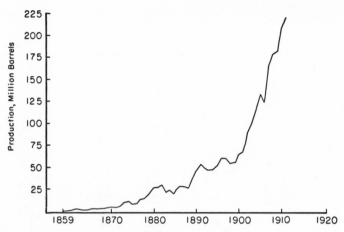

Figure 6-2. Oil production during the Standard Trust

The purpose of the chapter is to describe the industry and its corporate members, and to examine the economic incentives that have led to its present structuring. This empirical description will serve as necessary background in Chapter 9, where we will consider corporate policy toward the production and depletion of domestic reserves, and assess the applicability of each of the economic theories to the evidence available about the petroleum industry.

Major Oil Companies

What is a major oil company? Do we measure it by total sales revenue or by total sales of petroleum products? Mobil Oil owns Montgomery Ward, and Tenneco owns the Kern County Land Company and the Newport News Ship-building Company. Du Pont owns Conoco oil and coal, and United States Steel owns Marathon Oil. Should non-oil operations be included in consider-ing the size of oil companies? And what of solar, natural gas, coal, and nuclear operations? Or should the focus be on oil itself? Should size be measured by gasoline sales, or refinery output of all products, or crude oil production, or reserve ownership?

Fortunately, the problem is simple. Any of these measures of size would define essentially the same group of major companies because they are domi-nant in all these areas. Tables 6-1–6-4 give a comprehensive listing of the 20 largest petroleum companies ("the majors"). They are arranged in order of amount of revenue. (See Appendix 6-A for a discussion of these four tables.)

Most of these major oil companies are active in natural gas, nuclear power, and coal. The largest companies are also engaged in solar energy develop-ment. The past performance of the petroleum industry shows both advantages

[105]

Table 6-1. Major oil companies: Economics, 1981

Company	Revenue ($ billion)	Assets ($ billion)	Funds earned on operations ($ billion)	Net income ($ billion)	Stockholders' equity ($ billion)	Net income (as a % of stock. equity) end '81	beg. '81
1. Exxon	115.148	62.931	10.428	5.567	28.517	19.5	21.9
2. Mobil	68.587	34.776	4.333	2.433	14.657	16.6	18.6
3. Texaco	59.297	27.489	3.833	2.310	13.752	16.8	18.4
4. Standard/Cal.	46.609	23.680	3.607	2.380	12.703	18.7	21.5
5. Standard/Ind.	31.729	22.916	3.816	1.922	10.665	18.0	20.5
6. Gulf	30.461	20.429	2.885	1.231	9.984	12.3	12.8
7. Atlantic Rich.	28.747	19.733	3.589	1.671	8.665	19.3	22.5
8. Du Pont/Conoco	23.092	23.829	2.724	1.401	10.458	13.4	30.6
9. Shell	22.457	20.118	3.475	1.701	9.245	18.4	21.0
10. Phillips	16.288	11.264	1.954	.879	5.481	16.0	17.8
11. Sun	15.967	11.822	1.784	1.076	5.006	21.5	25.0
12. Tenneco	15.777	16.808	2.135	.813	5.045	16.1	19.5
13. Occidental	15.335	8.075	1.069	.722	2.864	25.2	35.2
14. Standard/Ohio	14.378	15.743	2.858	1.947	5.963	32.7	42.7
15. Getty	13.252	9.536	2.041	.857	4.774	18.0	20.7
16. Union/Cal.	11.296	7.593	1.812	.791	4.114	19.2	22.7
17. Marathon	9.815	5.994	.821	.343	2.063	16.6	17.8
18. Ashland	9.727	4.097	.266	.090	.972	9.3	9.8
19. Amerada Hess	9.444	6.322	.749	.213	2.491	8.6	9.0
20. Cities Service	8.643	6.049	.930	−.049	2.107	−2.3	−1.9
Totals	$566.049	$359.204	$55.109	$28.298	$159.526	17.7%	20.7%

and problems with respect to activity in other energy forms. These subjects—natural gas, coal, nuclear power, and solar energy—will be taken up in Chapters 8 and 10–14, and the significance of oil company participation in these industries will be examined.

The revenues of the 20 companies are larger than those of the U.S. government: their revenues of $566 billion exceed federal receipts of $424 billion.[4]

For the majors, assets of $359 billion are associated with employment of one million persons. With average assets per employee of $279,000, the petroleum industry is the most capital-intensive general manufacturing activity, and among all industries, second only to electric utilities in capital intensity. In comparison, all large manufacturing corporations (including petroleum) average $60,000 in assets per employee, while general retailing has $34,000 per employee.

The majors' profit ($28 billion) was large in absolute size in 1981, and the *profitability* of the industry was greater than for other industries. Petroleum's 17.7 percent return was higher than the 13.8 percent median for large manufacturing corporations. This is not typical of most of the post–World War II

[4]Excluding social security.

[106]

period; Figure 1-5 showed that the petroleum industry's profit rate has in the past always been about equal to or a little below overall industry returns.

It should be observed that conventional end-of-the-year reporting somewhat understates the rate of return on stockholders' equity. End-of-year equity includes that portion of the year's profit earnings which are retained. If rate of return is restated as the *net income* return on beginning-of-the-year equity, the major oil companies earned a 20.7 percent return rather than the end-of-year definition of 17.7 percent.

While the general statement about normal profitability is probably correct, it must be noted that federal and state income taxes and depreciation accounting distort profit statistics to a considerable degree. Chapters 11 and 12 will examine the tax implications of utility expansion and nuclear power generation, and it will be apparent that the revenue from a new investment can be

Table 6-2. Major oil companies: Petroleum, 1981

Company	Crude oil & liquids production (million barrels) U.S.	Worldwide	Refining (million barrels) U.S.	Worldwide	Petroleum sales (million barrels) U.S.	Worldwide
1. Exxon	274	1,386	406	1,415	473	1,679
2. Mobil	115	215	233	647	254	780
3. Texaco	139	1,114	306	807	329	969
4. Standard/Cal.	125	174	394	726	416	793
5. Standard/Ind.	160	290	299	345	303	370
6. Gulf	126	224	237	402	252	436
7. Atlantic Rich.	197	209	242	242	236	265
8. Du Pont/Conoco	51	137	106	134	121	192
9. Shell	188	196	316	316	301	301
10. Phillips	98	149	104	104	163	183
11. Sun	79	89	168	168	203	203
12. Tenneco	36	46	36	36	52	52
13. Occidental	2	96	–	–	–	–
14. Standard/Ohio	262	262	132	132	138	138
15. Getty	101	155	91	97	104	122
16. Union/Cal.	61	86	115	130	133	133
17. Marathon	61	86	140	154	167	184
18. Ashland	1	4	117	117	161	161
19. Amerada Hess	30	57	151	151	188	188
20. Cities Service	54	65	76	76	91	91
Totals, major companies	2,160	5,040	3,669	6,199	4,085	7,240
1981 U.S. total	3,135		4,554		5,840	
Majors, % U.S. total	69%		81%		70%	
1981 World total		20,398		20,398		20,398
Majors, % world total		25%		30%		35%

Table 6-3. Major oil companies: Petroleum reserves and natural gas production and reserves, 1981

Company	Petroleum Reserves (million barrels) U.S.	Worldwide	Natural Gas Production (billions of cubic feet) U.S.	Worldwide	Natural Gas Reserves (trillions of cubic feet) U.S.	Worldwide
1. Exxon	2,822	6,792	1,118.7	2,455.4	16.9	43.9
2. Mobil	898	2,996	684.0	1,089.2	6.3	17.6
3. Texaco	1,120	2,118	791.3	877.8	6.9	10.2
4. Standard/Cal.	1,237	1,635	438.4	510.6	5.3	7.2
5. Standard/Ind.	1,674	2,678	832.2	1,130.0	8.9	14.9
6. Gulf	865	1,912	615.0	692.4	4.2	5.9
7. Atlantic Rich.	2,549	2,632	425.2	473.0	13.4	13.7
8. Du Pont/Conoco	387	1,637	269.0	328.1	2.6	3.6
9. Shell	2,208	2,324	664.3	666.9	7.1	7.4
10. Phillips	476	916	324.9	506.6	3.4	6.8
11. Sun	716	812	380.3	419.4	3.3	3.8
12. Tenneco	246	392	437.2	446.0	3.4	3.6
13. Occidental	20	758	19.0	39.9	0.2	0.4
14. Standard/Ohio	3,419	3,419	46.5	46.5	6.7	6.7
15. Getty	1,322	2,072	319.5	328.4	2.5	2.8
16. Union/Cal.	533	731	413.7	440.3	5.4	7.1
17. Marathon	641	1,129	160.8	217.4	2.0	3.6
18. Ashland	6	17	10.2	10.2	0.2	0.2
19. Amerada Hess	276	816	133.8	228.0	1.1	2.3
20. Cities Service	564	584	301.0	320.0	2.7	3.1
Totals	21,979	36,370	8,385.0	11,266.1	102.5	164.8
1981 U.S. total	29,785		19,590.0		198.0	
Majors, % U.S. total	74%		43%		52%	
1981 World total		670,709		57,816.0		2,911.3
Majors, % world total		5%		19%		6%

wholly exempt from tax liability.[5] Sufficient tax credits also exist to shelter revenue from other facilities. Generally, *deferred taxes* are treated as current expense.

Funds from operations (i.e., *cash flow*) will often give a better indication of current short-run financial performance. On this basis, the majors did very well, earning $55 billion from operations.[6] For example, in Table 6-1 Exxon's profit is $5.6 billion for 1981. To this amount should be added (A) *retained earnings* for *depreciation* and *depletion* ($2.9 billion), (B) income taxes reported as paid on 1981 profit but actually deferred to the future ($1.7 billion), and (C) other items such as additions to employee pension funds ($0.2 billion). This gives a total amount of funds earned from operations of $10.4 billion. (Tables 7-4 and 7-5 show part of the financial transition for

[5]The present value of tax liability for new investment with normal expected profit is negative. It is negative during the construction period, negative in the first years of operation, and positive over most of the operating period. Chapters 11 and 12 study this subject in greater depth.

[6]Cash flow and other data sources for Tables 6-1–6-4 are in Appendix 6-A.

Table 6-4. Major oil companies: Windfall profit tax, employees, coal, nuclear, solar, non-energy activities, 1981

Company	Windfall profit tax paid ($ million)	Employees (thousands)	U.S. Coal Production (millions of tons)	Uranium; nuclear fuel	Solar	Major non-energy business
1. Exxon	2,118	180	13.900	X	X	
2. Mobil	936	206	X	X	X	X
3. Texaco	1,120	67	X	X	X	
4. Standard/Cal.	1,165	43	X	X	X	
5. Standard/Ind.	1,663.6	59	3.643	–	X	
6. Gulf	1,082	59	11.800	X	–	
7. Atlantic Rich.	1,392	53	15.835	X	X	
8. Du Pont/Conoco	206	177	38.500	X	X	X
9. Shell	1,214	37	3.522	X	X	
10. Phillips	409	35	X	X	–	
11. Sun	626	45	14.316	X	–	
12. Tenneco	225	103	–	X	–	X
13. Occidental	n.a.	48	19.645	X	–	X
14. Standard/Ohio	1,438	57	9.546	X	X	
15. Getty	666.7	19	2.770	X	–	
16. Union/Cal.	522.8	19	–	X	–	
17. Marathon	646.5	16	X	X	–	
18. Ashland	n.a.	34	14.500	–	–	
19. Amerada Hess	319	9	–	–	–	
20. Cities Service	313.8	20	X	–	–	
Totals	$16,063.4	1,286				

Exxon from 1972 to 1980.) Thus, it is accurate to say that the majors earn normal profit rates, but this understates their actual current funds, as it does for every major industry.

Although the tables show the 20 largest companies by revenues, they also include the 18 largest U.S. crude oil producers, and obviously include the U.S. parents, subsidiaries, or affiliates of the world's seven largest oil companies (as discussed in Chapter 5). Similarly, these companies import America's oil, handling about 80 percent of the country's petroleum imports.[7] The 19 largest refiners are also included, as are the 18 largest retailers of gasoline and other petroleum products.[8]

[7]The Financial Reporting System (FRS) used by the U.S. Department of Energy, Energy Information Administration in *Performance Profiles of Major Energy Producers 1979,* July 1981, uses 26 energy companies. These include all 20 major petroleum companies in Tables 6-1–6-4, and six other energy companies. Although these other six are generally not comparable in size to the majors, it is not possible to delete them from the FRS published data. FRS reports its 26 companies imported 88% of crude imports in 1979; I assume my text estimate of 80% for the 20 majors is, perhaps, a little low.

[8]Data on production, refining, and gasoline sales are from *National Petroleum News,* Fact Book Issue, 1982; *Oil and Gas Journal,* May 25, 1981; and U.S. DoE, Energy Information Administration, *Petroleum Refineries in the United States and U.S. Territories: January 1, 1981,* May 22, 1981.

Seventeen of the 20 major oil companies produce significant levels of natural gas. There are three exceptions. Occidental and Ashland import oil into the United States, and neither was a significant producer of oil or gas within the country. Standard of Ohio is a leader in natural gas reserves because of its Alaskan holdings, but is not currently a major producer. This natural gas ownership is inevitable for oil companies. On the average, 30 percent of proven gas reserves are dissolved in or associated with crude oil. An important exception is Alaska, where 84 percent of the gas reserves are found with oil.

The 20 also include 16 of the largest energy producers when the energy in coal, natural gas, and uranium is added to the energy of oil.[9]

All 20 companies have production, refining, and sales operations, and are usually *vertically integrated* in the United States. This means their operations in crude oil production, refining, and sales will be of comparable size. Exxon sells more than it refines and refines more than it produces in the United States, but it is balanced on a worldwide basis. Mobil and Texaco are in a position similar to Exxon in the United States, and are also balanced on a worldwide basis. Standard of Ohio/British Petroleum is balanced in quantity terms within the United States in refining and sales. But, because it can refine almost none of the Alaskan oil it produces, it is active in exchange and processing agreements and sales with other companies.

In 1981, the year for the data in Tables 6-1 and 6-4, Occidental was primarily an oil producer, and most of its operations were in foreign areas. Occidental sold crude to other majors and independents. With its 1982 acquisition of Cities Service, the new Occidental has a different character. It will have considerable domestic U.S. oil and gas operations, and a much higher degree of vertical integration.

Table 6-4 shows that *horizontal integration* is also common, for the majors have expanded into non-oil forms of energy. Eight of the ten largest oil companies have sales or research activities in solar energy. Nuclear fuel activities are common. As Chapter 11 will indicate, although this provides efficiency gains, it also creates serious problems for public policy. Major coal interests are common, and this, too, raises problems for policy, which are discussed in Chapter 10. As oil and gas reserves disappear, the economic incentive for increased activity in coal, nuclear power, and solar energy will accelerate. At present (in the mid-1980's) most oil companies are primarily oil companies in the sense that most of their revenues come from oil and gas activities. As noted in the tables, there are exceptions. Mobil and Tenneco

---

[9]American Petroleum Institute, *Market Shares and Individual Company Data for U.S. Energy Markets: 1950–1980* (Washington, 1981), p. 154. For 1980, Kerr-McGee, Peabody Coal, Utah International (subsidiary of General Electric), and Union Carbide displace Tenneco, Occidental, Ashland, and Cities Service. However, the newly formed Occidental/Cities Service combination will place the new oil company in the group of 20 largest energy producers.

were energy companies that expanded into non-energy areas. Conoco was acquired by Du Pont in 1981, and Marathon has been acquired by U.S. Steel in 1982. The new parent corporations (U.S. Steel and Du Pont) should now be considered to be major oil corporations. Mobil earned one-sixth of its revenue in retailing (Montgomery Ward) and packaging (Container Corporation of America). Tenneco earned one-third of its revenues in non-energy sectors. These Tenneco activities include construction and farm equipment (J. I. Chase Co), auto equipment (e.g., Monroe Muffler), agriculture (e.g., Sun Giant), packaging (Packaging Corporation of America), shipbuilding (Newport News), and insurance.

Du Pont earns one-half of its revenue from non-energy products such as vinyl chlorides, orlon, explosives, and biomedical products. U.S. Steel's major operations in addition to steel products and energy are in coal sales, plastics, chemicals, natural gas utilities, and real estate.

In 1981, most major petroleum companies continued to be primarily in the energy business. In fact, the rapid growth in oil and gas revenue and profit from 1979 to 1982 increased the relative importance of traditional oil and gas operations. But, as oil and gas resources disappear and profitability remains high, this diversification into non-energy areas will increase. Similarly, acquisition of major oil companies by companies in metals, chemicals, and other sectors of industry may also continue.

Tables 6-1–6-4, then, show a picture of the typical major American oil company as being large in terms of multi-billion dollar revenues, vertically integrated, international, and horizontally integrated into other fuels and solar energy. But it is still primarily in the energy business. The industry is capital-intensive in the sense of having a very high ratio of assets per employee. In the past, petroleum companies usually earned normal or below-normal reported profit rates. Since 1973, rates of return have periodically been much higher for petroleum companies than for other industries.

The tables also show considerable variance in certain aspects of this structure. A lesser major may be primarily a foreign producer, or a domestic retailer. A lesser major is also less likely to be horizontally integrated into other energy sources and solar energy development.

Notwithstanding the variations in individual company structure, the overall structure of major companies is quite stable. Table 6-5 shows variations in sales leadership at the beginning and end of the 27-year period, 1954–1981. Every major in the 1954 group is in the 1981 group. Three have changed their names, and four have been absorbed by other majors and two by companies from outside the industry. The basic structure is unchanged. The seven largest sellers of petroleum in 1981 were also the largest seven gasoline marketers in 1974, 1954, 1935, and 1926.[10] This stability is also evident in the relationship

[10]See Chapman, Flaim, et al., *Structure of the U.S. Petroleum Industry*, p. 8, or T. Flaim, p. 233.

Table 6-5. Change in sales leadership, U.S. oil companies

1954	1981
1. Standard Oil/New Jersey ——————→ Exxon	
2. Gulf ——————————————→ Mobil	
3. Socony-Vacuum ———————→ Texaco	
4. Standard Oil/Indiana ——————→ Standard Oil/California	
5. Texas Co.——————————→ Standard Oil/Indiana	
6. Shell ——————————————→ Gulf	
7. Standard Oil/California ————→ Atlantic Richfield	
8. Sinclair-------------------→ Conoco (Du Pont)	
9. Cities Service ———————→ Shell	
10. Phillips ——————————→ Phillips	
11. Sun Oil Co.————————→ Sun Co.	
12. Atlantic Refining ----------→ Tenneco	
13. Continental————————→ Occidental	
14. Tidewater-----------------→ Standard Oil/Ohio	
15. Pure Oil ——————————→ Getty	
16. Union Oil/California————→ Union Oil/California	
17. Standard Oil/Ohio ———————→ Marathon (U.S. Steel)	
18. Ohio Oil ——————————→ Ashland	
19. Ashland —————————→ Amerada Hess	
20. Richfield-------------→ Cities Service	

Note: Segmented lines denote mergers and acquisitions. Solid lines indicate changes in relative position.

of leading regional gasoline retailers to the territories allocated to the members of the original Standard Oil Trust.[11]

In fact, the imposition of stability upon such a complex geological, technological, and economic system is a major accomplishment of the petroleum industry. One important mechanism by which stability has been achieved is cooperation between major companies in various activities.

Cooperation in Oil Production

Cooperation begins in exploration and production. John Wilson prepared a summary of partnerships in bidding for federal leases in the Gulf of Mexico Outer Continental Shelf offshore from Lousiana.[12] According to his analysis, Texaco shared 32 bids with Tenneco. Shell had Standard of Indiana as a partner in 14 bids, and had three other partners in 79 bids. Standard of Indiana had Union and Shell among its partners in 321 joint bids. This is the general pattern for the majors. Four of the companies had partners in all their bids. Only Exxon, with 80 independent bids, had no partners.

[11]The geographic similarities between current marketing patterns and those of the original trust are described in this chapter's section on marketing, below.

[12]U.S. Senate, Committee on the Judiciary, *Petroleum Industry Competition Act of 1976*, Report to accompany S.2387, June 28, 1976, pp. 28–31. The period covered was 1970–1972.

The picture is the same for crude oil production leases held on the Louisiana offshore areas. All the majors have partners. The small companies engaged in offshore production have even more partnership arrangements than the majors. North Sea leases (Table 5-4) show the same pattern for the same major companies in Western Europe.

The reasons for this cooperative activity are economic in nature, and operate without regard to distinctions between public or private ownership. The motivations include reducing the risk of large financial losses and increasing the industry's geological knowledge of a general region.

Suppose you are responsible for managing bids for a hypothetical oil company. The federal government is offering 20 tracts in a field for bidding, and it so happens there are 19 other companies considering bids. Suppose any one tract has a 1 in 15 chance of producing usable oil, and every tract will require expensive exploration costs. You believe the field as a whole has a 3 in 4 chance of earning a profit. Your company has allotted you money for bidding which you believe would be sufficient to win one of the tracts.

If you offer an independent winning bid, your chances of losing money are 14 out of 15. On the other hand, if you go into partnerships to make winning bids on 10 of the tracts, your chances of making money rise to 1 in 2. If you can arrange winning partnerships on all 20 tracts, your chances of making money are now 3 out of 4, and the odds of losing money have fallen to 1 in 4. You have not changed the total profit expected from the field, but you have greatly reduced your chances of losing money for your company. This reduction in risk is one of the major motivations for partnerships in bidding. Indeed, if small companies could not form partnerships, the high risk might force them to drop out completely.

A second motivation for cooperation in these bids and leases is that it increases the geological knowledge available to each company by permitting the pooling of findings from each tract. This, in turn, permits more efficient production from the field and better knowledge about potential adjacent fields.

These circumstances dictate the logic of partnership in exploration for any kind of company. In Alaska, the British government is a partner with the majors through the participation of British Petroleum—owned primarily by their government. A farmers' cooperative in the oil business, a consumers' cooperative, a state or national corporation: all face the same problem of risk and can reduce it by partnership.

It would not be surprising to learn that the recent federal prohibition of joint bidding by the largest companies on federal Outer Continental Shelf leases has caused costs there to increase.

In the traditional oil-producing areas in the United States, cooperation is well established. Most oil is produced from wells held in partnership. Ashland, for example, owns more than 100 wells in partnership with *every* other major. Cities Service, Standard of California, and Standard of Ohio have the same level of partnership. This is probably more or less typical of the majors in their crude operations in the United States. Table 1 in Appendix 6-B to this chapter lists representative patterns of shared ownership in oil wells. Again, economic motivations provide the impetus for joint ownership. In the discussion of the Outer Continental Shelf, the advantages of risk avoidance and better geological information were cited. Now a third reason should be added, that of more efficient recovery. Suppose that a pool is being drilled by each of 10 companies. If competitive conditions exist, each company will attempt to extract as much oil as it can as fast as it can for its own use. The result is that the economically efficient flow of oil is left behind in the rush to "beggar thy neighbor." This was the way the industry worked from 1859 until the 1930s. The states began regulating production, and the concept of *unitized operation* emerged as a state-assisted method of promoting efficient cooperation between companies.

This is an important point. A useful illustration is given by Marathon:

> The most significant development in our domestic production operations in 1976 was the unitization of Yates Field in West Texas. . . . Efforts to obtain a higher MER [maximum efficient rate of production] . . . had long been hampered by disagreement among the operators as to the proper allocation of production. And, with individual operation of each of the many tracts in the field, it was impossible to conduct a needed pressure maintenance program . . . We developed a plan for unitized operation and began . . . to secure the necessary agreement among the hundreds of interest owners. . . . Through a pressure maintenance program . . . recoverable oil reserves will be increased by an estimated 200 million barrels. . . . The program . . . increased our net production at Yates by about 20,000 barrels per day.[13]

Operation as a unit means each pool or field is to be managed as a single production unit, and the annual production is based upon physical and economic considerations. One result is that there is much less waste in production and more efficient use of reserves. But a second consequence is less widely recognized: economic efficiency has dictated that cooperation replace competition between companies, and state governments have set guidelines.

But if these kinds of cooperation are economically sensible, what implication does this have for our concept of competition? Is the classical notion of competition, described in the first two chapters, irrelevant? In general, the cooperative activities described here are economically efficient and lead to lower production costs. In the succeeding examination of other aspects of the energy industry, it will become apparent that many kinds of cooperation and affiliation are efficient.

[13]Marathon Oil Company, *1976 Annual Report*, pp. 9, 11.

Refining

In the discussion of concentration and cooperation in crude oil production no reference was made to regional variations within the United States. While regional patterns do exist, there is no publicly available information to describe these patterns. Refinery ownership, however, is publicly reported.

In the average state that had refineries in 1981, the largest refiner owned 63 percent of the capacity, and in 13 states there was only one company. The states with companies in the refining business had, on the average, the four leading companies owning 90 percent of the capacity.[14]

This extensive concentration within regions is much higher than is suggested by national averages. Nationwide, Exxon had only 8 percent of the total U.S. refining capacity, compared to the 63 percent held by the largest refiner in the average state. The four largest American refiners owned 29 percent of total national capacity, but in the average state their percentage was 90 percent. These and other data are summarized in Table 6-6.

The economy of scale in refining has been a primary influence in bringing about this high degree of regional concentration. New refineries are believed to be efficient at a capacity on the order of 75 million barrels per year, and a refinery of this size costs about one billion dollars.[15]

The marketing area of an individual refinery is the result of many factors. These include refining costs, crude oil availability and cost, volume of refining products, regional patterns of consumption of products, marketing territories of refining companies and other oil companies, the extent of processing and exchange agreements between companies, and transportation costs. It is obvious that marketing areas will not coincide with state boundaries. Therefore, the actual regional concentration in petroleum supply must, of course, differ from concentration calculated from state sales. It is impossible to determine whether concentration defined on a marketing area basis would be uniformly higher or lower than that suggested by state data.

However, the very large size of efficient refineries and the existence of regional concentration motivate cooperative behavior between companies. Costs are reduced for Company A if it supplies its gas stations near a refinery owned by B with gasoline from one of B's refineries. Or A and B may exchange crude oil for similar reasons. The net result is that both can have larger, more efficient refineries and lower costs. An *exchange agreement* for finished products and crude oil enables them to avoid the higher transportation costs that would otherwise follow. This trading in crude oil, gasoline, and other products is widespread. Probably less than one-half of the gasoline

---

[14]From U.S. DoE, *Petroleum Refineries in the United States.* In states with one, two, or three refiners, "four leading companies" means the one, two, or three companies.

[15]$800 million in 1979 dollars. *National Petroleum News,* May 1979, pp. 73, 74.

Table 6-6. Concentration levels in the U.S. petroleum industry

Stage of industry	Percentage of national or state totals controlled by majors	Number of companies (year)
Crude Oil		
Net ownership of U.S. proved oil reserves	74	20 (1981)
Control of crude oil production	79[a]	20 (1981)
Net ownership of Alaskan oil	93	3 (1979)
Crude oil shipment, majors' pipelines	92	16 (1973)
Refining		
National refinery capacity	72	20 (1981)
National refinery capacity, Exxon, Std./Cal., Std./Ind., Shell	29	4 (1981)
Refinery capacity in each state with one or more refiners, 1 to 4 companies	90	1 to 4 (1981)
Refined products	81	20 (1981)
Marketing		
Sale of gasoline	71	20 (1980)
Sale of petroleum products	70	20 (1981)
State leaders, gasoline	100	9 (1977)
Four leading gasoline retailers, all states	93	20 (1977)
Retail gasoline sales in average state, four leaders	44	4 (1977)
National gasoline sales, Std./Ind., Shell, Exxon, Texaco (or Gulf)	27[b]	4 (1981)

Sources: Company Annual Reports; U.S. Senate, *Petroleum Industry Competition Act; National Petroleum News;* U.S. DoE, *Monthly Energy Review* and *Petroleum Refineries in the United States.*
[a]See discussion of ownership, control, and royalty oil in the section on concentration ratios in this chapter.
[b]Texaco and Gulf each marketed 5.7% of U.S. gasoline sales in 1981.

purchased by customers was refined by the company selling the gas.[16] Perhaps one-fourth or less of the gasoline sold by the average major company was refined by that company from crude oil produced and owned solely by that company.

This ability to trade products is completely dependent upon the homogeneous nature of gasoline, home heating oil, jet fuel, and other products. It is essential that Exxon's gasoline be basically identical to Texaco's gasoline. Pipeline transportation reinforces the need for standardized products. Many companies take products from a single pipeline. Additives may be used according to the taste of the customer. It is somewhat similar to food coloring. As one executive said, "We have additives of our own, our uniquely devel-

[16]Chapman, Flaim, et al., *Structure of the U.S. Petroleum Industry,* p. 59; and T. Flaim, pp. 203, 223.

oped additives that we develop in our own research labs—and these are added after it comes out of the pipeline."[17]

Marketing

As I pointed out earlier, gasoline is homogeneous, and much gasoline advertising is simply foolish. But advertising does have one economic function. It produces customer loyalty, and thereby creates predictable stability, enabling a company to manage its production, transportation, refining, and marketing system at maximum capacity and efficiency. This stability has been gained by eliminating serious economic competition in terms of pricing. Such competition would lead customers to move in large numbers from company to company, thereby creating instability and a sort of intercompany cycling of low- and high-capacity utilization. This, in turn, would raise costs and prices. Again, it is apparent that economic competition would lead to inefficiency and higher prices.

Table 6-6 also summarizes the position of the majors in marketing in the United States. The same structural features of concentration and cooperation noted in production and refining are evident in marketing. The geographic pattern of gasoline marketing reflects these influences. On a national basis, the 20 majors have about 70 percent of the sales. In individual states, the four leading retailers handle on the average 44 percent of the gasoline sold. The largest national retailer in 1981 was Standard Oil of Indiana, with only 7 percent of the market on a national basis. However, the leading retailer in each state averages 16 percent of the state's sales.

Each company does not make an equal attempt to sell products in all parts of the country. Four companies are the leading sellers in 44 states. Figure 6-3 portrays this regional pattern for 1977. It has not changed significantly in succeeding years. Exxon is the leading seller in 12 states in the East and South as well as in the District of Columbia. Standard of California leads in 11 states in the West and Southeast, Mobil in 5 states in the Northeast, and Standard of Indiana in 13 states in the Midwest.

One interesting aspect of this pattern is its similarlity to the 1911 pattern of marketing territories held by the ancestors of these four companies when they belonged to the original Standard Oil Trust. Compare the Trust in 1911 in Figure 6-4 to the pattern in 1977 in Figure 6-3. The Standard Oil Company of New Jersey has become Exxon. The Vacuum Oil Company (a member of the Trust) merged with Standard Oil of New York, becoming Socony Vacuum and later Mobil. Standard of Kentucky is now a subsidiary of Standard of

---

[17]U.S. Senate, Committee on the Judiciary, Subcommittee on Antitrust and Monopoly, Hearings, *Vertical Integration, Part I*, 1975, p. 427.

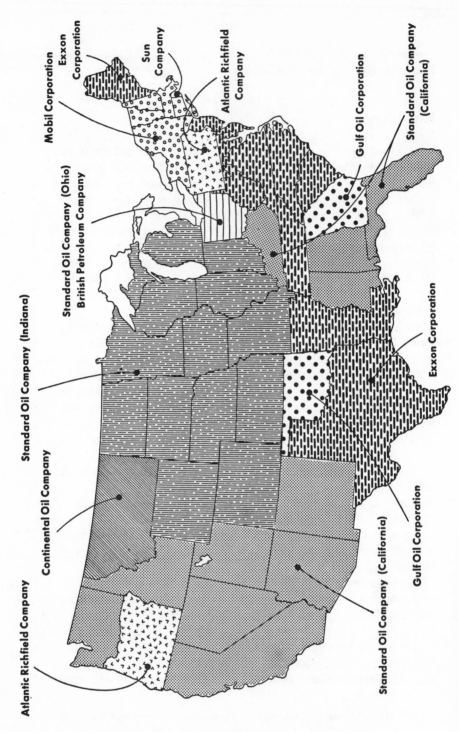

Atlantic Richfield Company

Continental Oil Company

Standard Oil Company (Indiana)

Standard Oil Company (Ohio) British Petroleum Company

Mobil Corporation

Exxon Corporation

Sun Company

Atlantic Richfield Company

Gulf Oil Corporation

Standard Oil Company (California)

Exxon Corporation

Standard Oil Company (California)

Gulf Oil Corporation

Figure 6-3. Leading marketers, 1977

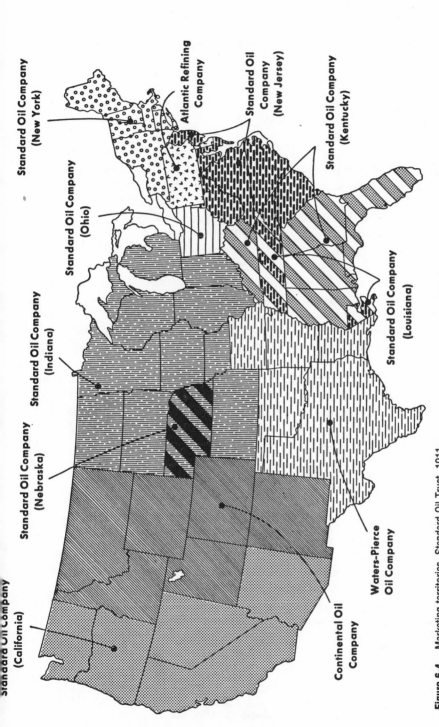

Standard Oil Company
(California)

Standard Oil Company
(New York)

Standard Oil Company
(Ohio)

Standard Oil Company
(Indiana)

Standard Oil Company
(Nebraska)

Atlantic Refining
Company

Standard Oil
Company
(New Jersey)

Standard Oil Company
(Kentucky)

Standard Oil Company
(Louisiana)

Continental Oil
Company

Waters-Pierce
Oil Company

Figure 6-4. Marketing territories, Standard Oil Trust, 1911.
Adapted, with the permission of the publisher, from Ralph W. Hidy and Muriel E. Hidy, *Pioneering in Big Business, 1882–1911*, (New York: Harper, 1955).

California; Standard of Nebraska was acquired by Standard of Indiana; and Standard of Louisiana was acquired by Standard of New Jersey.

Three other Standard Trust descendants are the leading marketers in four states originally held by members of the Trust. Continental Oil (Conoco) continues to lead in Montana, Atlantic Richfield in Pennsylvania (and now in Oregon, too), and Sohio in Ohio.

The most complex genealogy is surely that of Atlantic Richfield. The Waters-Pierce Oil Company (the Standard Trust company in the Southwest) became the Pierce Company and was purchased by Sinclair. Sinclair merged with the Atlantic Richfield Company in 1969. The Atlantic Refining Company (another original Standard Trust member) had merged with Richfield in 1966. Richfield had itself once been a Sinclair subsidiary. But—the Waters-Pierce territory has been assumed by Exxon!

The similarities in Figure 6-3 and 6-4 are remarkable. The patterns coincide in 40 states. In some states only fractions of a percentage point separate the leader from another company. Yet this pattern has endured for more than 66 years. Another interesting aspect of it is the domination of sales leadership by old Standard companies. Forty-eight of the 51 states (including the District of Columbia) had an old Standard Trust member as their leading retailer. Shell and Texaco are two leading sellers nationwide, but did not lead in any one state.

The stability of this geographic pattern of corporate dominance over a span of two or three human generations is difficult to explain. Considering the reluctance of the majors to compete continuously on a price basis with each other, one might suggest that this regional leadership pattern serves to provide a set of regional pricing patterns, each primarily determined by the regional leaders.

It should be emphasized here that these leadership patterns are based upon retail sales only. They exclude from A's total the gasoline A refines for B which B then sells. Standard of California's Kentucky subsidiary serves as an illustration of such disposition in Table 6-7. Our Cornell study of the petroleum industry found that in 1973 this division of Socal sold 63 percent of its gasoline to retail customers.[18] The remaining 37 percent went to other companies. The division obtained one-half of its home heating oil and other distillates from other companies through processing, exchange, and purchasing agreements. Products were exchanged with 28 other companies. Companies trading with Socal's division include Ashland, Atlantic Richfield, British Petroleum, Cities Service, and nine other majors. This trade is simply not evident from data on final sales or market concentration ratios.

---

[18]Some of this 63 percent of gasoline handled by Standard of California/Standard of Kentucky distributors probably went to other companies' stations. And some of the 37 percent taken by other companies probably went to Standard of California stations.

Table 6-7. Standard Oil of California: Gasoline and distillate oil sources and disposition for Standard Oil of Kentucky and the Chevron-Southern Division

	Source				
Product	Socal refinery	Processing receipts	Exchange receipts	Trade purchase	Intracompany purchases; inventory
Gasoline	60%	11%	19%	7%	3%
Distillate oil	46%	9%	31%	10%	4%
	Disposition				
	Socal marketing	Processing deliveries	Exchange deliveries	Refinery sale	Intracompany sales; inventory
Gasoline	63%	11%	25%	1%	0%
Distillate oil	62%	1%	25%	1%	11%

Source: Chapman, Flaim, et al., Structure of the U.S. Petroleum Industry, p. 69. Key terms here are explained in the Glossary at the end of this book.

Similar information on marketing for other petroleum products on a state basis is not publicly available. However, the Cornell study concluded that concentration within states is higher for the sale of jet fuel, residual oil, and heating oil than it is for gasoline.[19] Trade between oil companies in these finished products is also common, but probably on a smaller scale.

One major reason for the apparently lesser concentration in gasoline markets is this intercompany trade. Suppose Company A supplies 100 percent of the gasoline in an area. It sells one-third as its regular brand, A Plus Gasoline. It sells one-third through its self-service subsidiary, Independent Gas, and the remaining third is sold or traded to other oil companies. Company A supplies 100 percent of the gasoline, but its concentration ratio is only 33 percent. So the absence of data on supply and intercompany trading in actual marketing areas means that concentration ratios based upon final sales may understate concentration in supply. The apparent situation is that the majors are numerous and no one or small group of them commands national markets. The reality, however, is significant concentration in supply in most regional areas because of widespread trading in standardized products between companies.

The role of independent retailers in this system is complex. In general, independents sell the same gasoline at a lower price. In fact, about one-half of their gasoline is purchased from the majors, and about one-half is acquired from independent refiners.[20] (Recall, however, that independent refiners ob-

[19]Chapman, Flaim, et al., Structure of the U.S. Petroleum Industry, pp. 77–81.

[20]U.S. Senate, Committee on Government Operations, Permanent Subcommittee on Investigations, Investigation of the Petroleum Industry, July 12, 1973, pp. 9–11.

tain crude from, or in partnership with, the majors, and sell products to them as well.)

But the typical customer at a major brand station is buying more than gasoline. The customer is buying whatever quality security is associated with buying major brand gasoline. He or she is also buying the assurance that the gasoline might contain one or another additive, that the windshield might be washed, and that an attendant may fill the tank. And use of a credit card adds 3 to 5 percent to the cost. The typical independent customer pays in cash and seeks a lower price. The services and assurances available at the branded outlet are unimportant to this buyer. The independent customer may also have less income than the brand customer.

A major will exploit this situation by selling its gas at its own stations at one price, and selling gas to independents to enable them to sell at a cheaper price. And a major can have its "independent" brand as well, as has Mobil with Big-Bi, Socal with Calco, Exxon with Alert, Du Pont/Conoco with Kayo, and so on.

In our illustration here, Company A may retail its A Plus gas at $1.40 a gallon with full service and at $1.35 a gallon with self-service. It may compete with itself through its sales to Independent Gasoline, which is selling at $1.30 a gallon. And other major companies may also be selling A's gas.

The result is that this differentiation in type of retailing permits greater volume. As gas prices rise, however, customers will shift to the self-service and low-price stations and there will be a reduction both in the number of major brand stations and in the total number of retail gas stations. In fact, the number of gas stations has already declined by 80,000 since the peak in 1972; in 1982 there were 147,000. This trend will continue.

These structural characteristics of the industry permit the apparent percentage share of the majors in gasoline sales on a national basis to fall while concentration of supply in states may be increasing.

Alaska

As inexpensive oil and gas in the United States are depleted, Alaska assumes growing importance for industry structure and resource use. In 1982, Alaskan oil reserves are one-third of the country's 30 billion barrels of proven reserves of crude oil. In the 1980s, Alaskan annual production should be about 600 million barrels, 20–25 percent of total U.S. production. In the 1990s, Alaskan oil production will probably be declining, as will production in the lower 48 states.

Two companies manage the Prudhoe Bay field on the North Slope of Alaska. Sohio/BP (Standard Oil of Ohio/British Petroleum) operates the west side of the field, while Atlantic Richfield operates the east side. Both operators coordinate their plans for maximum efficiency. The oil itself is owned essentially by three companies. Sohio/BP owns 53 percent, and Atlantic Richfield and Exxon together own 40 percent on the east side. Mobil, Phillips, Socal, Getty, and Amerada Hess individually own small shares of from ½ percent to 2 percent.[21]

Alaska's North Slope provides a picture of many aspects of the U.S. petroleum industry's future. It has significant implications for future concentration of crude oil production, cooperative activities, costs, prices, and public/private interaction.

Suppose U.S. oil production outside Alaska continues to decline by 5 percent each year (see Figure 5-3). U.S. production outside Alaska would be about 1.5 billion barrels in 1990, and total production about 2.0 billion barrels. North Slope production would thus be 20–25 percent of total U.S. oil production. A simple guess is that the three companies in the North Slope would then own about 25–30 percent of total U.S. production. They presently own 23 percent.

The economics of cooperation is also apparent. The virtues of unitized operation, summarized by Marathon in this chapter's section on production, are equally applicable to the North Slope. Two companies (Atlantic Richfield and Sohio/BP) can operate an integrated management plan for maximum efficient extraction of the oil in place. Competitive production with 17 owners each producing as much as possible as soon as possible would be utterly foolish.

Cooperation is equally practical in refining and marketing. Sohio's refineries are in Ohio and Pennsylvania, and they receive little if any Alaskan oil. Sohio/BP trades or sells its Alaskan crude to other majors or to independents, and presumably receives products from these same companies. The basic picture of trading shown in Table 6-7 above for Socal will apply on a larger scale in the future.

Similarly, with U.S. gas stations closing, it makes little sense for the three major Alaskan companies to undertake major programs for gas station construction. They will expand their sales to other majors and independents.

These cooperative activities reduce the costs of producing Alaskan oil and refining and marketing its products. A competitive approach (competitive in the economic sense) would probably be so inefficient as to render North Slope production impossible.

It is surprising to learn that Alaskan oil may have caused the British government to become the largest single controlling interest in proved U.S. reserves. Recall that the British government owns 39 percent of British Petroleum and

[21]This information is based upon Annual Reports and private communications. There are also other minor interests besides those listed above. These latter shares (each for less than one-twentieth of 1%) are held by Louisiana Land and Exploration, Marathon, Placid, and five separate Hunt family interests.

British Petroleum owns 53 percent of Standard of Ohio. Sohio/BP reserves in the United States are 3.4 billion barrels; Exxon follows with 2.8 billion barrels. There is a certain irony in the absence of federal or state petroleum corporations in the United States, while the British government becomes the largest owner of U.S. oil.

Concentration Ratios

The *concentration ratios* in Table 6-6 define the proportion of business activity that is owned or controlled by a specific number of companies. As we have seen, cooperative activities between companies make such statistics of little value in understanding the industry.

The Appendix Table 6-B-1 shows partnerships in production. As a consequence of these partnerships, net ownership data understate actual economic influence. In Alaska, for example, Atlantic Richfield has net ownership of about 20 percent of the oil. Yet that company operates half the field, and Atlantic Richfield and BP/Sohio jointly manage the field. Noting its percentage holding clearly understates Atlantic Richfield's influence.

Another factor to be considered in interpreting concentration ratios is *royalty oil*. This is the amount of oil which may be assigned through lease agreements. For example, if individual A or company B holds mineral rights to a producing property, then company C, operating the field, may provide either dollars or oil to individual A or company B. Company C, then, directs the production of this royalty oil, and probably takes it directly into its own *gathering system*.

Consider this illustration. Total production at Xerxes Field is 10,000 barrels per day, of which 1,000 barrels is "royalty oil," dedicated to the landowner. All the actual production at Xerxes Field is done by company Z, which is a partner in the field with four other companies. Z gathers and refines the oil for the other companies under processing and exchange agreements. Z is the largest owner, holding 3,000 barrels in its own name; Z also buys all the royalty oil from the landowner. Thus, although the largest firm's "concentration ratio" is Z's 30 percent ownership, Z also buys the 10 percent royalty oil and refines the 60 percent belonging to the other companies. Z manages 100 percent of the production.

One estimate of royalty oil is that it is one-eighth of total production.[22] In Table 6-6, *control* means *net ownership* plus this assumed royalty oil.

The concentration ratios for control still understate economic influence since they do not include the oil that the companies participate in controlling through partnerships with other oil companies.

[22]See Mulholland and Webbink, p. 35. Table 6-6 shows the majors with 79 percent "control" of crude production. This is the 69 percent net figure in Table 6-2 multiplied by 8/7 to adjust for royalty oil.

Table 6-8. Comparative concentration ratios, 1972

Industry	Percent accounted for by	
	largest 4 companies	largest 8 companies
Primary aluminum	79	92
Flat glass	92	na
Motor vehicles	93	99
Primary copper	72	100
Tires and inner tubes	73	90
Aircraft	66	86
Industrial gases	65	81
Alkalines and chlorine	72	91
Synthetic rubber	62	81
Blast furnaces and steel mills	44	65
Industrial trucks and tractors	50	66
Semiconductors	57	70
Weaving mills (synthetic)	39	54
Ship building and repairing	47	63
Construction machinery	43	54
Lubricating oils and greases	31	44
Fertilizers	35	53
Petroleum refining	31	56
Weaving mills (cotton)	31	48

Source: William A. Johnson, et al., Competition in the Oil Industry (Washington: Energy Policy Research Project, 1976), p. 4.

Concentration in U.S. crude oil production is increasing. The majors controlled 56 percent in 1955, and this has risen regularly in the succeeding 26 years, to 79 percent in 1981.[23] This increase will inevitably continue.

The importance of regional patterns has been noted in this chapter in the sections on marketing and on Alaska. Table 6-6 shows some of these data on regional control or influence. Suppose we count the number of leading state retailers. The maximum possible number would be 51, one for each state. But only 9 companies are leading retailers; they represent 100 percent of the state leaders. Similarly, if we count the four leading retailers in each state, the maximum number would be 204. But the 20 majors listed in Tables 6-1–6-4 occupy 93 percent (109) of these "top four" slots.

The use of national concentration ratios seems to obscure as much as it reveals of the petroleum industry. Why, then, have they been considered useful? On a theoretical basis, a high level of concentration may create the structural conditions necessary for the existence of a profit monopoly. If three or four firms were to control 80–90 percent of an industry's production and sales, they could choose to restrict production, force prices up, and reap excessive monopoly profits. This is why national concentration ratios have

[23]From U.S. Senate, Judiciary Committee, Petroleum Competition Act, p. 21, and Table 6-6.

traditionally been computed, and on the basis of this theory, the petroleum industry is competitive. In Table 6-8, petroleum is shown to be one of the least concentrated industries.

However, economic theory provides little information on the significance of cooperative activities between large companies. I shall defer a discussion of the problems of interpreting this conflicting evidence on competition, growth, and monopoly until Chapter 9.

Future Prices

As domestic U.S. production declines, Alaskan and imported production accelerates. Inexpensive oil from large, old producing areas costs $5–$6 per barrel to produce, but is now probably less than one-third of U.S. production and less than 15 percent of American oil supplies. As this is written (autumn 1982) growing instability in crude oil pricing is now reducing the price of imported oil to $30–$35 per barrel. Table 6-9 shows how oil costs affect gasoline prices.

Formerly, refining was the most expensive stage, and this is still the case for old, inexpensive domestic oil. In contrast, gasoline from Alaskan oil now carries costs from pipeline and tanker transportation that are equivalent in magnitude to refining cost.

Imported oil is another story. Nearly half of what is paid for it will be pure monopoly profit for the major Middle East producers. Actual production costs may be as low as 50¢ per barrel, as shown in Table 5-5 (approximately 1.2¢ per gallon). Automobile drivers in the United States are paying about 74¢ "tax" to Middle East governments on each gallon of gasoline costing $1.35.

In the past the industry has avoided price differentials based upon crude oil costs. Gasoline sold by Shell or Exxon in neighboring states has had essen-

Table 6-9. Illustrative costs of importing, refining, and marketing Middle East oil (1982)

	Cost for each stage	
	per gallon	per barrel
Crude oil		
royalty profit	$ .74	$31.08
production cost	.02	.84
Transportation, tanker	.04	1.68
Refining	.20	8.40
Distribution	.07	2.94
Retail gas station	.11	4.62
Federal, state, and local taxes	.17	7.14
Total cost to consumer	$1.35	$56.70

Sources: U.S. DoE, Monthly Petroleum Product Price Report and Monthly Energy Review.

tially the same price regardless of how expensive the crude oil. Uniform pricing has meant that the greater profit from low-cost sources was used to pay for oil from high-cost sources. It remains to be seen whether the industry and government will continue this general policy.

For 114 years throughout the Growth Era the companies obtained increasing supplies at declining cost, and supplied growing consumption levels at declining real prices. In the 1970s, the industry and government evened out the very large differentials in crude oil cost by averaging these cost differentials throughout the industry.[24] In the 1980s, the American oil industry confronts several new economic factors: (1) the absence of price control for the first time since OPEC crude oil prices were increased; (2) major differences in costs of production from domestic sources; (3) the fact that the largest owner of U.S. oil is now a foreign company controlled by the British government; and (4) periods of decline in petroleum use in response to high prices, recession, and conservation.

I make no prediction about the near-term movement of gasoline and petroleum prices in the mid-1980s. The long-term prediction is clear: declining resources, higher prices, and lower consumption.

The crude oil *windfall profit tax* amounted to $16 billion in 1981 for the 20 major petroleum companies (Table 6-4 above), equivalent to 9¢ per gallon on all the domestic sales by these companies. The $16 billion is about 3 percent of the majors' total revenues, and about 8 percent of the retail cost of petroleum products.

The future impact of the crude oil windfall profit tax depends upon the global price of oil, the levels of domestic U.S. consumption and production, and legislative amendment to the tax. It is impossible to hold a confident opinion about its future. Together with several other problems in public policy, the windfall profit tax is discussed again in Chapter 15.

Summary

(1) John D. Rockefeller's Standard Oil Trust secured a clear monopoly position in control of the U.S. petroleum industry. Railroads gave the Trust volume discounts on Trust shipments, and also placed a tax on the shipments of independent producers, and paid this tax to the Trust. Throughout the era of the Trust (to 1911), petroleum prices declined and consumption grew at exponential rates. The Trust, by eliminating major economic competition, was able to consolidate the economies of scale inherent in the industry.

[24]Formally, this was termed the entitlement program. See U.S. DoE, Energy Information Administration, *Monthly Energy Review,* January 1977, "Crude Oil Entitlements Program." The combination of cost-averaging through entitlements and price control meant that there was no apparent monopoly profit in the 1974–78 period.

(2) The 20 major petroleum companies have revenues that exceed the receipts of the federal government. There has been little change in the membership of this group in three decades. In gasoline sales, the seven leaders have remained essentially unchanged since 1926.

(3) Cooperation, partnerships, and unitization increase efficiency by eliminating competition. Total physical production can thus be increased. Most oil wells are owned by partnerships of major companies. (In fact, cooperation is so extensive that major oil companies can disagree on the number of wells each thinks it owns with the other. As Appendix 6-B-2 shows, for example, Cities Service throught it owned 1,301 wells in partnerships with Sohio, but Sohio believed it owned 2,669 wells in partnerships with Cities Service!)

(4) In order to achieve economies of scale in production and refining, companies must engage in considerable trading in crude oil and refined products. On a nationwide basis, Exxon owned only 8 percent of refinery capacity in 1981, but in the average state with one or more refineries, the largest refiner in that state owned 63 percent of capacity.

(5) Nationally, the 20 majors control[25] about 85 percent of the reserves, 79 percent of crude production, 72 percent of refinery capacity, refine 81 percent of the oil refined, and sell 70 percent of the petroleum products. In other words, control by the majors is about 75 percent at each stage of the industry.

(6) Four companies (Standard of California, Standard of Indiana, Exxon, and Mobil) are regional leaders in gasoline sales. Each was a member of the Standard Trust in 1911. Marketing territories of the original trust are similar to the areas of regional leadership of the trust's modern corporate descendants.

(7) Gasoline and other petroleum products are homogeneous, and are widely exchanged between companies to lower costs and prices.

(8) Alaskan oil will continue to dominate American crude oil production. Sohio/BP, Atlantic Richfield, and Exxon will control 25–30 percent of U.S. oil production in the late 1980s.

(9) Because the British government owns a 39 percent interest in British Petroleum, and British Petroleum owns 53 percent of Sohio, and Sohio owns more U.S. oil than any other company, the British government may become the major interest-holder in the largest producer of U.S. crude oil. BP/Sohio is the largest owner of proved U.S. oil reserves.

(10) The necessity of cooperation in production is even greater in Alaska than elsewhere in the United States. Two companies (Atlantic Richfield and Sohio/BP) jointly manage the North Slope.

(11) Concentration ratios show that on a national basis the petroleum industry is less concentrated than the large majority of other American industries.

---

[25]Remember that control includes net ownership plus royalty oil. See the section on concentration ratios in this chapter.

However, the patterns of cooperation that exist greatly reduce the significance of this fact.

(12) In earlier years, crude oil was inexpensive and refining was the most expensive stage. For Alaskan oil, transportation is as costly as refining. For Middle Eastern oil, the royalty profit is the highest cost. In the 1980s, when a customer pays $1.35 for a gallon of gasoline from Middle East oil, one-half of this will go to the Middle Eastern governments as pure monopoly profit.

(13) The windfall profit tax on domestic oil is equivalent to 9¢ per gallon of petroleum products or an 8 percent sales tax. It is very complex in application, and its future is uncertain.

Several questions remain unanswered by this description of the American petroleum industry's economic structure, and new ones have arisen. What is the nature of the ownership and management systems of major oil companies? Do these management systems operate efficiently—or independently? Where do the Rockefeller family interests fit into today's petroleum industry?

How about natural gas? Is it, too, nearing depletion? Have oil companies held production back?

Finally, can some conclusions be reached on these perplexing questions of growth, monopoly, and profitability? Are efficiency and monopoly contradictory?

If some conclusion of this matter of industry behavior is possible, how can this help to illuminate the theory of resource use and depletion? In addition, how much is really left, and how long will it last?

In the end we cannot complete our investigation of these subjects without taking up questions of public and private ownership, and the problems of regulation. Perhaps the author's perspective is increasingly apparent. I think narrow economic criticism of the performance of petroleum companies and utilities is usually inaccurate; and more important, it is misleading—or at least irrelevant to the basic issues. As our understanding of energy use increases, we should begin to see that we should not be seeking to criticize the management of energy corporations for their performance as managers. The serious questions are harder: what should U.S. energy goals be—or perhaps more modestly—how should goals be decided? And what kind of structure and organization is necessary?

Appendix 6-A. Notes on Tables 6-1–6-4

Sources: 1981 Company Annual Reports, 10-K Financial Reports, and Statistical Supplements; *Oil and Gas Journal* (December 28, 1981 and March 8, 1982); U.S. Department of Energy, Energy Information Administration, *Monthly Energy Review* (June 1982). Sources often differ on definitions, interpretations, and amounts. Conoco became a wholly owned subsidiary of Du Pont on September 30, 1981; data reported are from the Du Pont *Annual Report,* and reflect this merger. Marathon Oil was acquired by the United States Steel Corporation and Occidental acquired Cities

Petroleum, Natural Gas, and Oil Companies

Service in 1982; data reported are from the 1981 Annual Reports, and do not reflect the mergers. Ashland data are for the year ending September 30, 1981; all other data are for the year ending December 31, 1981. Tables were prepared by Lucrezia Herman.

Table 6-1: Revenue includes sales, other operating revenues, excise taxes, interest income, and equity in earnings of affiliates. Assets include total current assets plus investments, advances, net properties, plant, and equipment. Net income includes extraordinary items such as property sales. Data are for worldwide operations, with the exception of Standard of Ohio and Shell, which are U.S. subsidiary affiliates of two major international firms, British Petroleum and the Royal Dutch/Shell group of companies. The data for U.S. subsidiaries are for their operations alone, and exclude the parents' international operations. (Chapter 5 describes these international relationships.)

Table 6-2: Production data combine crude oil, condensate, and natural gas liquids, and are reported on a net ownership basis (excluding royalty oil owned by some other company, but including oil received under agreements with foreign governments). Refining data, where possible, are for crude oil refined by the company. The petroleum sales data often include petrochemicals. Refining data for Sun include a small amount of crude oil refined in Canada. Petroleum sales data for Sun, Tenneco, Union/California, and Ashland include small amounts of products marketed outside the United States. Occidental markets a small amount of petroleum products refined by other companies, its own refining activities having ceased in mid-1978. The company primarily sold crude oil and natural gas liquids.

Table 6-3: All petroleum and natural gas reserve data are for net proved developed and un-developed reserves, including proportional interest in proved reserves of equity companies and supplies available under long-term agreements with foreign governments. Natural gas production data for Atlantic Richfield and Du Pont/Conoco are reported on an "as sold" basis; data for other companies are for net production.

Table 6-4: Windfall profit taxes paid by Occidental and Ashland in 1981 are insignificant, because of their very low production of domestic crude oil.

For coal, "X" means reserve holdings or exploration, but no reported production for 1981. Data for Atlantic Richfield, Shell, and Ashland are for coal sold, and may include purchased coal. The Ashland total includes 100 percent of the coal produced by Arch Mineral, a 50 percent-owned subsidiary.

For uranium and nuclear fuel, "X" means mining, milling, reserves, exploration, or fuel fabrication. (These fuels are discussed in Chapter 11.) Texaco and Shell reported uranium activity in 1980, with no evidence of discontinuation in 1981.

For solar energy, "X" means sales or research activity. (See Chapter 14.) Exxon and Texaco reported solar-related research in 1980, with no evidence of discontinuation in 1981. Gulf's research on solar energy was discontinued at the close of 1980.

Only four majors earned more than 10 percent of their revenues in non-energy activities in 1981. In the 1980s, U.S. Steel and Du Pont will each have about one-half of their revenues from non-energy activities. Chemicals are energy related because of the energy-intensive manufacturing process and the frequent use of oil and gas as a feedstock.

Appendix 6-B-1. U.S. oil wells jointly owned among major and non-major[1] petroleum firms in 1973

Reporting company	Total wells in which reporting company owns an interest	Total wells owned with others	Major petroleum firms with whom wells are owned		Wells jointly owned with each major as a percent of reporting firm's total joint wells
			Company name	Number of gross wells	
SEC. A: MAJOR PETROLEUM FIRMS[a]					
Ashland Oil, Inc_____	6,471	4,867	Amerada Hess Corp_____	226	5
			Atlantic Richfield Co_____	1,149	24
			Cities Service Co_____	496	10
			Continental Oil Co_____	539	11
			Exxon Corp_____	936	19
			Getty Oil Co_____	796	16
			Gulf Oil Corp_____	483	10
			Marathon Oil Co_____	112	2
			Mobil Oil Corp_____	619	13
			Phillips Petroleum Co_____	482	10
			Shell Oil Co_____	101	2
			Standard Oil of California__	390	8
			Standard Oil of Indiana____	1,024	21
			Standard Oil of Ohio_____	742	15
			Sun Oil Co_____	409	8
			Tenneco, Inc_____	342	7
			Texaco, Inc_____	1,001	21
			Union Oil of California_____	184	4
Cities Service Co.[a]_____	14,715	10,818	Amerada Hess Corp_____	2,228	21
			Ashland Oil, Inc_____	473	4
			Atlantic Richfield Co_____	6,514	60
			Continental Oil Co_____	3,886	36
			Exxon Corp_____	3,676	34
			Getty Oil Co_____	5,116	47
			Gulf Oil Corp_____	3,719	34
			Marathon Oil Co_____	1,898	18
			Mobil Oil Corp_____	4,069	38
			Phillips Petroleum Co_____	4,433	41
			Shell Oil Co_____	2,440	23
			Standard Oil of California___	837	8
			Standard Oil of Indiana____	3,877	36
			Standard Oil of Ohio_____	1,301	12
			Sun Oil Co_____	3,763	35
			Tenneco, Inc_____	938	9
			Texaco, Inc_____	5,120	47
			Union Oil of California_____	1,764	16
Occidental Petroleum Corp_	359	294	Exxon Corp_____	13	4
			Gulf Oil Corp_____	22	7
			Shell Oil Co_____	18	6
			Standard Oil of California___	27	9
			Sun Oil Co_____	1	(*)
			Tenneco, Inc_____	1	(*)
Standard Oil of California__	11,168	3,298	Amerada Hess Corp_____	1,125	34
			Ashland Oil, Inc_____	14	(*)
			Atlantic Richfield Co_____	1,355	41
			Cities Service Co_____	1,039	32
			Continental Oil Co_____	1,587	48
			Exxon Corp_____	1,113	34
			Getty Oil Co_____	1,781	54
			Gulf Oil Corp_____	1,190	36
			Marathon Oil Co_____	1,001	30
			Mobil Oil Crop_____	1,463	44
			Occidental Petroleum Corp_	23	1
			Phillips Petroleum Co_____	1,417	43
			Shell Oil Co_____	389	12
			Standard Oil of Indiana____	1,620	49
			Standard Oil of Ohio_____	1,253	38
			Sun Oil Co_____	1,298	39
			Tenneco, Inc_____	1,025	31
			Texaco, Inc_____	1,664	50
			Union Oil of California_____	1,610	49

(*continued*)

[131]

Reporting company	Total wells in which reporting company owns an interest	Total wells owned with others	Major petroleum firms with whom wells are owned		Wells jointly owned with each major as a percent of reporting firm's total joint wells
			Company name	Number of gross wells	
Standard Oil of Ohio	6,664	6,292	Amerada Hess Corp	1,413	22
			Ashland Oil, Inc	1,232	20
			Atlantic Richfield Co	3,425	54
			Cities Service Co	2,669	42
			Continental Oil Co	2,340	37
			Exxon Corp	2,210	35
			Getty Oil Co	3,108	49
			Gulf Oil Corp	2,947	47
			Marathon Oil Co	1,821	29
			Mobil Oil Corp	2,920	46
			Phillips Petroleum Co	2,655	42
			Shell Oil Co	685	11
			Standard Oil of California	829	13
			Standard Oil of Indiana	2,429	39
			Sun Oil Co	3,028	48
			Tenneco, Inc	1,304	21
			Texaco, Inc	3,485	55
			Union Oil of California	1,016	16
SEC. B: NON-MAJOR PETROLEUM FIRMS [4]					
American Petrofina, Inc	6,696	5,910	Amerada Hess Corp	349	6
			Ashland Oil, Inc	4	(*)
			Atlantic Richfield Co	404	7
			Cities Service Co	51	9
			Continental Oil Co	78	1
			Exxon Corp	115	2
			Getty Oil Co.[5]	[5]209	4
			Gulf Oil Corp	537	9
			Marathon Oil Co	41	7
			Mobil Oil Corp	556	9
			Phillips Petroleum Co	345	6
			Shell Oil Co	822	14
			Standard Oil of California	306	5
			Standard Oil of Indiana	527	9
			Standard Oil of Ohio	69	1
			Sun Oil Co	333	6
			Tenneco, Inc	42	1
			Texaco, Inc	624	11
			Union Oil of California	132	2
Apco Oil Corp.[6]	4,612	4,488	Amerada Hess Corp	64	1
			Atlantic Richfield Co	786	18
			Continental Oil Co	77	2
			Exxon Corp	292	7
			Getty Oil Co	861	19
			Gulf Oil Corp	1	(*)
			Mobil Oil Corp	578	13
			Shell Oil Co	445	10
			Standard Oil of California	231	5
			Tenneco, Inc	19	(*)
			Texaco, Inc	52	1
			Union Oil of California	253	6
Crown Central Petroleum Corp.	1,231	125	Ashland Oil, Inc	2	2
			Atlantic Richfield Co	3	2
			Cities Service Co	7	6
			Continental Oil Co	2	2
			Exxon Corp	7	6
			Phillips Oil Co	10	8
			Standard Oil of Indiana	1	1
			Sun Oil Co	1	1
			Texaco, Inc	5	4
			Union Oil of California	7	6
Diamond Shamrock Oil & Gas Co.	975	294	Amerada Hess Corp	1	(*)
			Ashland Oil, Inc	2	1
			Continental Oil Co	1	(*)
			Exxon Corp	2	1
			Getty Oil Co	1	(*)
			Gulf Oil Corp	1	(*)
			Phillips Petroleum Corp	7	2
			Shell Oil Co	1	(*)
			Standard Oil of Indiana	2	1
			Union Oil of California	3	1
El Paso Natural Gas Co.[7]	46	[7]36	Atlantic Richfield Co	35	97
			Exxon Corp	1	3
			Getty Oil Co	1	3
			Standard Oil of California	1	3
			Sun Oil Co		3

(continued)

Reporting company	Total wells in which reporting company owns an interest	Total wells owned with others	Major petroleum firms with whom wells are owned — Company name	Number of gross wells: Forest-operated wells	Number of gross wells: Partner-operated wells [8]	Wells jointly owned with each major as a percent of reporting firm's total joint wells
Forest Oil Corp.[8]	2,642	2,503	Amerada Hess Corp	1	180	7
			Atlantic Richfield Co	0	1	(*)
			Cities Service Co	104	0	4
			Continental Oil Co	100	76	7
			Exxon Corp	34	57	4
			Getty Oil Co	15	90	4
			Gulf Oil Corp	0	2	(*)
			Marathon Oil Co	0	2	(*)
			Mobil Oil Corp	0	156	6
			Phillips Petroleum Co	68	81	6
			Shell Oil Co	0	10	(*)
			Standard Oil of California	19	68	4
			Standard Oil of Indiana	101	423	21
			Standard Oil of Ohio	0	96	4
			Sun Oil Co	108	29	5
			Tenneco, Inc	2	17	1
			Texaco, Inc	103	584	27
			Union Oil of California	0	2	(*)

Reporting company	Total wells in which reporting company owns an interest	Total wells owned with others	Company name	Number of gross wells: Unitized wells	Number of gross wells: Non-unitized wells [9]	Wells jointly owned with each major as a percent of reporting firm's total joint wells
General American Oil Co. of Texas.[9]	6,793	5,878	Amerada Hess Corp	2,139	6	36
			Ashland Oil, Inc	568	1	10
			Atlantic Richfield Co	3,176	16	54
			Cities Service Co	2,157	4	37
			Continental Oil Co	1,300	20	22
			Exxon Corp	3,273	34	56
			Getty Oil Co	3,010	16	51
			Gulf Oil Corp	2,981	19	51
			Marathon Oil Co	914	2	16
			Mobil Oil Corp	2,870	28	49
			Occidental Petroleum Corp	0	3	(*)
			Phillips Petroleum Co	2,260	0	38
			Shell Oil Co	2,084	2	35
			Standard Oil of California	18	12	1
			Standard Oil of Indiana	2,684	10	46
			Standard Oil of Ohio	670	6	12
			Sun Oil Co	2,001	25	34
			Tenneco, Inc	2,250	87	40
			Texaco, Inc	2,256	17	39
			Union Oil of California	521	1	9

Reporting company	Total wells in which reporting company owns an interest	Total wells owned with others	Company name	Number of gross wells	Wells jointly owned with each major as a percent of reporting firm's total joint wells
General Crude Oil Co	3,012	2,034	Ashland Oil, Inc	125	6
			Atlantic Richfield Co	318	16
			Cities Service Co	235	12
			Exxon Corp	148	7
			Gulf Oil Corp	250	12
			Mobil Oil Corp	116	6
			Standard Oil of California	265	13
			Standard Oil of Indiana	116	6
			Standard Oil of Ohio	32	2
			Sun Oil Co	326	16
			Texaco, Inc	376	18
			Union Oil of California	51	3
Lone Star Gas Co	2,277	2,123	Amerada Hess Corp	64	3
			Atlantic Richfield Co	78	4
			Continental Oil Co	150	7
			Exxon Corp	10	(*)
			Getty Oil Co.[10]	[10] 11	1
			Gulf Oil Corp	38	2
			Marathon Oil Co	1	(*)
			Mobil Oil Corp	3	(*)

(continued)

Petroleum, Natural Gas, and Oil Companies

Appendix 6-B-1 (Continued)

Reporting company	Total wells in which reporting company owns an interest	Total wells owned with others	Major petroleum firms with whom wells are owned		Wells jointly owned with each major as a percent of reporting firm's total joint wells
			Company name	Number of gross wells	
			Shell Oil Co	269	13
			Standard Oil of California	1,261	59
			Standard Oil of Indiana	31	1
			Standard Oil of Ohio	7	(*)
			Sun Oil Co	35	2
			Tenneco, Inc	4	(*)
			Texaco, Inc	1	(*)
Louisiana Land & Exploration Co.	458	429	Amerada Hess Corp	132	31
			Exxon Corp	26	6
			Marathon Oil Co	119	28
			Texaco, Inc	153	36
			Union Oil of California	48	11
Pennzoil Co	8,168	2,502	Atlantic Richfield Co	199	8
			Continental Oil Co	244	10
			Exxon Corp	49	2
			Getty Oil Co	39	2
			Gulf Oil Corp	217	9
			Marathon Oil Co	79	3
			Mobil Oil Corp	202	8
			Phillips Petroleum Co	100	4
			Shell Oil Co	316	13
			Standard Oil of California	46	2
			Standard Oil of Indiana	80	3
			Standard Oil of Ohio	56	2
			Sun Oil Co	2	(*)
			Tenneco, Inc	27	1
			Texaco, Inc	263	11
			Union Oil of California	1	(*)
Tesoro Petroleum Corp	706	390	Ashland Oil, Inc	39	10
			Atlantic Richfield Co	41	11
			Gulf Oil Corp	107	27
			Standard Oil of California	70	18
			Standard Oil of Indiana	99	25
			Sun Oil Co	99	25
			Tenneco, Inc	70	18
			Texaco, Inc	175	45
			Union Oil of California	43	11
Total Leonard, Inc	37	37	Shell Oil Co	1	3

*Less than 0.5 percent.
[1] Major refers to those reporting companies among the 20 largest petroleum firms by sales for 1974. Non-majors are all firms excluding majors.
[2] Texaco, Inc. and Sun Oil Co. both listed all other majors as joint owners but did not specify the number of wells owned with each. All references to Shell Oil also include the Royal Dutch Shell group of companies.
[3] Cities Service Co.'s joint wells represent unitization agreements and do not reflect original lease agreements.
[4] Southwestern Oil & Refining Co. and Texas Oil & Gas Corp. listed joint owners but none were among the majors.
[5] Both Getty and its subsidiary Skelly were listed as partners in 103 wells in PAD 3; if these wells are the same then the total wells for Getty should be 106 and not the 209 listed here.
[6] Apco's joint wells are for the operators of partner-operated wells only.
[7] El Paso reported 31 joint wells in PAD 3 and 1 in PAD 4 for 1973. However, they also reported 35 wells owned with Atlantic Richfield in PAD 3 for 1973. Consequently, the total joint wells were assumed to be 36 for 1973. El Paso also reported separate data for its subsidiary Odessa Natural Gas Corp. which owned 112 gross wells and 34 joint wells in 1973. Data for Odessa were excluded from these calculations because of discrepancies in the data reported for years prior to 1973.
[8] Forest separated wells into Forest-operated and partner-operated. The identity of joint owners on partner-operated properties, other than the partner-operator, was not reported.
[9] General American Oil was the only company which separated wells into unitized and nonunitized wells.
[10] 6 wells were jointly owned with Skelly Oil Co. and 5 with Getty.

Source: Chapman, Flaim, et al., *Structure of the U.S. Petroleum Industry*, pp. 42–45.

Appendix 6-B-2. Discrepancies in reported jointly owned U.S. oil wells among firms for 1973

Reporting company and partner	Number of oil wells listed by reporting company as jointly owned with partner	Number of oil wells listed by partner as jointly owned with the reporting company
SEC. A: MAJOR PETROLEUM FIRMS[1]		
Ashland Oil, Inc.:		
Cities Service	496	473
Standard (California)	390	14
Standard (Ohio)	742	1,232
Cities Service Co.:[2]		
Standard (California)	837	1,039
Standard (Ohio)	1,301	2,669
Occidental Petroleum Corp.: Standard (California)	27	23
Standard Oil of California: Standard (Ohio)	1,253	829
SEC. B: NON-MAJOR PETROLEUM FIRM[3]		
Apco Oil Corp.:[4]		
Cities Service	0	877
Standard (California)	231	39
Standard (Ohio)	0	40
American Petrofina, Inc.:		
Ashland	4	226
Cities Service Co.	51	1,092
Standard (California)	306	213
Standard (Ohio)	69	1,118
Occidental	0	4
Crown Central Petroleum Corp.:		
Ashland	2	0
Cities Service Co.	7	200
Standard (Ohio)	0	165
Diamond Shamrock Oil & Gas Corp.: Ashland	2	0
El Paso Natural Gas Co.:[5]		
Cities Service	0	56
Standard (California)	1	0
Standard (Ohio)	0	215
Forest Oil Corp.:[6]		
Cities Service	104	356
Standard (California)	87	56
Standard (Ohio)	96	95
Occidental	0	1
General American Oil Co. of Texas:[7]		
Ashland	569	132
Cities Service	2,161	1,181
Occidental	3	2
Standard (California)	30	48
Standard (Ohio)	676	482
General Crude Oil Co:		
Ashland	125	61
Cities Service	235	503
Standard (California)	265	989
Standard (Ohio)	32	733
Lone Star Gas Co.:		
Cities Service	0	465
Standard (California)	1,261	989
Standard (Ohio)	7	530
Louisiana Land & Exploration Co.: Standard (California)	0	2
Pennzoil Co.:		
Cities Service	0	152
Standard (California)	46	1
Standard (Ohio)	56	109
Occidental	0	22
Tesoro Petroleum Corp.:		
Ashland	39	45
Cities Service	0	50
Standard (California)	70	

[1] Major refers to those reporting companies among the 20 largest petroleum firms by sales for 1974.
[2] Cities Service Co.'s joint wells reflect unitization agreements.
[3] Nonmajors are all firms excluding the top 20 firms by sales.
[4] Apco's joint wells are for the operators of partner-operated wells only.
[5] Data for the Odessa Natural Corp. (a subsidiary) are excluded because of data discrepancies.
[6] The identity of joint owners on partner-operated properties, other than operators, is not known.
[7] Includes both unitized and non-unitized wells.

Source: Chapman, Flaim, et al., *Structure of the U.S. Petroleum Industry,* pp. 46–47.

7

Ownership, Management, and Finance

As a matter of fact, nine-tenths of the stockholders of the Standard Oil Company are now and always have been Republicans. Within my knowledge there are but two Democrats who have ever been stockholders in the Company.

—Senator Oliver Payne, 1888

If "socialism" is defined as "ownership of the means of production by the workers" . . . then the United States is the first truly "Socialist" country. Through their pension funds, employees of American business today own at least 25 percent of its equity capital, which is more than enough for control.

—Peter Drucker, 1976

Few subjects stimulate more contradictory responses than does the question of ownership, management, and control of petroleum companies.[1] Perhaps this is because the subject is so closely linked to the economic theories introduced in Chapters 1 and 2. Those theories are in turn often used to criticize or defend particular characteristics or actions of the industry. Followers of the profit monopoly theory have sought and found evidence that the owners and managers of major oil companies are affiliated with those of others. Adherents of the competitive theory can argue that the major companies are in clear conformity with antitrust law and regulation, and might also advance the proposition that major oil companies—in general—give great weight to antitrust law and potential conflict of interest in formulating their executive management structure. Proponents of the growth monopoly view can adopt by incorporation each of the preceding views, and add an additional emphasis to the significant general separation of ownership and management.

[1]Senator Payne was quoted in Ida M. Tarbell, *The History of the Standard Oil Company*, 2 vols. (New York: McClure, Phillips, 1904), 2:118. Peter F. Drucker's observation is from his *The Unseen Revolution: How Pension Fund Socialism Came to America* (New York: Harper & Row, 1976), p. 1.

One critic of the industry reviewed the data on the subject of this chapter and concluded, simply, "The basic approach should be to break up the control relationships which make joint action possible.[2] John Blair, an important industry critic in the middle years of this century, shared this concern, ending his assessment by arguing that: "Through interlocking corporate relationships and joint ventures of every conceivable form, the opportunities for substituting collective for individual judgment are legion."[3] Perhaps John Wilson put the critics' case most persuasively:

> They must work together to further their joint interests. . . . But it is, most assuredly, not the kind of institutional setting within which a free market economy can be expected to function efficiently. Real economic competition is made of tougher stuff. . . . In order to function both efficiently and in the public interest, free markets must be competitive. This means that the participants must be structurally and behaviorally independent of each other. That precondition, quite apparently does not apply to the petroleum industry.[4]

As will be explained in this chapter, however, the empirical evidence can be interpreted not only to strengthen the Wilson-Blair position; it also provides support for interpretations which oppose that position.

Table 7-1 shows 1980 data for stock control by the six largest *shareholders* for America's largest oil companies. Several important points are evident. First, there are forty-two investment positions: seven oil companies, six largest investors. But in only two cases does a single investor hold as much as 5 percent of the stock. These two cases are Standard Oil of California, where the Crocker National Corporation has 10.7 percent of the voting stock, and Gulf Oil, where the Mellon National Corporation holds 6.2 percent. This is significant because economists have generally concluded that 5- to 10-percent shareholding is necessary for a shareholder to have the potential for controlling the company.[5] The research done in 1974 for our *Structure of the U.S. Petroleum Industry*, which surveyed all the major petroleum companies, had similar findings. Our survey identified just four companies, shown in Table 7-2, in which a single interest held 10 percent or more of the stock.

One might roughly generalize as follows: two of the largest American companies—Socal and Gulf—could have significant potential for being influenced by their largest shareholder. Two of the other largest majors—Sohio

---

[2]Norman Medvin, *The American Oil Industry: A Failure of Anti-Trust Policy* (New York: Marine Engineers' Beneficial Association, 1973), p. 121.

[3]John M. Blair, *The Control of Oil* (New York: Pantheon, 1976), p. 136.

[4]Cited in Blair, p. 136.

[5]According to Robert J. Larner, *Management Control and the Large Corporation* (New York: Dunellen, 1971), p. 779; Gerald L. Salamon and E. Dan Smith, "Corporate Control and Managerial Misrepresentation of Firm Performance," *The Bell Journal of Economics*, 10:1 (Spring 1979), 332; and Miron Stano, "Monopoly Power, Ownership Control, and Corporate Performance," *The Bell Journal of Economics*, 7:2 (Autumn 1976), 678.

Petroleum, Natural Gas, and Oil Companies

Table 7-1. Six largest institutional shareholders in largest American petroleum corporations, percent of common stock voting shares owned or managed, 1980

Exxon		Mobil	
Chase Manhattan	1.7%	J. P. Morgan	2.4%
Mfg. Hanover	1.3	Nat'l Detroit Corp.	1.6
J. P. Morgan	1.1	Bancoklahoma	1.3
TIAA/CREF	1.0	Chase Manhattan	1.2
Fayez Sarofim	0.9	Fayez Sarofim	1.0
Citicorp	0.8	Prudential Ins.	1.0
Total	6.8%	Total	8.5%
Texaco		**Standard Oil–California**	
Union National Bank	1.0%	Crocker National	10.7%
Nat'l Detroit Corp.	0.9	Chase Manhattan	1.3
TIAA/CREF	0.9	Fayez Sarofim	1.0
Fayez Sarofim	0.8	Mfg. Hanover	0.9
Continental Ill.	0.7	Wells Fargo	0.8
Mfg. Hanover	0.6	J. P. Morgan	0.7
Total	4.9%	Total	15.4%
Gulf Oil		**Standard Oil–Indiana**	
Mellon National	6.2%	Fayez Sarofim	1.2%
TIAA/CREF	1.4	Citicorp	1.1
Prudential Ins.	1.3	Chase Manhattan	1.1
J. P. Morgan	1.3	Harris Bank Corp	1.1
First Tulsa Bancorp	1.1	Nat'l Detroit Corp.	1.0
Fayez Sarofim	0.8	First Chicago Corp.	0.9
Total	12.1%	Total	6.5%
Atlantic Richfield			
Security Pacific Corp	2.9%		
Citicorp	2.4		
Mfg. Hanover	2.0		
Marsh & McLennan	1.3		
Prudential Ins.	1.0		
Calif Pub. Emp. Ret. Sy.	1.0		
Total	10.6%		

Source: U.S. Senate, Committee on Governmental Affairs, *Structure of Corporate Concentration,* Committee Print, December 1980, Vol. 1, pp. 69–71.

Table 7-2. Large individual shareholding (over 10 percent) in major U.S. petroleum companies, 1974

Company	Shareholder, and percentage held
Amerada Hess	Leon Hess, 20%
Getty Oil	J. Paul Getty, 64%
Occidental Petroleum	Large but unidentified interest held through the New York Stock Exchange
Sun Oil	Glenmede Trust Company, 39% representing Pew Memorial Trust

Source: Duane Chapman, Theresa Flaim, Kathy Cole, Jan Locken, and Silvio Flaim, *The Structure of the U.S. Petroleum Industry: A Summary of Survey Data,* Committee Print, U.S. Senate, Committee on Interior and Insular Affairs, Special Subcommittee on Integrated Oil Operations, 1976, pp. 14–16.

and Shell—are subsidiaries of European oil companies, as explained in Chapter 5. The four smaller majors listed in Table 7-2 may have large holdings by individuals or families which are substantial enough to guarantee considerable responsiveness by management. This leaves 12 of the 20 which are probably management-controlled.

Management control does not mean that management is unresponsive to shareholders large and small. It does imply that shareholders are one of many interest groups. For a management-controlled petroleum company, shareholders as a group are simply near the front of a crowd of contending interests which includes OPEC, employees, public attitudes, debt holders, and government.

It is evident in Table 7-1 that financial institutions are the major shareholders. All 42 entries are banks, investment companies, pension funds and their managers, or insurance companies. Shareholding actually takes many forms. It may be direct ownership, or management of stock which is owned by trusts, pension plans, estates, individuals, or corporations. The bank itself may be a nominee in managing the stock, or it may ask another organization to act as nominee.

It may be surprising to learn that these 42 large owners are generally not represented on the boards of directors of the companies. On the average, only two of the six largest shareholders are represented on the board, and usually one of these two directors is also on the board of the very largest shareholder.

The final point about the Table 7-1 ownership data is its interconnectedness. All seven companies share at least one major shareholder with each of the other companies. Table 7-3 summarizes much of the data on ownership and management. On shareholding, for example, Table 7-3 shows that Exxon and Mobil share three large stockholders (i.e., S3 for the Exxon/Mobil pair). (The three largest shareholders, of course can be identified in Table 7-1.)

In fact, the 1980 study done for the U.S. Senate Governmental Affairs Committee found that thirty-six investors held stock in all seven oil companies.

No information was available on large individual owners for the 1980 study. Because of the role of the Rockefellers in the economic history of petroleum, it is interesting to note John Blair's report that in 1938 the Rockefeller family averaged a 15 percent equity position in Exxon, Mobil, and the Standards of Indiana and California.[6] In the 1974 Cornell Study, however, no evidence of such large interests was found.

The puzzle of affiliated ownership interests grows more complex when one examines the second level: ownership of the bank corporations which are the major owners of the petroleum companies. Table 7-1 shows that the Chase Manhattan Corporation is the largest shareholder in the largest oil company.

[6]Blair, p. 149.

Table 7-3. Summary of ownership and management affiliations, largest oil companies

	Mobil	Texaco	Stnd./Cal.	Gulf	Stnd./Ind.	At. Rich.	Royal Dutch/ Shell	Brit. Pet. Stnd./Ohio
Exxon	S3,D9,T2	S3,D2,T2	S4,D6,A,T	S3,D1,T3	S3,D8,A,T2	S2,D5,T2	D1,A	D2,T
Mobil		S2,D4,T2	S3,D12,T	S3,T3	S3,D5,T	S1,D2,T3	D2,T	D2,T
Texaco			S2,D2,T	S2,T2	S2,T2	S1,T2		T
Stnd./Cal.				S2,T3	S2,A,T3	S1,D12,T3	A,T3	D1
Gulf					S1,T2	S1,A,T5	D1,T2	
Stnd./Ind.						S1,D6,T2	D1,A,T	
At. Rich.							D1,T3	D1,T2
Royal Dutch/Shell								D1,A

S = shared institutional shareholders among largest 6 in each company. Thirty-six large investors held stock in all 7 American oil companies, 1980. Data on 1980 shareholders was not available for Royal Dutch Shell and British Petroleum/Standard Oil of Ohio.

D = the same firm is represented on the boards of directors of a pair of oil companies, 1980.

A = shared accountants, 1980.

T = shared debtholders among largest 10 for each oil company for debt with at least five years original term, as of Jan. 1, 1974.

Sources: U.S. Senate, Structure of Corporate Concentration; Chapman, Flaim, et al., Structure of the U.S. Petroleum Industry; Theresa Ann Flaim, "The Structure of the U.S. Petroleum Industry: Concentration, Vertical Integration, and Joint Activities," Ph.D. thesis, Cornell University 1977, and "The Structure of the U.S. Petroleum Industry: Joint Activities and Affiliations," Antitrust Bulletin, 24:3 (Fall 1979), 555–572; and Company Annual Reports.

But Chase Manhattan's largest shareholder is another banking corporation, Citicorp. Citicorp itself is a large shareholder in Exxon, Standard of Indiana, and Atlantic Richfield. As the pattern becomes evident, it is no surprise to learn that the largest shareholder in Citicorp is J. P. Morgan, and J. P. Morgan is in turn the largest shareholder in Mobil.[7]

It should be noted that the banking companies owning oil companies are completely or almost completely bank holding companies. National banks are prohibited from direct investment of their funds in stocks of companies outside of banking or finance. However, a bank may be owned by a bank holding company, and this second kind of company can in general own a maximum of 5 percent of the stock of a nonfinancial corporation.[8]

While the evidence is amenable to several interpretations, it seems clear that major petroleum companies are not controlled by blocks of large private shareholders.

Supposing this conclusion, why, then, is it common for *large* investors to have positions among several petroleum companies? One reason may be an economy of scale. Petroleum is complicated, and much effort is needed to gain a confident understanding of a single company and its role in the industry. Once a financial institution has done so, it will find it easier to do so for the second company, and still easier for the third. At the same time, the investment policies of the major financial institutions give petroleum company management some signals about how that petroleum company is viewed by investors and serve to influence management direction. Generally, stock market prices for an individual company may be influenced by three factors: the general state of the stock market, the relative financial strength of petroleum industry stock, and the comparative position of the individual oil company with respect to other oil companies. When a large investor sells stock in oil company A and buys stock in oil company B, the managements of A and B each learn something about how their financial strengths are perceived.

A second reason for multiple holdings is simply size itself. Large financial institutions have sizable assets, and petroleum companies absorb much capital.

We may conclude that, for most of the major oil companies, the largest shareholders are financial institutions. These leading institutions commonly own small percentages of several oil companies, and of each other.

[7]The shareholding data on bank corporations are also from U.S. Senate, Committee on Governmental Affairs, *Structure of Corporate Concentration*, Committee Print, December 1980, Vol. 1, p. 33.

[8]See Pauline B. Heller, *Handbook of Federal Bank Holding Company Law* (New York: Law Journal Press, 1976), p. 221.

Management

Peter Drucker's assertion about pension fund ownership now becomes relevant. While Tables 7-1 and 7-3 have summarized institutional shareholding in the largest American oil companies, the largest blocks of stock in six of those companies are actually held by employees of the companies. In Exxon, Mobil, Socal, Standard/Indiana, Atlantic Richfield, and Texaco, employee investment plans have more stock than any other shareholder.[9]

I think it a reasonable speculation that management is a major participant in these employee stock plans.

Since the major companies have their own stock plans as largest shareholders, and, in general, single blocks of shares do not exceed 10 percent of voting stock, it seems a fair conclusion to assume that the largest major oil companies are controlled by management rather than owners. In arriving at this conclusion we are making a specific determination of a general question. Larner's detailed analysis of 1963 data concluded that management control was typical for the 500 largest nonfinancial corporations.

The 20 largest oil companies in Larner's 1963 study are, as indicated in Table 6-5, essentially the same group as the 20 largest of 1954 or 1979.[10] Larner had concluded that 15 were management-controlled and 5 owner-controlled. However, the 5 owner-controlled companies included two that were subsidiaries of management-controlled firms. Shell was identified as owner-controlled. But as Chapter 5 shows, Shell (U.S.A.) is a subsidiary of Royal Dutch Shell. In 1963, Richfield was also identified as owner-controlled. The two major owners were Cities Service and Sinclair. Since 1963, both Sinclair and Richfield have become part of Atlantic-Richfield.

The question "Does it matter? Does owner control versus manager control matter?" remains for the concluding section of this chapter. It seems likely, however, that the very largest oil companies should be viewed as management-controlled, and that some of the lesser majors (and Gulf) are more likely to be owner-influenced. I would suppose that as the size of oil companies decreases below a billion dollars in annual sales, owner influence increases.

Figure 7-1 portrays the typical management structure of a major oil company. The board of directors will have approximately 15 members, about one-half of whom are *inside directors* and one-half of whom are *outside directors*. The inside directors are management executives in the company, and the outside directors are not company employees. While the shareholders have the formal responsibility of electing directors, the nominating committee of

[9]U.S. Senate, *Structure of Corporate Concentration*, vol. 1, p. 22. Crocker National in Table 7-1 is probably representing the Standard/California employee investment plan.

[10]Larner, Appendix A. Although Larner used assets as a measure of size in 1963, the companies in Tables 6-1–6-4, are identical for 1954, except Larner has Sunray DX rather than Occidental as one of the 20 largest. Sun Oil is now the formal owner of Sunray.

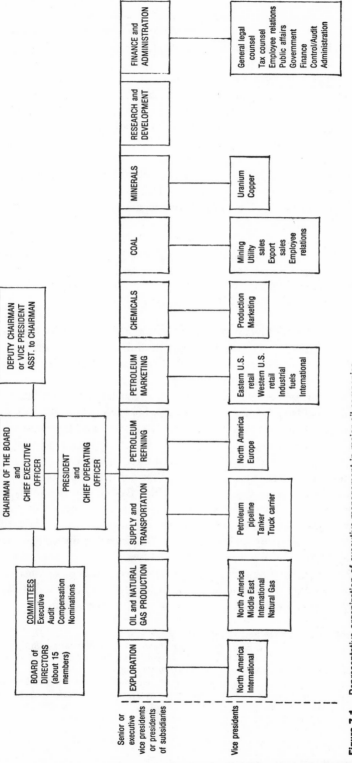

Figure 7-1. Representative organization of executive management in major oil companies

the board has the larger role. In this limited sense, boards may be viewed as self-selecting.

Formally, the responsibility of a board of directors of a major oil company is quite large. Its domain includes investment decisions, product determination, employee wages, executive salaries, dividend payments, and debt authorization. In practice, I would think it conservative to assert that corporate managements dominate many of these decisions. And many observers share a similar view.[11]

Table 7-3, in addition to shareholding affiliations, reports interlocking directorates among the largest majors. Twenty-three interconnections are shown between the 9 companies. Is this legal? The Clayton Antitrust Act asserts:

> . . . no person at the same time shall be a director in any two or more corporations . . . if such corporations are or shall have been . . . competitors, so that the elimination of competition by agreement between them would constitute a violation of any of the provisions of any of the antitrust laws.[12]

These 23 multiple director affiliations do not in any way violate the Clayton Act. They consist of two persons from a third firm serving on the boards of two oil companies. For example, in 1980, William J. DeLancey and Martha Peterson were on the board of Metropolitan Life. DeLancey was also on Sohio's board, while Peterson served on Exxon's board. John Place, also from Metropolitan's board, was on Arco's board. Superficially, this would seem to be in contradiction of the Clayton Act. Here are Ms. and Mr. A, B, and C, all on the board of one of the largest investors in major oil companies, and A, B, and C are also on the boards of three oil companies.

But these affiliations must be viewed in the context of the economic structure described in preceding chapters. Sohio, Exxon, and Arco must work closely together in Alaska, as Chapter 6 indicates, because this is the only way to achieve maximum recovery of the oil in place. In the early 1970s, Sohio shared ownership with Exxon in 2200 producing oil wells, and with Arco in 3400 wells.[13] Given this kind of partnership in production, and similar cooperative arrangements in transportation, refining, and marketing, the three directors can create no potential for restraint of trade that had not already come into existence because of the cooperative activities.

[11]Peter F. Drucker, "The Bored Board," *The Wharton Magazine*, 1:1 (Fall 1976), 19; Harry G. Henn, *Handbook of the Law of Corporations and Other Business Enterprises* (St. Paul: West Publishing, 1961), pp. 337–339; William E. Knepper, *Liability of Corporate Officers and Directors*, 3d ed. (Indianapolis: Allen Smith, 1978), pp. 5–16.

[12]Cited in U.S. House of Representatives, Committee on the Judiciary, *The Antitrust Laws: A Basis for Economic Freedom*, A Staff Report to the Antitrust Subcommittee, 1965, p. 8.

[13]See Chapter 6, Appendix B.

It seems likely that these multiple directorate affiliations are analogous to the multiple ownership affiliations described in the preceding section, and have similar economic motivations and functions.

The significance of multiple director affiliations is reduced still further by the fact that directors play a relatively minor role in the management of large corporations.

In one important area it is evident that the companies are very sensitive to potential conflicts of interest for their directors. There is no single person who is on one of the petroleum company boards, and also serving on the board of directors for an automobile company, or a major supplier of equipment to the company. The motivation for eliminating such ties is to remove the potential conflict whereby, for example, a director would work to secure a special contract between an equipment manufacturer of offshore drilling rigs and a major oil company. Such special contracts could benefit the director personally, but create financial and other problems for the oil company. One problem this sensitivity creates, however, is to make cooperation difficult even when it might be economically efficient. More direct association between petroleum company and automobile company management might have eased the transition to smaller cars using less gasoline with lesser air pollution emissions.

In Figure 7-1, the main line of authority is seen to run through the chief operating officer and the chairman of the board. The *board of directors* is given a prominent position, but off to the side. The board chairman will be termed the *chief executive officer,* and this person's attention will be oriented toward the board, and to the president, who is the *chief operating officer.* We might say that the chief operating officer worries about running the refineries and whether to build new ones or close old ones. The chief executive officer worries about the chief operating officer, how to resolve the financial and governmental problems of those new refineries, and how to present the problems and policies to the board.

The upper level of Figure 7-1 is fairly standard for the majors: a board of directors with several committees, a chief operating officer, a chairman of the board with a top-level executive as assistant to the chairman. However, the organization of operating activities varies considerably from one company to another. Some companies give first emphasis to functions such as exploration, production, oil refining, coal, nuclear fuel, finance, law, etc. Others emphasize geographic areas. To some extent this delineation of functional versus geographic organization can be overemphasized, since middle-level management in each company probably is responsible for one or two functions in a specific geographic area.

The formal titles held by operating executives also vary considerably. Exxon, for example, has vice presidents of functions and presidents of major divisions and affiliates such as Esso Middle East, Esso Europe, and Exxon Chemical. Texaco has presidents of subsidiary corporations and senior vice presidents. Conoco has executive vice presidents. Sun Oil and Getty have group vice presidents.

Subsidiary corporations have a significant role for most of the companies. Appendix 7-A shows British Petroleum's more important subsidiaries. Note that Sohio, America's largest oil producer, is merely one entry at the end of the third column.

Figure 7-1, then, should be viewed as a picture of the organization of a composite major oil company in the early 1980s. It might be recalled from Chapter 6 that the average company in the group of 20 majors had 64,000 employees and $28 billion revenues in 1981. Exxon's $115 billion revenues were, of course, the largest of any corporation in the world.

Finance

For most of its history, the petroleum industry financed much of its expansion from the banking, insurance, and investment business. Exxon's lenders held $2.6 billion of Exxon debt in 1972. This increased $2.1 to $4.7 billion in 1980. However, this must be contrasted to capital stock, which declined from $2.6 billion to $1.7 billion, and retained earnings, which increased from $9.6 billion to $23.7 billion. In 1980, Exxon debt increased by only $100 million. This tendency for debt to lessen in importance will probably continue in the 1980s. Given the high level of funds from operations that are being earned in the 1980s, it is inevitable that the petroleum industry consider major expansion into non-energy fields. Table 6-4 will need significant revision.

Table 7-4 summarizes some basic financial statistics for Exxon in 1972 and 1980.

Debtholding indicates the same pattern of multiple affiliations described for ownership and boards of directors. In Table 7-3, "T2" for Mobil and Texaco means that the 1974 Cornell survey reported two financial institutions were major *debtholders* for both Mobil and Texaco. Metropolitan Life Insurance was the leader: it was a large debtholder for 13 of the 20 largest companies.[14]

Table 7-3 shows that 30 of the pairs of companies had affiliated large debtholders. The economic incentives for this pattern are presumably similar to those incentives described above for multiple affiliations in ownership and boards of directors.

Auditing is an important function linking finance and management. Basically, the independent accountants who audit the petroleum companies are expected to guarantee that financial and operating data are accurately reported

[14]Chapman, Flaim, et al., *Structure of the U.S. Petroleum Industry*, p. 19.

Table 7-4. Basic financial data, Exxon, 1972 and 1980, in billion dollars

	1972	1980
Income and funds		
Revenue	$22.4	$110.4
less Expenses	−20.9	−104.7
Net income	$ 1.5	$ 5.7
Add to net income:		
Depreciation, depletion	+1.1	+2.3
New deferred income taxes	+0.2	+1.8
Other additions	+0.1	+1.0
Total funds from operations	$ 2.9	$ 10.8
Financial transactions	+0.1	−1.1
Net increase in debt	+0.5	+0.1
Total funds available and utilized	$ 3.5	$ 9.8
Capital items		
Long-term debt	2.6 (16%)	4.7 (12%)
Deferred income taxes	0.8 (5%)	6.2 (16%)
Annuity reserves, etc.	0.6 (4%)	2.0 (5%)
Capital stock	2.6 (16%)	1.7 (4%)
Retained Earnings	9.6 (59%)	23.7 (62%)
Total Capital Items	16.2 (100%)	38.3 (99%)
Shareholders' equity		
Amount, beginning of year	$11.6	$ 22.6
Amount, end of year	$12.3	$ 25.4
Net income, % return, beginning of year	13%	25
Operating funds, % return, beginning of year	25%	48%

Source: Annual Reports, 1972 and 1980.

and fairly presented, and that a company's position with respect to oil reserves and other assets is reasonably described. The complexity and size of the industry creates an economy of scale that leads some few accounting firms to provide auditing services for several major oil companies. Five accounting firms act as independent auditors and accountants for all 20 major companies and the two international parents, Royal Dutch Shell and British Petroleum.[15] Sharing of accounting firms among the largest majors is also shown in Table 7-3. For example, Exxon and Royal Dutch Shell both have Price-Waterhouse as accountants.

The question that arises here is what incentives may exist to prevent a single firm from becoming the auditor for all major companies. I speculate that the motivation is a desire on the part of management to avoid placing its auditors themselves in a position of potential conflict of interest, as they might be if they controlled the formal publication of all major company data.

[15]These five firms are Arthur Andersen & Co., Arthur Young & Co., Coopers & Lybrand, Ernst & Whinney, and Price-Waterhouse & Co.

As in many other aspects of the industry, economies of scale and affiliation are considerable, but are nevertheless limited.

As the 1980s begin, the American petroleum corporations have unique opportunities and problems in terms of the industry's financial situation. On the positive side, revenues have continued to increase regularly. Profit levels have increased above Growth Era levels, but not regularly. The industry is well situated in coal and nonfuel minerals. The basic problems have been enumerated in other chapters: (1) declining petroleum and natural gas resources in the United States; (2) unstable sales growth in the United States and around the world; (3) an uncertain future for nuclear power, where the petroleum industry is strong (see Chapter 11); (4) an uncertain political environment for expansion out of energy minerals.

In Table 7-4, the major differences between the last year of the Growth Era (1972) and the first year of the 1980s are evident. The outstanding change is size: revenue grew five times as a result of OPEC price increases.

Profit can be measured in many ways. Two common measures are net income and *funds earned from operations*. The latter concept defines cash income after expenses, while net income subtracts allowances for depreciation of plant and equipment, and depletion of oil, gas, and coal resources. Net income also subtracts deferred income taxes. Consequently, funds from operations are greater than net income.

In 1972, Exxon earned a 13 percent rate of return in terms of net income as a percentage of shareholders' equity, and a 25 percent return for funds from operations. In 1980, the rates of return are much higher, being 25 percent for net income and 48 percent for funds from operations. (Recall from Chapters 1 and 2 that a net income return of 15 percent is considered normal.)

Rising profitability in the 1970s has led to a rearrangement of capital structure. For Exxon, retained earnings rose in value as well as in proportion to other selected capital items. Capital stock held by shareholders actually declined in amount and percentage because Exxon bought back stock over the 1970s. The interaction of rising profit and increased tax incentives led to a growth in deferred taxes which had contributed $6 billion to Exxon's capital in 1980.

Growth in profit also led to a reduction in the significance of long-term debt over the 1970s, as it declined from 16 percent to 12 percent of capital investment.[16]

The negative entry of −$1.1 billion in 1980 for financial transactions reflected a financial loss caused by a significant growth in Exxon's petroleum inventories that year, a result of lower product sales.

---

[16]Standard of Ohio has an unusual amount of debt for U.S. petroleum companies in the 1980s. Because of its Alaskan investments, its long-term debt was nearly half of its capital. This is declining rapidly, however, because of Sohio's profit levels.

Table 7-5. Exxon's application of funds in 1980

	$ billion	Percent
Shareholder dividends	2.4	24
Investment in U.S. oil and gas exploration and production	1.8	18
Investment in foreign oil and gas exploration and production	1.9	19
Investment in foreign refining and marketing facilities	0.8	8
Investment in U.S. refining and marketing facilities	0.2	2
Investment in transportation	0.2	2
Other property, plant, and equipment	1.7	17
Purchase of Exxon shares by company	0.4	4
Increase in cash and securities	0.4	4
Total funds available and used	9.8	98%*

Source: Exxon Annual Report 1980.
*Rounding is responsible for 98% rather than 100% sum.

Table 7-5 shows how Exxon used its funds in 1980. Dividend payments are significant, as are continued investment in foreign and U.S. oil and gas production facilities. It should be noted, however, that investment in U.S. refining, marketing, and transportation is not large.

As opportunities for profitable oil and gas development continue to decline in proportion to the industry's revenue and profit, major changes must be anticipated in the kind of picture that is shown in Table 7-5. Future investments must necessarily be in non-energy areas and in other energy and mineral resources.

The financial status of the OPEC national oil companies is quite different. According to the estimates shown in Table 5-5 and Figure 5-5, petroleum is costing less than $1.50 per barrel to produce, and is being sold to the international companies at prices between $30 and $40 per barrel. In the late 1970s, OPEC governments were dividing their revenue from the United States between investment in the United States and the purchase of American goods and services.[17] About one-fifth of their revenues was spent on capital goods, a high proportion of which was probably for use in their petroleum operations. About one-seventh of the petroleum revenue was used to purchase military goods from private U.S. contractors and from the federal government. Almost half the revenue was reinvested in financial capital in the United States in Treasury bills and notes, in corporate debt, in purchasing stock in U.S. corporations, and in commercial bank deposits. Since electric utilities were important in capital expansion in the 1970s, it seems nearly certain that much OPEC revenue from petroleum sales in the United States was invested in

[17]See Christopher L. Bach, "OPEC Transactions in the U.S. International Accounts, 1972–77," Survey of Current Business, 58:4 (April 1978), 21.

[149]

electric utilities, and much of this, in turn, was probably used in nuclear power development.

The interdependency of international institutions is underlined by this sequence. When an American purchased a dollar's worth of gasoline from a private U.S. oil company, part of the dollar went to an OPEC government oil agency, and some fraction of that sum was reinvested in the United States in electric utilities and nuclear power.

Does It Matter?

Do multiple affiliations between oil companies create a problem? This chapter has described those extensive affiliations with major shareholders, boards of directors, debtholders, and auditors. Were Wilson and Blair correct in arguing that these affiliations are injurious to economic competition? The position taken here is that these multiple affiliations contributed to efficiencies in management during the Growth Era. Perhaps one casualty of this empirical view, however, will be our concept of economic competition, which as outlined in Chapters 1 and 2 requires competitors to be independent. (This theoretical question of competition, monopoly, and growth will be addressed in Chapter 9 where all the evidence is summarized.)

Three other observations are relevant. First one should note the international nature of multiple affiliations in ownership and boards of directors. The pattern applies to European and Japanese corporations as much as to American oil companies.[18]

Second, it is not clear how the degree of ownership influence affects the actual management of the companies. Robert Larner's research in the late 1960s concluded that the degree of owner influence did not affect profitability. However, Miron Stano found that owner-controlled firms performed significantly better in the stock market, and Gerald Salamon and E. Dan Smith find that management-controlled firms are more likely to use accounting policies to misrepresent firm performance.

Casual inspection of rate of return shown in Table 6-1 yields an amusing result. The highest rate of return is that of Standard of Ohio, the company owned by British Petroleum, which is the company controlled by the British government. Apparently, the public/private ownership question did not affect profitability. The 4 owner-influenced oil companies (Gulf, Amerada Hess, Occidental, Sun) have an average rate of return similar to that of the 15 privately owned management-controlled oil companies. We cannot reach a

[18]Empirical data in Europe and Japan are discussed in P. S. Johnson and R. Apps, "Interlocking Directorates among the UK's Largest Companies," *Antitrust Bulletin*, 24:2 (Summer 1979), 357–369; and in F. M. Scherer, *Industrial Market Structure and Economic Performance*, 2d ed. (Boston: Houghton Mifflin, 1980), pp. 51, 52.

logical conclusion based upon the slim evidence that is available. The kind of control (public/private, management/owner) is not, in the 1980s, a reliable guide to the performance of the company.

The third observation is that similar patterns exist in publicly owned and privately owned companies. The organization of British Petroleum and CFP (the French Petroleum Company) is as well reflected by Figure 7-1 as is that of any privately owned company. Both BP and CFP have governments as their largest shareholders. The subsidiary listing in Appendix 7-B is similar in form to that for the large American majors.

Finally, a warning, to be expanded upon in Chapter 9. The two publicly known efforts to create profit monopolies in energy have both involved government corporations or agencies. OPEC, of course, is the most familiar. Less widely known is the effort of one or more American oil companies to work with Canadian, South African, and Australian governments to establish a uranium cartel in the early 1970s.[19] Whatever the potential for economic power created by the incentives for affiliations in ownership, management, and finance, it cannot be supposed that public ownership in and of itself will provide a remedy to the problem.

[19]Described in Geoffrey Rothwell, "Market Coordination by the Uranium Oxide Industry," *The Antitrust Bulletin,* 25:1 (Spring 1980), 233–268. See Chapter 11.

Appendix 7-A. Representative International Subsidiaries: The British Petroleum Company

This list contains the names of the more important subsidiary and associated companies of the group at 3 December 1980 indicating group percentage of equity capital (to nearest whole number). Those held directly by th parent company are marked with an asterisk. The percentage owned being that of the group unless otherwi indicated.

Subsidiary companies

	%	Country of Incorporation	Principal Activities
International			
BP Chemicals International	100	England	Holding company
BP Coal	100	England	Coal production overseas
BP Exploration	100	Scotland	Holding company
BP Gas	100	England	Natural gas
*BP International (formerly BP Trading)	100	England	Integrated oil operations
BP Minerals International	100	England	Minerals
BP Nutrition	100	England	Nutrition
BP Oil International	100	England	Integrated oil operations
BP Shipping (formerly BP Tanker Company)	100	England	Oil transportation
Scicon Consultancy International	100	England	Computer software
*Selection Trust	99	England	Mining finance
*Tanker Insurance	100	England	Insurance
Europe			
UK			
Alexander Duckham	100	England	Lubricants
Amari	99	England	Manufacturing
Bäkelite UK	100	England	Chemicals
Border Chemicals	100	England	Chemicals
BP Capital	100	England	Finance
BP Chemicals	100	England	Chemicals
BP Nutrition (UK)	100	England	Nutrition
BP Oil	100	England	Refining and marketing
BP Oil Development	100	England	Oil production
BP Oil Trading	100	England	Marketing
BP Petroleum Development	100	England	Oil and gas exploration and production
*BP Properties	100	England	Property
Bristol Composite Materials	100	England	Manufacturing
BXL Plastics	100	England	Chemicals
Forth Chemicals	100	England	Chemicals
AUSTRIA			
*BP Austria	100	Austria	Marketing
BELGIUM			
*ABP	100	Belgium	Processing and supply
*BPNV (parent 20%)	100	Belgium	Marketing and chemicals
Tensia	100	Belgium	Detergents
DENMARK			
*BP Olie-Kompagniet	100	Denmark	Marketing
FINLAND			
*BP Petco	100	Finland	Marketing
*Suomen BP	100	Finland	Marketing
FRANCE			
BP Chimie	79	France	Chemicals
*Société Française des Pétroles BP	79	France	Refining, marketing and oil transportation
Streichenberger	59	France	Marketing
GERMANY			
*Deutsche BP	100	Germany	Refining and marketing
Fanal	100	Germany	Marketing
Gelsenberg	100	Germany	Holding company
Oelwerke Julius Schindler	100	Germany	Refining
Stromeyer	100	Germany	Marketing

Subsidiary companies

	%	Country of Incorporation	Principal Activities
Europe			
GREECE			
*BP of Greece	100	England	Marketing
ITALY			
*Britannica Petroli (BP)	100	Italy	Marketing
NETHERLANDS			
BP Nutrition Holdings	100	Netherlands	Holding company
British Petroleum BV	100	Netherlands	Holding company
British Petroleum(Overzee)	100	Netherlands	Holding and finance company
*British Petroleum Maatschappij Nederland	100	Netherlands	Holding company
British Petroleum Raffinaderij Nederland	100	Netherlands	Refining
British Petroleum Benzine en Petroleum Handel Maatschappij	100	Netherlands	Marketing
British Petroleum Exploratie Maatschappij Nederland	100	Netherlands	Exploration
Hendrix Fabrieken	100	Netherlands	Nutrition
Noordzee Selection	99	Netherlands	Natural gas
Trouw	100	Netherlands	Nutrition
NORWAY			
*BP Norge	100	Norway	Marketing
*BP Petroleum Development of Norway	100	Norway	Oil exploration
PORTUGAL			
*Companhia Portuguesa dos Petróleos BP	100	Portugal	Marketing
REPUBLIC OF IRELAND			
BP Ireland	100	Ireland	Marketing
SPAIN			
*BP Española de Petroleos	100	Spain	Marketing
SWEDEN			
*Svenska BP	100	Sweden	Marketing
BP Raffinaderi (Göteborg)	78	Sweden	Refining
SWITZERLAND			
*BP (Schweiz)	100	Switzerland	Marketing
BP Chemicals (Suisse)	100	Switzerland	Chemicals
Geldner Stromeyer	100	Switzerland	Marketing
TURKEY			
*BP Petrolleri	100	Turkey	Marketing
*BP Overseas Refining	100	England	Refining
Middle East			
*BP Arabian Agencies	100	England	Marketing
BP-Japan Oil Development	55	Scotland	Marketing of crude
Africa			
*BP Africa Medwest	100	England	Marketing
BP Southern Africa	100	South Africa	Holding and marke company
Société des Pétroles BP d'Afrique Occidentale	100	Senegal	Marketing

Appendix 7-'A (Continued)

Subsidiary companies

	%	Country of Incorporation	Principal Activities
Far East			
HONG KONG			
*BP Oil Hong Kong	100	Hong Kong	Oil products supplier
INDONESIA			
*BP Petroleum Development of Indonesia	100	England	Oil exploration
JAPAN			
*BP Far East	100	England	Group services
MALAYSIA			
*BP Malaysia	100	Malaysia	Marketing
SINGAPORE			
*BP Refinery Singapore	100	Singapore	Refining
*BP Singapore	100	Singapore	Marketing
Australasia			
AUSTRALIA			
British Petroleum Company of Australia	100	Australia	Holding company
BP Australia	100	Australia	Integrated oil operations
BP Coal and Minerals Australia	100	Australia	Holding company
Clutha Development	100	Australia	Coal production and marketing
Kwinana Nitrogen	87	Australia	Chemicals
Seltrust Holdings	78	Australia	Minerals
FIJI			
*BP South-West Pacific	100	Fiji	Marketing
NEW ZEALAND			
BP Chemicals NZ	100	New Zealand	Chemicals
*BP New Zealand Holdings	100	New Zealand	Holding company
BP New Zealand	100	New Zealand	Marketing
BP (Oil Exploration) New Zealand	100	New Zealand	Oil and gas exploration and production
Europa Oil (NZ)	100	New Zealand	Marketing
PAPUA NEW GUINEA			
BP (Papua New Guinea)	100	Papua New Guinea	Marketing

	%	Country of Incorporation	Principal Activities
Western Hemisphere			
BRAZIL			
BP Petroleum Development Brazil	100	Bermuda	Oil exploration
CANADA			
*BP Canadian Holdings	100	Canada	Holding company
BP Canada	65	Canada	Integrated oil operations
Selco	99	Canada	Minerals
USA			
Amselco Natural Resources	99	USA	Minerals
BP Alaska Exploration	100	USA	Oil exploration
BP Chemicals Americas	100	USA	Chemicals
BP North America	100	USA	Holding company
*BP North American Finance	100	USA	Finance company
BP North America Trading	100	USA	Marketing
BP Pipelines	100	USA	Pipeline company
Sohio/BP Trans Alaska Pipeline Capital	68	USA	Finance company
The Standard Oil Company (Sohio)	53	USA	Integrated oil, coal, chemicals and mineral operations

Associated companies

	%	Country of Incorporation	Principal Activities
International			
Irano-British Shipping	50	Bermuda	Oil transportation
Sub Sea International	50	USA	Deep-sea diving service
Europe			
UK			
Associated Octel	37	England	Manufacture and marketing
Orobis	50	England	Chemicals
Rockwool	50	England	Insulation material
Sub Sea Offshore	50	England	Deep-sea diving service
BELGIUM			
*Petrogaz (parent 15%)	50	Belgium	Marketing,
*SIBP	50	Belgium	Refining
FRANCE			
Distugil	50	France	Chemicals
Gerland	26	France	Other marketing
Naphtachimie	39	France	Chemicals
Raffinerie de Strasbourg	26	France	Refining
GERMANY			
Erdölchemie	50	Germany	Chemicals
Ruhrgas	25	Germany	Gas distribution
Middle East			
Abu Dhabi Gas Liquefaction	16	Abu Dhabi	Natural gas liquefaction
Abu Dhabi Marine Areas	37	England	Marketing of crude oil
Abu Dhabi Marine Operating	15	Abu Dhabi	Oil exploration and production
Abu Dhabi Petroleum	24	England	Marketing of crude oil
Abu Dhabi Onshore Operating	9	Abu Dhabi	Oil exploration and production
Africa			
Central Chemical Investments	24	South Africa	Chemicals
Consolidated Petroleum	50	England	Marketing
Shell and BP South African Petroleum Refineries	50	South Africa	Refining
Société Ivoirienne des Pétroles BP	50	Ivory Coast	Marketing
Unisel Gold Mines	34	South Africa	Minerals
Far East			
Singapore Refining Company	30	Singapore	Refining
Australasia			
AUSTRALIA			
CS BP & Farmers	33	Australia	Marketing of fertilisers
NEW ZEALAND			
New Zealand Refining	24	New Zealand	Refining
USA			
Alyeska Pipeline Service	33	USA	Pipeline company

Source: The British Petroleum Company, *Annual Report & Accounts, 1980,* pp. 39–40.

8

Natural Gas

> I feel that it is unreasonable that natural gas be promoted for home and business heating on a national basis, since supplier public announcements . . . have already announced gas curtailment in some areas and predicted more to come.
>
> —W. E. (Trez) Lee, 1974

> A consensus seems to have developed in the energy community that natural gas will be the most abundant, most accessible and surest alternative to imported oil for at least a decade.
>
> —New York Times, 1979

These two wholly contradictory statements illustrate the dilemma we face in understanding the U.S. position in natural gas.[1] Is there shortage or abundance? In Chapter 3 we examined natural gas in the context of its relative efficiency, substitutability, and cost when compared to other forms of energy. Chapter 4 showed that natural gas prices are rising rapidly, but, on a Btu basis, remained considerably below comparable energy forms in the mid 1980s. In Chapter 6, the major oil companies were seen to be large producers of natural gas.

There are several subjects to be covered in this chapter: reserves, production, and consumption in the United States; the organization of the natural gas industry and the positions of petroleum companies and electric utilities in the industry; and the natural gas shortages of the 1970s. American and world reserves will be compared, and the potential for natural gas imports in the United States clarified.

As a preview of the conclusion of this chapter, it can be said that both the above statements are correct in the context in which they were made. What is necessary is to gain a sufficient understanding of the economics of natural gas to place these views in perspective.

[1]Trez Lee's statement is taken from Duane Chapman, Timothy Mount, John F. Finklea, et al., *Power Generation: Conservation, Health, and Fuel Supply,* U.S. Federal Power Commission National Power Survey, March 1975, p. 178. Lee is on the staff of the American Public Power Association. The energy community's consensus was reported in the *New York Times,* July 23, 1979.

Table 8-1. U.S. and world natural gas resources in 1981, trillion cubic feet

	U.S.	World
Cumulative production to Jan. 1, 1981	603	1,086
Proven reserves	191	2,639
Possible additional potential resources	913	4,701
Original-in-place (illustrative)	1,707	8,426
Population, million	229	4,481
Original endowment, per capita, million cubic feet per person	7	2
Annual production, 1980	20.1	52.8
Reserves to production ratio, years remaining		
Proven reserves	10 years	50 years
Additional potential plus proven reserves	85 years	139 years

Sources: National Academy of Sciences, *Mineral Resources and the Environment,* prepared by the Committee on Mineral Resources and the Environment (COMRATE), Commission on Natural Resources, National Research Council, 1975; *Oil and Gas Journal,* December 29, 1980; U.S. Department of Energy, Energy Information Administration, *World Natural Gas* and *Monthly Energy Review;* American Gas Association, *Gas Facts: 1980;* Colorado School of Mines, *Potential Supply of Natural Gas in the United States,* May 1981.

U.S. Natural Gas Production and Consumption

Natural gas is found in geological formations which are similar to oil-bearing formations. About 30 percent of natural gas is associated with oil, but in Alaska the proportion is a much higher 84 percent. Natural gas may be produced from underground reservoirs in which the pressure is thousands of pounds per square inch. The gas itself in original form may have significant amounts of water vapor and sulfur, and may also contain valuable amounts of complex hydrocarbons. These hydrocarbons can be liquefied, and provide gasoline, kerosene, propane, and other products. As a resource associated with natural gas, these hydrocarbons are termed *natural gas liquids.*

After the gas has been cleaned and the valuable hydrocarbons removed, the natural gas is transformed into a homogenous product. It is primarily methane, is standardized at a pressure of 14.73 pounds per square inch,[2] and now averages 1,026 Btu per cubic foot.

Natural gas consumption in the United States has grown very rapidly, increasing fivefold from 1946 to 1973. In part, this is the result of the economic influence of price elasticity. Figure 1-3 showed the decline of natural gas prices over 35 years to their historic minimum in 1970. During this period of growth, the price of natural gas on a Btu basis was always below that of competing types of energy. (In the section on price control later in this chapter, the problem of the interaction of supply, demand, price control, and profit is explained.)

[2]The pressure is actually expressed as "PSIA," and is the absolute pressure difference from a perfect vacuum. It is defined in the context of a 60° F temperature. When natural gas is at 14.73 pounds per square inch PSIA, it is pressurized at twice the normal atmospheric pressure.

Estimated reserves in the United States and the world at the beginning of 1981 are indicated in Table 8-1. It is evident that in terms of original endowment America was very well provided. With 1980 populations, the U.S. original endowment is about 7 million cubic feet per person, three or four times the world average. But our high levels of use create a different future prospect. When proven reserves are examined, the United States has a 10-year reserves/production ratio, while the world average is about 50. When potential reserves are considered as well, the American figure rises to 85 years, and the world average is somewhat less than 150 years of proven plus potential reserves at present production levels.

These terms have the same meaning they did when introduced in Chapter 5. The uncertainty involved in this kind of analysis can be illustrated by contrasting part of Table 8-1 with another source, the U.S. Geological Survey.[3] The entry for U.S. potential resources (913 Tcf) was estimated by the Potential Gas Committee, a joint effort of the Colorado School of Mines and the American Gas Association. This 913 Tcf can be separated into two components: probable natural gas associated with existing fields, and undiscovered resources. For these two components, the Potential Gas Agency estimates amounts of 193 and 720 Tcf.

The U.S. Geological Survey estimates the latter amount (undiscovered resources) at a mean value of 594 Tcf. This, obviously, is a much more pessimistic conclusion. The Geological Survey applies probability analysis to its estimates, and gives a one-in-twenty probability that these undiscovered resources may be as high as 739 Tcf. In other words, the Geological Survey sees less than a 5 percent chance that the mean estimate of the Potential Gas Agency will be correct.

The significance of the difference is that if the Survey's mean estimate is correct, the United States has 50 years rather than 85 years of remaining natural gas resources. The important similarity is that both organizations offer ultimate estimates of American total remaining resources (proven and potential) that would be wholly exhausted in less than 100 years at current production levels.

Natural gas proven reserves (191 trillion cubic feet in Table 8-1) are not evenly distributed throughout the country. Although more than 40 states produce natural gas, 6 states have 85 percent of the reserves. American

[3]See U.S. Department of the Interior, Geological Survey, *Estimates of Undiscovered Recoverable Resources of Conventionally Producible Oil and Gas in the United States, A Summary*, Open File Report 81-192, 1981; and Colorado School of Mines, Potential Gas Agency, *Potential Supply of Natural Gas in the United States* (Golden, Col., May 1981). Other definitions and estimates of reserves are in U.S. Department of Energy, *U.S. Crude Oil, Natural Gas, and Natural Gas Liquids*, annual; American Petroleum Institute, *Market Shares and Individual Company Data for U.S. Energy Markets: 1950–1980*, 1981; American Gas Association, *Gas Facts: A Statistical Record of the Gas Utility Industry*, annual; and the *Oil and Gas Journal*, end-of-year issues.

natural gas is concentrated in Louisiana, Texas, Alaska, New Mexico, Oklahoma, and Kansas. Louisiana and Texas together have 50 percent of the proven reserves, and Alaska has 16 percent.

Proven reserves in the United States are usually consumed more rapidly than new reserves are being found. Each year from 1970 through 1980 experienced declining proven reserves. Reserves increased slightly in 1981, and the 198 trillion cubic feet of proven reserves at the beginning of 1982 exceed the value in Table 8-1. However, they are two-thirds of the 1970 level. The Prudhoe Bay reserves were formally incorporated into these estimates in 1970, which was thus the last year in which new field discoveries accounted for most of the new reserves. For the last several years, new reserves have been found primarily in existing fields and reservoirs.

A departure from the 1970–80 trend occurred in 1981, when new proven reserves roughly equaled production. It is probable, however, that 1970 will be seen as the historic year of maximum proven reserves in the United States. If the Alaskan gas is not actually available to users in the rest of the United States, then the resource situation, of course, will be considerably worse. Alaska's proven reserves of about 32 trillion cubic feet are one-sixth of the country's total.

The Hubbert prediction, incidentally, held that the peak year of U.S. production would be 1978, with 21 trillion cubic feet produced. The actual peak was 1973, with 22.6 trillion cubic feet. Notwithstanding the economic discussion of supply and demand in a subsequent section of this chapter, it is very likely that American production of natural gas from underground reservoirs will have diminished considerably by the end of the century. A stable level of production may be anticipated for much of the 1980s, and a rapid decline in the 1990s.

Theoretically, this decline could be avoided if production and consumption utilized significant amounts of domestic unconventional gas resources, synthetic gas, or imported gas. Unconventional gas exists in very large quantities. One well-known unconventional gas source is in a particular geologic formation known as Devonian shale. In fact, America's first natural gas well in 1821 in Fredonia, New York, produced gas from this formation. Estimates of gas resources in the ground made by the National Petroleum Council range up to 2 quadrillion cubic feet of Devonian gas, equivalent to 100 years at present consumption levels.[4] The problem is economic. Because the gas cannot move through the solid shale, the normal recovery rate is only about 2.5 percent. Increasing the recovery rate might require strip mining, or underground heating.

There is also the possibility of coal gasification, analyzed in Chapter 14. But that too will be much more costly than the conventional large gas fields in

[4]See "Natural Gas from Devonian Shales," in *Potential Supply of Natural Gas.*

use in the latter half of this century. Table 8-1 should be seen as an approximation of gas resources that may be obtainable under known technologies with predictable costs. Unconventional gas resources or imported gas may both be available in large quantities, but at much higher cost. This, as will be seen in following sections of this chapter, has a major influence on future supply and demand interactions in the remainder of the century.

The Natural Gas System and Oil Companies

The three major sectors of the natural gas supply system are (1) production from underground reservoirs and processing, (2) transportation by pipeline, and (3) sale by a utility to customers. Pipelines are used to transmit gas within a state as well as for long-distance transmission over hundreds of miles. Most natural gas is sold by a utility to the user, but some is sold directly by the natural gas producer to users, and some natural gas is used by gas producers or gas pipeline operators as energy for production, processing, and transportation.

Two useful summary statistics help provide perspective on the relative importance of these various facets of the industry: more than half of the marketed natural gas passes through interstate gas pipelines, and three-fourths of the total production is sold by utilities.

Table 8-2 gives national cost data for these three major sectors. Note that prices paid to natural gas producers are increasing rapidly, and pipeline charges to utilities are rising almost as fast. Prices to customers increase

Table 8-2. Representative natural gas prices ($/MBtu)

	1972	1981	?1990?
Price at natural gas (or oil) well paid to producer	$0.18	$1.68	$ 8.50
Price charged to utility by pipeline company	$0.43	$2.63	$11.00
Prices charged by utilities to customers			
Residential	$1.19	$4.35	$15.25
Commercial	0.91	4.00	14.00
Industrial	0.45	3.25	11.40
Average	0.73	3.85	13.50

Sources: The 1972 data are from American Gas Association, *Gas Facts: 1975.* The 1981 values have been estimated by the author from information in U.S. DoE, *Monthly Energy Review.* The 1990 speculative prices are calculated on the basis that, in early 1981, heating oil cost a residential customer $1.25 per gallon, or $9.05/MBtu, that this oil price will increase by one-half by 1990 and that the price of decontrolled natural gas will rise to meet it. Remember that one cubic foot of natural gas usually has 1,026 Btu, so a price of $15.25 per MBtu is equivalent to a price of $15.57 per 1,000 cubic feet.

significantly, but proportionately less than the prices utilities pay for the gas. Because 1972 is the last year before the first oil crisis, it is the benchmark year for price increases. Recall that during the Growth Era, real natural gas prices to consumers declined.

America's major petroleum companies are a significant part of the natural gas industry. The 20 major oil companies studied in earlier chapters had 43 percent of the country's natural gas production, as shown in Table 6-3. Three of those companies (Occidental, Amerada Hess, and Ashland) have relatively little oil or gas production in this country, but they are substantial oil importers. Standard of Ohio/British Petroleum owns very large Alaskan gas reserves, but these are not in production in the mid-1980s. The remaining 16 major oil companies are all in the group of the 20 largest natural gas producers in Table 8-3. Superior Oil, El Paso, Pennzoil, and Panhandle Eastern are the four new companies which are added to constitute the group of 20 largest gas producers. Each of these gas producers of course also produces oil.

As I have suggested, there are important technological reasons why the oil industry is so central in natural gas. The first and most obvious is the common association of natural gas with petroleum in underground reservoirs. The second and related reason is that, even for natural gas disassociated from petroleum, the technologies are similar for extraction from underground wells, collection in field pipeline systems, processing, and transportation by pipeline.

Although information on cooperative activities among gas producers is not publicly available, it may be assumed that affiliations of the kind existing in

Table 8-3. Twenty largest U.S. natural gas producers

1980 rank and company	1980 production billion ft^3	1980 rank and company	1980 production billion ft^3
1. Exxon	1,235	15. Superior Oil*	296
2. Texaco	958	16. El Paso*	286
3. Standard/Indiana	823	17. Conoco (Du Pont)	285
4. Mobil	714	18. Pennzoil*	214
5. Gulf	662	19. Marathon (U.S. Steel)	164
6. Shell	652	20. Panhandle Eastern*	147
7. Standard/California	450	Total	9,508
8. Atlantic Richfield	449		
9. Union/California	426		
10. Tenneco	408	U.S. natural gas production,	
11. Sun	384	excluding NGL	19,602
12. Phillips	345		
13. Getty	313	Largest producers' share of	
14. Cities Service (Occidental)	297	U.S. total	49%

*Not one of the 20 major oil companies in Chapter 6. Parentheses indicate new parent owners since 1980. Data are intended to represent production within the United States.
Sources: Annual Reports 1980, and U.S. DoE, *Monthly Energy Review.*

the petroleum industry are also characteristic of the natural gas industry. These include partnerships and unitization agreements in gas production, and exchange and processing agreements in processing plants and natural gas liquids. Similarly, including *royalty gas* would increase concentration ratios, and it is likely that concentration in gas-producing regions is higher than concentration calculated on a national basis. On this latter point, the Federal Trade Commission surveyed concentration in three major regions in 1974. The 20 largest companies produced an average 92.5 percent in each region. Unfortunately, however, the commission's study examined only companies in its own survey. The surveyed companies in the three regions produced 73 percent of the gas in the four states that include the regions. The 20 largest producers had 92.5 percent of the surveyed gas, so it can be estimated that these 20 companies accounted for 68 percent of total gas production in the four states. If royalty gas and gas produced in the four states outside the three regions are considered, the true concentration ratio is probably between 68 percent and 80 percent rather than the 92.5 percent shown in the FTC survey.[5]

Branded Natural Gas, Pipeline Companies, and Utilities

Why is it impossible to buy brand-name natural gas similar to Exxon, Arco, or other brands of gasoline and heating oil? In physical terms, the logic is comparable. Both crude oil and raw natural gas must be processed and standardized into homogenous products, and both are produced by companies that produce the other. Yet gasoline is branded and natural gas is not. A hint at the solution to the paradox is offered by bottled gas, also known as *liquefied petroleum gas*. It is a natural gas product, but is sold under brand names by petroleum dealers who also market heating oil. Bottled gas, then, is not sold by utilities, and this is the key. Bottled gas is a natural gas product that can be marketed by vertically integrated oil companies and by independent dealers. Natural gas itself is generally sold by utilities, and these utilities usually are not vertically integrated. Therefore bottled gas is branded and natural gas is not.

Most of the country's natural gas is transported by pipeline companies, and the largest 20 account for delivery of 70 percent of the total natural gas consumption. Appendix 8-A lists these companies. The major pipeline companies are essentially independent of both the major gas producers and the utilities. Only two of the largest pipeline companies are petroleum majors, and only four are major gas producers.

The companies operate more than 250,000 miles of pipeline, in every state in the country. In 1979, the pipeline companies paid $26 billion for purchased

[5]See U.S. Federal Trade Commission, Bureau of Economics, *Economic Structure and Behavior in the Natural Gas Production Industry*, February 1979, p. 35 and Appendix A-13, 14, 15; and American Gas Association, *Gas Facts: 1975* (Arlington, Va., 1976), p. 25.

gas, had $7 billion in other net expenses, and earned a net income of $3 billion from gas revenues of $36 billion.

Utilities and Consumption Patterns

The natural gas utilities are the final stage in the supply system. These companies purchase natural gas from the pipelines, and sell it to the users. Table 8-2 indicates that about $1.25 of the $3.85/MBtu which natural gas cost in early 1981 was for gas utility costs. In the future, it is certain that the gas utility will have an even smaller share of customer revenue. This is, as should now be evident, because of the rising field price paid for natural gas, a subject to be taken up at the end of the chapter.

The typical gas utility selling to final customers is privately owned and oriented toward the sole activity of distributing natural gas to its customers. There are, naturally, a bewildering assortment of exceptions to this generalization. There are some public gas utilities. Transmission pipeline companies often sell gas to final customers. Electric utilities often market natural gas. Some gas utilities are attempting to enter the gas production business. And some owners of gas utilities have participated in the development of shopping centers, condominiums, citrus fruit, and other activities to generate tax credits to shelter income from their profitable gas operations.[6] Nevertheless, the typical gas customer buying from a gas utility is buying from a company that does not have major outside interests.

The major categories of natural gas use are the same as those for electricity: residential, commercial, and industrial. Gas utilities have traditionally favored industrial users with lower prices than those charged to commercial and residential customers. The logic for this is economy of scale. The average industrial customer uses about 40,000 MBtu annually, while the average residential customer uses 120 MBtu. It is much less expensive on a cost per Btu basis to provide distribution pipelines for customers using large volumes than for those using small volumes. However, as gas producers gain a growing proportion of natural gas revenue, the price advantage held by industrial customers will decline.

As Chapters 3 and 4 explained, natural gas is substitutable for and competitive with electricity and petroleum in a variety of uses.

Table 8-4 shows typical annual natural gas usage for common appliances in two regions as well as national averages. Notice that there is little regional variation in the amount used for cooking, or clothes drying—or for gas lamps, for those having them. Water heating shows some variation. However, home heating shows considerable difference, an obvious result of climate dif-

[6]Described in California Public Utilities Commission, Finance Division, Tax Unit, "Report On Income Taxes of Southern California Gas Company and Pacific Lighting Service Company," February 22, 1978. Also, see Annual Reports of the latter company, 1977–80.

Table 8-4. Typical residential natural gas consumption, in M̄Btu per year

	New England	Pacific Coast	United States average
Cooking Range in Home	10.1	10.8	10.5
Water Heater	28.4	35.7	32.5
Clothes Dryer	8.6	8.2	7.3
Home Heat	137.2	75.5	106.9
Light	17.8	18.8	18.4
Average residence total, 1980	90.1	81.6	111.4
Percent of residences with gas heat	64%	96%	88%

Source: American Gas Association, *Gas Facts: 1980,* pp. 135, 69, 76. Since many households do not have all these gas appliances, the actual residential averages are less than the sum of the appliance averages.

ferences. The average New Englander uses more gas, but is also less likely to have natural gas heating, because New England is the longest distance from gas producing regions and the most costly to serve. Alaska and California residents paid less than $2.50/M̄Btu in the late 1970s; the New England average exceeded $4.50/MBtu.

In Chapter 3 we noted the greater efficiency and lower cost of natural gas heating in the late 1970s and early 1980s. But while economic cost minimizing should mean customers would always choose natural gas heating, such is not the case. Throughout the latter half of the 1970s, one-half of the new houses and two-thirds of the new apartments and condominiums were built with electric heating.[7] Although these proportions were somewhat lower in the North than in the South or West, the period since the end of the Growth Era has meant increased installation of electric heating.

Several reasons have been noted in other chapters. Probably foremost has been the concern with supply dependability. Most American have felt electricity to be more reliable than declining natural gas resources. In fact, in the middle 1970s, more than half of the country's residences had to face formal gas restrictions.[8] In many states new connections were prohibited. This problem was exacerbated by the price control system and the disagreement between the federal government and the oil and natural gas industry on the proper pricing policies. (More on this in the next section.)

Although concern about reliability is probably the major reason for dramatic increases in electric heating installation, other factors are relevant also. The combination utilities that sell natural gas as well as electricity have historically had greater motivation to promote electric heating, for two reasons. One is, again, reliability: utility management in the 1970s believed nuclear power and coal power to be more dependable than gas supplies. A second

[7]American Gas Association, *Gas Facts: 1979.*
[8]American Gas Association, *Gas Househeating Survey,* annual.

factor is investment. New power plant investment goes into the rate base and earns more income for shareholders as well as gaining considerable tax credits (a point to be discussed in Chapters 12 and 13). Higher gas revenues, on the other hand, usually mean simply larger payments to the pipeline companies that supply the gas. Finally, as was discussed in Chapter 3, electric heating costs less to install, providing an incentive for the contractor to build with electric heat if the buyer is indifferent.

Even with these many influences working against new gas heating, there are instances where buyers of homes with electric heating or appliances are replacing them with natural gas units.

Supply, Demand, and Regulation

Table 8-5 summarizes much of the information available on the natural gas supply system and the consumption of natural gas. The Growth Era can generally be considered to end in 1972 or 1973. Total U.S consumption was at its historic maximum in 1972, and that peak year is compared with 1980,

Table 8-5. Natural gas supply and use: A summary of the data

	1972	1980
Gas supply		
Gas wells, trillion ft³	19.0	17.8
Gas from oil wells, trillion ft³	5.0	3.7
less repressuring, venting, flaring	−1.5	−1.4
Equals marketed production, trillion ft³	22.5	20.1
Energy content, Btu/ft³ (approx.)	1,100	1,092
Marketed production, or "wet gas" production, Q	24.8	22.0
Add net imports, Q	1.0	1.0
Total gas supply available, Q	25.8	23.0
Gas consumption and use		
Change in storage, unaccounted, rounding	0.4	0.3
Natural gas liquids	2.7	2.3
Pipeline used for pumping, etc.	0.8	0.6
Total residential and commercial	7.9	7.6
Gas utility residential	(5.1)	(4.8)
Gas utility commercial	(2.3)	(2.4)
Total industrial	9.9	8.5
Producer sales or uses	(4.0)	(2.6)
Gas utility sales	(5.9)	(5.9)
Total electric utility use	4.1	3.7
Producer sales to electric utilities	(1.2)	(1.2)
Gas utility sales or supply	(2.9)	(2.5)
Total gas consumption and use, Q	25.8	23.0

Sources: Estimated from American Gas Association, *Gas Facts: 1979* and *1980,* and U.S. DoE, *Monthly Energy Review.* There are small amounts of gas sold to government customers which are not indicated separately in the table. Total gas utility sales were 17.1 Q in 1972 and 15.4 Q in 1980.

by which time total gas consumption has declined only slightly. Since net imports are unchanged, the decline in consumption is matched by the decline in *marketed production*. This is in spite of the magnitude of price increases shown in Table 8-2, which might have been expected to increase production, or considerably reduce demand.

Gas production from oil wells has played a decreasing role in natural gas supply in the lower 48 states. However, the high level of Alaskan associated gas will change this trend if Alaskan gas becomes available.

The net imports are from Canada: about one trillion cubic feet a year. Small amounts of exports of natural gas to Canada and Mexico, and liquefied natural gas exports from Alaska to Japan have little impact on the supply situation.

The largest class of gas use shows the greatest reduction. Industrial use declined by 1.5 Q, and this one sector accounts for more than 50 percent of the reduction in total national use. While individual consumers may consider that their own efforts have conserved gas, it is evident that, as a class, the industrial sector contributed most to the reduction in gas use. The industries which have contributed the most to reduced gas consumption are three of the biggest users: paper, chemicals, and petroleum refining. These three industries by themselves have made more than one-fourth of the total American reduction in natural gas consumption.[9]

On the whole, the phenomenon of small decreases in production and consumption (Table 8-5) and large increases in prices (Table 8-2) seems to constitute a logical problem. Why hasn't domestic production increased in response to the 900 percent increase in producer prices? Why hasn't demand changed in response to the 500 percent increase in customer prices and the removal of most restrictions on gas use? The answer lies in the interaction of supply and demand relationships with regulatory policy and remaining reserves.

The regulatory history of natural gas can be summarized according to its impact on each of the three major sectors of the industry. State regulation of the gas utility distributors began early in this century, and by the late 1960s 48 state commissions existed, regulating the price of natural gas sold to customers by utilities operating within their jurisdictions.[10]

As interstate pipelines became important in bringing natural gas from the producing areas to the rest of the country, it became apparent that state commissions could not determine the possible existence of monopoly profit in

[9]Based upon data in *Gas Facts: 1979*, p. 93.

[10]Information on the history and pattern of natural gas regulation is in: Charles F. Phillips, Jr., *The Economics of Regulation: Theory and Practice in the Transportation and Public Utility Industries* (Homewood, Ill.: Richard D. Irwin, 1969); Alfred E. Kahn, *The Economics of Regulation: Principles and Institutions*, 2 vols. (New York: John Wiley, 1970 and 1971); Robert Stobaugh and Daniel Yergin, eds., *Energy Future: Report of the Energy Project at the Harvard Business School* (New York: Random House, 1979); National Association of Regulatory Utility Commissioners, *1979 Annual Report on Utility and Carrier Regulation*, 1980.

the natural gas industry. Consequently, the Natural Gas Act of 1938 created the Federal Power Commission (FPC) to regulate the interstate transportation of natural gas. In 1954, the Supreme Court required this commission to regulate prices charged by producers to pipeline companies. The logic was simple: the FPC could not regulate pipeline prices unless it regulated the prices which pipeline companies had to pay for natural gas.

For perhaps 10 years, from, roughly, 1954 to 1965, the situation was stable. Natural gas prices were regulated at each of the three sectors, gas production increased, consumption increased, real prices to consumers declined, and profit rates were generally satisfactory for oil/gas companies, pipeline corporations, and gas utilities. However, in the middle 1960s, a problem began to emerge. Producer prices for *intrastate* gas were unregulated, while *interstate* prices were subject to rigid control.

The result was the slow emergence of differentiated markets in intrastate and interstate sales. The end of the Growth Era created additional pressure. Rising oil prices led many large and small customers to seek the lower-price gas alternative. The picture in the middle 1970s is illustrated by Figure 8-1. The market price that would lead gas producers to supply Q_e and customers to

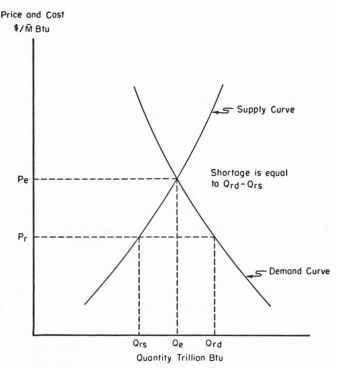

Figure 8-1. The effects of interstate price control in the 1970s: The breakdown of regulation

buy Q_e is the equilibrium price P_e. However, the regulated price P_r is much lower. It leads producers to offer Q_{rs} and customers to demand Q_{rd}. The result is a theoretical shortage, the difference between the two, $Q_{rd} - Q_{rs}$.

In reality, a shortage is precisely what developed. We have already noted the development of widespread rationing restrictions on gas use in the 1970s. The divergence between regulated price and equilibrium market price also became real: in 1976, the average regulated producer price for interstate gas was 48¢/M̄cf while producers were selling intrastate at $1.60.

In 1978, the Natural Gas Policy Act implemented a complicated phase-out of federal regulation of producer prices. Passing over much of the intricate detail, we can roughly summarize the act as being intended to allow much of natural gas production to rise to market-clearing prices by the late 1980s. (This does not do justice to the 61 pages of fine print covering 46 sections and the cornucopia of equations and definitions contained in the act.[11])

The act also terminated the Federal Power Commission, which was succeeded by the Federal Energy Regulatory Commission. FERC is no longer a wholly independent agency, but has been placed within the Department of Energy. In the 1980s, either the Department of Energy or FERC (or both) may be reorganized or terminated, a spectrum of possibilities that simply illustrates the continuing uncertainty in national natural gas policy.

The 1980s may well see the complete termination of producer price control. Whether or not this happens, prices have already risen significantly and rationing has ended.

It is now possible to summarize the net effect of these relationships. Figure 8-1 explains the mid-1970s, when regulation ceased to function in the sense that significant shortages appeared. What is necessary is to understand the economic circumstances that existed before and after Figure 8-1. Why did regulation work in the late 1950s and 1960s, and why does deregulation work in the 1980s?

The upper part of Figure 8-2 shows a hypothetical representation of regulation in the early 1960s. All three sectors of the natural gas industry are regulated according to fair rate of return principles, so price is set equal to average cost. Producers, pipelines, and utilities accept regulation and earn a fair profit. The arrows on the demand curve indicate that the curve itself shifts each year. The quantity produced in regulated markets, Q_r, is increasing each year because growing demand is interacting with portions of the average cost curve which have declining costs. Everyone appears satisfied: real prices decline, costs decline, proven reserves grow, production and sales increase each year, and profit is fair but not excessive.

---

[11]Explained, more or less, in U.S. Department of Energy, Energy Information Administration, Energy Policy Study, Vol. 3, *Pricing Provisions of the Natural Gas Policy Act of 1978*, October 1979.

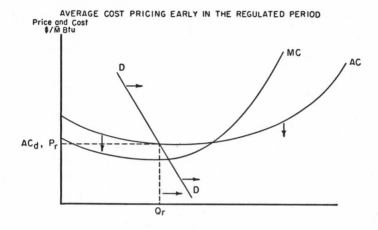

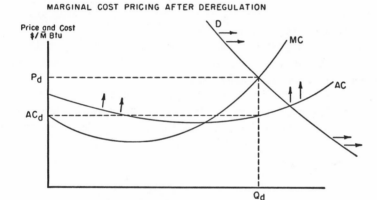

Figure 8-2. Natural gas pricing before and after regulation

Yet it is evident that this market will enter a disturbed period when the demand function shifts into a region where marginal costs of production grow rapidly. Gas producers become unwilling to sell interstate gas at average cost when intrastate gas is sold at the much higher marginal cost. The result, of course, was severe restrictions in interstate gas in the mid-1970s.

The lower half of Figure 8-2 represents the current period. Demand has shifted into a region where marginal cost is steeply rising. Although pipelines and utilities are still regulated, producers can increase real prices according to the rules of the 1978 act. The price P_d defines a market-clearing price for consumption Q_d. No rationing is necessary. However, note that P_d exceeds AC_d; the price in the deregulated period is considerably higher than average cost. Over a 20-year period, the gas producing industry has, on the whole,

[167]

shifted from regulated average cost pricing into a path leading to deregulated marginal cost pricing.

One important question remains. How can natural gas markets work in such a way that real prices grow rapidly while consumption and production are stable or decline slowly? An explanation is seen in Figure 8-3. Since 1973, we have seen rapid increases in the costs of fuels which compete with natural gas. This is true for consumers, as was shown, in Figure 4-2. Chapter 4 also indicated the cost advantage that natural gas had in home uses in the early 1980s is considerable. These factors have also been important for industry. In 1973, electric utilities paid 34¢/MBtu for natural gas and 79¢/MBtu for residual oil. By early 1981, the oil cost was $5.73/MBtu, $3.12/MBtu more than the gas cost. The result is that demand curves for natural gas have been shifting upward since 1973.

Supply curves are shifting backward, however. As low-cost gas reserves are exhausted, as other fuel prices grow, as deregulation allows gas producers rather than FERC to set wellhead prices, these factors interact to lead producers to offer less gas each year for old prices.

The net effect of these simultaneous and opposite shifts in supply and demand is that market prices rise rapidly, and production and consumption remain stable or decline slowly.

Since the lower half of Figure 8-2 is associated with Figure 8-3, these circumstances mean the creation of monopoly profit in the sense of a profit above normal levels. Further, because the major petroleum companies are

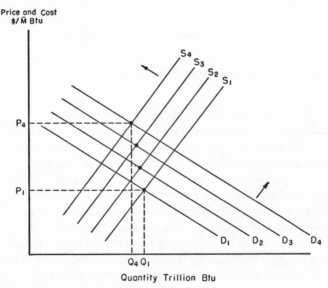

Figure 8-3. Supply and demand with deregulation

[168]

shown in Table 8-3 to be the major interest in natural gas production, these monopoly profits are accruing in large part to the petroleum companies.

Conclusions

To sum up, the history of natural gas use may be that of a modified Hubbert cycle. Figure 8-4 shows the usual characteristics during the Growth Era. Slow growth in the first stage is followed by accelerated production to a peak in the

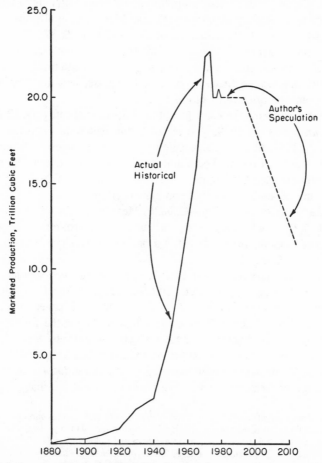

Figure 8-4. The natural gas cycle: Historical and projected.
Sources for the actual historical: Sam H. Schurr and Bruce C. Netschert, *Energy in the American Economy, 1850–1975* (Baltimore: Johns Hopkins University Press, 1960), pp. 492–493; *Historical Statistics of the United States*, p. 595; American Gas Association, *Gas Facts: 1979*, p. 23; U.S. DoE, *Monthly Energy Review*, May 1981, p. 50.

early 1970s. However, a period of arrested decline, the subject of the preceding analysis, has appeared in the late 1970s. My speculation is that the near future in natural gas will resemble the near past. Production and consumption will be stationary while real prices rise very rapidly. Marginal cost of new production grows rapidly, but monopoly profit does also. Sometime in the mid-1990s, I speculate, domestic reserves will begin to be seriously depleted, and natural gas prices will increase even more rapidly while production declines.

Future imports of natural gas will be of two kinds. The most significant will continue to be pipeline gas imported from Canada and, increasingly, from Mexico. Canada is at present supplying about one trillion cubic feet a year, and Mexico essentially nothing. Both countries have significant proven reserves: Mexico holds 65 trillion cubic feet and Canada 87 trillion. In each country, future exports to the United States will depend upon price agreements. Canada and Mexico want Americans to pay a Btu price comparable to that which the United States pays for oil imports, and the American purchasers want the price to be closer to cost.

A similar problem affects the potential of the second category, LNG (*liquefied natural gas*). In 1979, 250 billion cubic feet were imported, all from Algeria. In 1981, however, arrangements collapsed. Algeria, also influenced by the Btu price of oil, wanted a higher price. El Paso Natural Gas estimates its loss on attempted LNG imports to be $550 million.[12] Whether or not LNG has a future depends primarily upon agreed pricing rules.

For each import possibility—Canada, Mexico, LNG—the best near-term forecast for the middle 1980s is a continuation of present relationships: steady but limited imports from Canada, no near-term Mexican imports, and no significant LNG supplies.

Many observers believe that other sources of gas will supplant existing natural gas sources. It is expected that LNG, or *synthetic gas*, or Alaskan gas will become available. These subjects are explored in Chapter 14 and in work by the California Energy Commission, Resources for the Future, and the National Academy of Sciences.[13] My own view is that these technologies will have production costs as high as $10/MBtu in 1990 dollars. It is premature to evaluate their role in the overall natural gas system. I think we can be sure that one or more of these new technologies will be developed, that gas produced

---

[12]See El Paso Natural Gas Company, *1980 Annual Report*, p. 30, and Securities and Exchange Commission Form 10-K Report for the same company for 1980, pp. 1–3. Gas utilities in California are considering LNG imports from Alaska and Indonesia.

[13]California Energy Commission, *Energy Tomorrow: Challenges and Opportunities for California 1981*, Biennial Report, January 1981, chap. 6; Sam H. Schurr et al., Resources for the Future, *Energy in America's Future: The Choices before Us* (Baltimore: The Johns Hopkins University Press, 1979), pp. 256–268; National Academy of Sciences, *Energy in Transition 1985–2010*, Final Report of the Committee on Nuclear and Alternative Energy Systems, 1980, pp. 173–178.

will be very costly, and that the cycle of growth and decline in natural gas from underground reservoirs will continue its own path without much influence from these new technologies.

In 1990, the situation may be rather different. For the present, caution is a better bet.

Appendix 8-A. Sales of Large Interstate Natural Gas Pipeline Companies, 1980

Rank	Company	billion ft³
1	El Paso Natural Gas Company	1,181
2	Tenneco, Inc.	1,113
3	Columbia Gas Transmission Corporation	1,068
4	United Gas Pipeline Company	1,021
5	Natural Gas Pipeline Company of America	946
6	Texas Eastern Transmission Corporation	931
7	Transcontinental Gas Pipeline Corporation	891
8	Northern Natural Gas Company	755
9	Michigan-Wisconsin Pipeline Company	675
10	Panhandle Eastern Pipeline Company	665
11	Texas Gas Transmission Corporation	633
12	Consolidated Natural Gas Corporation	628
13	Southern Natural Gas Company	597
14	Enserch Corporation	543
15	Trunkline Gas Company	485
16	Cities Service Gas Company	376
17	Colorado Interstate Corporation	366
18	Midwestern Gas Transmission Company	328
19	Northwest Pipeline Corporation	322
20	Pacific Gas Transmission Company	309
		13,833

Source: U.S. DoE, Energy Information Administration, *Statistics of Interstate Natural Gas Pipeline Companies: 1980,* October 1981, Appendix G. Total gas utility sales were about 15 trillion cubic feet in 1980.

9

The Evidence: Competition, Monopoly, Profit, and Growth

There is no evidence that the large vertically integrated oil companies are presently exercising monopoly power in any of the four major stages of the oil business: the production of crude oil, the transportation of crude oil and refined products, the refining of crude oil, and the marketing of refined oil products. Indeed, all available evidence supports the opposite inference.
—Richard Mancke and the American Petroleum Institute, 1976

On the basis of the voluminous record discussed above, the Committee finds a number of conclusions inescapable. First, a handful of large integrated companies have effectively monopolized the supply of domestic crude oil. With their control over crude, these firms have been able to restrict entry into refining and limit the competitiveness of independent refiners. This, in turn, has enabled them to force on the consuming public a wasteful, inefficient, and extremely costly system of marketing.
—U.S. Senate Judiciary Committee, 1976

The two statements that introduce this chapter are in complete opposition. Each sees the petroleum industry as having four stages, and each sees the vertical integration of the majors as a significant characteristic. However, the Senate Judiciary Committee perceived a system that was an effective monopoly which created waste and excessive cost for petroleum consumers.[1] In contrast, Professor Mancke and the American Petroleum Institute conclude that there is no evidence of monopoly power at any stage of the industry.[2]

Does the endorsement of Mancke's analysis by the American Petroleum Institute mean that his conclusion should not be accepted? The answer is a qualified no. Suppose Mancke has asserted that the profit rates for oil companies were below average before 1973, as he did in the same analysis cited

---

[1]U.S. Senate, Committee on the Judiciary, *Petroleum Industry Competition Act of 1976,* Report to accompany S.2387, June 28, 1976, p. 147.

[2]Richard B. Mancke, "Competition in the Petroleum Industry," in *The Future of American Oil* (Washington: American Petroleum Institute, 1976), pp. 130, 131.

[172]

here.[3] Should this assertion be rejected because of his relationship to the industry? The answer is suggested by Figure 2-5. Petroleum industry profit rates were not above average, and Mancke's assertion is factually correct. If his statement were to be denied because of his relationship to the industry, then we would be rejecting a statement that is apparently true.

Bias has several meanings. The most widely understood meaning is that an individual or group has a clear financial self-interest in a particular interpretation of events. As the preceding example indicates, it is illogical to reject statements just because they are associated with an interested party. Obviously, the self-interest of the American Petroleum Institute is such that statements of fact and conclusion must be subjected to scrutiny before acceptance or rejection. It should be clear, however, that the position taken here is that the industry's view of the economic world should be one factor in an independent evaluation of that world.

The statistical meaning of bias is rather different. It implies a structural inability to estimate the true nature of the system being studied. It means, literally, that the investigator's estimates are always predictably high or low. In terms of pure statistics, there is no basis for preferring either positive or negative bias.

This should be the goal in interpretation of theory and evidence: factual accuracy, neither too high nor too low in its view of the industry.

The economic theories introduced in Chapters 1 and 2 have several contrasts. The theory of social optimality emphasizes public welfare in an economic sense, searching for national efficiency and the maximum possible net social benefit. The focus of this chapter does not include the social optimum. The objective is smaller but still sufficiently difficult. The behavioral theories will be examined for their ability to be reconciled with the empirical characteristics of the industry that have been described in the preceding chapters.

A brief summary is in order. The competitive theory assumes the industry is motivated by the maximum profit rate on its invested capital. The competitive structure has thousands of independent producers, and no single producer or group of producers can intentionally influence aggregate industry prices or production levels. The pricing rule is price equals marginal cost.

The profit monopoly theory postulates the same goal: maximum profit rate. However, the industry structure is such that a small number of producers can determine industry production and price levels. The pricing rule which characterizes this theory is marginal revenue equals marginal cost. Monopoly permits marginal revenue to be differentiated from price.

The growth monopoly concept is also characterized by a monopolisitc structure in the sense that a group of producers can determine industry price and production levels. The goal of corporations in this theory is maximum

[3]Mancke, p. 122.

[173]

possible sales rather than maximum profit rate. The pricing rule is price equals average cost.

The predicted consequences of the theories of behavior are quite different. Growth monopoly has the highest production and consumption levels with no excess profit and the lowest consumer prices. The profit monopoly theory predicts the highest profit rate and price level, and the lowest consumption level. A competitive industry would have price, profit, and consumption at levels between those of profit monopoly and growth monopoly. The competitive theory has the virtue of giving the socially optimal results if there are no external effects in public health or environmental quality.

Oligopoly is a term developed to suggest circumstances intermediate between profit monopoly and competition. An oligopoly with a small number of dominant firms approaches profit monopoly; an oligopoly with many firms who do not coordinate pricing and production decisions will come close to a competitive industry in behavior.

The Evidence

Most of the evidence available on the petroleum industry is ambiguous in that it logically supports more than one theory. But in three areas, the evidence points in one direction alone, to the competitive view. National concentration ratios are the most commonly applied tests for determining whether the structural conditions exist for a competitive industry. These were examined at every stage of the petroleum industry, and the industry was seen in Chapter 6 to be far less concentrated than other major industries.

A second area where empirical statistics support the competitive industry interpretation is in the large number of independent firms in the oil business. There are more than 10,000 oil/gas producers, 176 refiners, and 200,000–300,000 individual companies engaged in marketing petroleum products.[4]

Finally, the real competition in exploration and leasing is a clear argument favoring the competitive view.

In other areas the evidence is considerably more ambiguous in its implications. Profit rates were clearly normal for the industry in the 1946–72 period and again in 1975–78. However, in 1973–74 and again in 1979–81, petroleum industry profit rates have been very high. For the Growth Era, this evidence clearly contradicts the profit monopoly theory, and supports either the competitive theory or the growth monopoly concept. But the wide fluctuations in profit rates since the end of the Growth Era give no clear picture of the current state of the industry.

Real prices for petroleum products have been shown to decline throughout the history of the industry to 1972. This evidence supports the competitive

[4]These estimates were made by persons at the American Petroleum Institute in mid-1981.

position or the growth monopoly theory. The price increases of 1973–74 were followed by four years of stable real prices, then major increases in 1979 and 1980. Again, the picture after the end of the Growth Era is unclear.

The accelerated use of petroleum products throughout the 1859–1973 period has been noted, indicating that consumption increased 100-fold from 1900 to 1973. Again this is consistent with two of the theories, but not with profit monopoly. It is most improbable that a profit monopoly would have pursued such an expansionary policy, with that policy coupled with declining real prices and profit rates which were wholly normal.

Since 1973, U.S. petroleum consumption has ceased to grow, and in the early 1980s it is actually less than in 1973. This new pattern—the absence of exponential growth since 1973—is least compatible with the growth monopoly perspective.

The history of OPEC (Chapter 5) actually supports either the competitive or growth monopoly positions, and contradicts profit monopoly prior to 1973. The logic is as follows. It is clear that, in 1973 and thereafter, OPEC has exercised profit monopoly power in arranging pricing and production levels which guarantee monopoly profit. However, it should be equally clear that the international oil companies held similar power before 1973. During their period of maximum power, the eight international companies held between 70 and 90 percent of crude oil production outside North America and Europe.[5] The companies did not exert their control to attain the high prices achieved by OPEC. Indeed, it should be recalled that OPEC was formed in order to work against the companies' policies of lower prices. The actual behavior of the international companies prior to 1973, in conjunction with their considerable control, argues against the profit monopoly interpretation.

The high degree of international concentration held by the companies in the Growth Era is compatible with either theory of monopoly. We know that production decisions were carefully coordinated, and that prices were as well. International behavior in this period, then, supports either competition or growth monopoly. However, the international structure at that time was more consistent with either of the monopoly theories.

In three areas of interest, the evidence can probably be viewed as consistent with any one of the three theories. Regional concentration as described in Chapter 6 gives a clear indication that, in individual states, the leading four companies hold a much higher proportion of sales or refining than is evident in national statistics. However, the degree of such regional concentration is not sufficiently high to be clearly negative in its implications for the competitive view.

The importance of major oil companies in natural gas production is also subject to several interpretations. Chapter 8 shows that the 20 largest gas producers include 16 major oil companies. This structural association of

[5]See the section on world petroleum marketing in Chapter 5.

majors and gas production is compatible with either monopoly theory. Alternatively, the physical association of natural gas and oil and the technological similarities of their production make these dual activities absolutely inevitable for most oil companies. Indeed, an oil company that attempted to avoid natural gas production would be not only wasting this resource, but imposing considerable financial loss upon itself. A competitive industry would require every petroleum corporation to make efficient use of its natural gas opportunities.

Vertical integration is the third facet of the industry compatible with all theories. As Chapter 6 suggested, the interaction of economies of scale and stability in production, refining, and marketing results in more efficient production practices. The existence of competing vertically integrated companies must be accepted as a real possibility. Obviously, however, vertical integration is itself compatible with either of the monopoly theories.

Monopoly and the Evidence

Ten distinct characteristics of the industry were discussed in the preceding section, and each was deemed either supportive of or compatible with the competitive theory. Several areas remain which are consistent with one or both monopoly theories.

Cooperative activities between companies were described extensively in Chapter 6. Partnerships in wells and leases, unitization agreements, processing arrangements at refineries, and trading in finished products are all efficient in the sense that they increase ultimate recovery or reduce costs, or both. Yet this extensive network of efficient cooperation does not easily blend with the theory of independent companies competing with one another.

Similarly, sound economic reasons underlie the multiple affiliations among oil companies in ownership, management, and finance described in Chapter 7. Yet the existence of these affiliations is not easily reconciled with the competitive theory.

In Chapter 6 I also noted the considerable stability in industry leadership in two areas. First, the leading state and regional retailers in the current period are descended from the companies that held similar territories 70 years ago in the Standard Oil Trust. Second, the identity of the leading seven companies and the leading 20 companies has remained quite stable over periods in excess of 50 years. This kind of stability is not an obvious characteristic of a competitive industry, and this evidence more readily fits the two monopoly theories.

The nature of the organization of the Alaskan oil holdings also more nearly fits the implications of monopoly rather than competition. It will be recalled that Prudhoe Bay covers more than one-third of American proven crude oil reserves, and that this field is essentially owned by three companies (Sohio/ BP, Exxon, Atlantic Richfield) and operated by two (Sohio/BP and Atlantic Richfield). Again, it must be kept in mind that unitized operation is essential for maximum recovery from the Prudhoe Bay sectors. Yet this degree of dominance, however efficient, is not readily made a competitive influence. Since Prudhoe Bay did not become a major producing area until the mid-1970s, the significance of Alaskan arrangements is greater for the present period than for the Growth Era.

Price controls in natural gas and oil are relatively clear in their implications. They presume some degree of monopoly power, and seek to lower prices and profits below the levels that would be attained in their absence. As Chapter 8 indicated, during the period 1954–78 consistent attempts were made to regulate natural gas prices at each of the three major sectors of the natural gas industry: production, pipeline transmission, and utility distribution. While it appears that elements of marginal cost pricing were partially present in some producing-area price regulation, the system essentially worked on the basis of average cost pricing. This, as we have seen, is the outstanding characteristic of the growth monopoly theory.

Petroleum price regulation was basically absent throughout most of the Growth Era. It did not begin until wage and price stabilization programs were enacted in 1971. By the late 1970s, OPEC-induced international price increases were met by a variety of domestic U.S. regulations. These laws and regulations defined prices for several sources of crude oil and product types. Additional regulations affected refineries, jobbers, and retailers.

One specific program, the "Entitlements Program," had as its explicit goal the equalization of costs to refiners. Companies with access to low-cost old oil had to provide financial assistance to companies without the same access.[6]

Overall, the objective was to lower prices to consumers by means of a set of regulatory policies that brought about a situation considerably closer to average cost pricing than to marginal cost pricing or profit monopoly pricing. The implication of this set of policies is that the federal government assumed the existence of some degree of monopoly, and attempted to control its pricing policies. One study ended as follows: "The general conclusion, then, is that the oil regulations, taken as a whole, result in generally lower prices for

[6]These price regulations were the subject of considerable analysis in the late 1970s and 1980. Some of these analyses are: U.S. Federal Energy Administration, National Energy Information Center, *Crude Oil Entitlements Program*, January 1977; Scott Harvey and Calvin T. Roush, Jr., *Petroleum Product Price Regulations: Output, Efficiency and Competitive Effects*, Staff Report of the Bureau of Economics to the Federal Trade Commission, February 1981; U.S. Department of Energy, Energy Information Administration, *The Impact of the Entitlements Program on the Market for Residual Fuel Oil on the East Coast*, February 1980, and *Price Controls and International Petroleum Product Prices*. February 1980. Also from DoE, EIA, are: Energy Policy Study, Vol. 1, *Effects of Oil Regulation on Prices and Quantities: A Qualitative Analysis*, October 1979, and Energy Policy Study, Vol. 2, *The Effect of Legislative and Regulatory Actions on Competition in Petroleum Markets*, October 1979.

petroleum products and hence greater quantities used than would be true in a free competitive market."[7]

Chapter 2 above reported the results of the Flaim and Mount quantitative analysis of the predictive ability of the three behavioral theories.[8] Neither the profit monopoly nor the competitive theory was able to account for the high levels of consumption in the 1960s and early 1970s. Nor could these two theories explain the actual price levels, because those levels were much lower than would be expected from the competitive and profit monopoly theories. Their study concluded that only the growth monopoly theory gave an accurate explanation of consumption and price levels.

But an important qualification is necessary: Flaim and Mount found that the growth monopoly theory ceased being accurate in 1974; the 1974–77 period did not fit the growth monopoly predictions.

The author's view has been much influenced by the empirical evidence as it has been reviewed and by the Flaim-Mount study. The long period 1859–1973 seems to fit the growth monopoly concept. The implication, then, is that both the structure and behavior of the industry in this era were oriented toward efficient production, low prices, normal profits, and accelerating consumption. Internationally, the industry sought to attain imported oil at the lowest cost, and passed these cost savings along in the form of lower product prices. These advantages to the U.S. economy as cited here are obvious. The disadvantages are less obvious, but may be put simply. American oil and gas resources have been depleted more rapidly than they might have been, given a different orientation for the industry, and the existence of large, growing, stable supplies of gasoline at declining real prices contributed to America's dependence upon automobile and air travel.

For the period since 1973, interpretation is more difficult. The weight of the empirical data no longer supports the growth monopoly view. OPEC dominates the private U.S. companies, British Petroleum is the largest holder of American oil, and Royal Dutch Shell is one of the country's leading gasoline retailers. The combination of declining U.S. resources and lesser influence over foreign countries' pricing decisions is a new factor in the global petroleum market. Almost any view could be supported for the post-1973 period. Perhaps it is a period in which the U.S. petroleum industry passes from its century-long pursuit of growth to a phase of mixed oligopolistic competition and profit monopoly.

Whether or not these views are compatible with the concept of workable competition will be left to the conclusion of the book.

As with Chapter 7, we may ask here if it matters which theory best fits the petroleum industry. Consider the implications of these theories for other

[7]U.S. DoE, *Effects of Oil Regulation on Prices and Quantities*, p. 82.
[8]Figure 2-3 in Chapter 2, and discussion.

energy sources. First, as shall be seen, the oil industry is important in coal, nuclear fuel, and solar energy. If the petroleum industry is seen as a growth monopoly, then the logical prediction is premature development of high levels of these three energy sources. Alternatively, if the profit monopoly view is held, then we expect the petroleum industry to seek to retard the efficient development of coal, nuclear fuel, conservation, rail and public transportation, and solar energy. Finally, if the industry is perceived to be competitive, then it can be assumed that the oil companies will, by following normal economic incentives, work toward proper levels of production and consumption with normal profits and efficient prices.

The Growth Era

This phrase first appeared in Chapter 1 and has made several subsequent appearances. With the preceding eight chapters as a foundation, it can be given a first definition here.

Examining the economic history of the oil/natural gas industry makes it apparent that several economic factors have operated in a parallel or congruent pattern. As I have observed repeatedly, the period from 1859 to 1973 is characterized by ever-declining real prices and exponential growth in consumption. Domestic production of oil and natural gas has a historical profile that resembles the Hubbert cycle, and peaks in the early 1970s. All industrial sectors of oil and gas generally earned profits at normal, or somewhat below normal, rates. This is true for petroleum corporations, natural gas pipeline companies, and natural gas utilities. American corporations participated in foreign concessions which controlled almost all the crude oil in world trade, and obtained this oil for domestic customers at low cost. In addition, the major oil companies and the other elements of the oil and gas industries established an effective network of cooperative activities in exploration, production, refining, transportation, and marketing. These cooperative activities made a substantial contribution to the stable, efficient production of oil and gas at low costs and prices.

These several facets of the pre-1973 period are all contained in the Growth Era concept. As we have seen, all of these characteristics except cooperative activity are unstable after the Growth Era.

This, then, is a first definition as relevant to American oil and natural gas use. Succeeding chapters investigate coal, nuclear power, electric utilities, environmental analysis, alternative technologies, and the macroeconomic context of energy use. In the last chapters, the Growth Era concept is to be examined again in this more complex context.

IV

COAL, NUCLEAR POWER, AND UTILITIES

10

Coal

Resources for the Future (RFF), a nonprofit, independent research organization, recently published a study that includes comparative energy reserve data.[1] It is shown in Table 10-1 in a simplified form.

Much of this chapter might be considered as the placing in context of Table 10-1. With 1.8 trillion tons of recoverable reserves and 38,000 Q in those reserves, coal has an estimated 93 percent of America's conventional and nuclear energy resources. Proven reserves alone in Table 10-1 are 260 billion tons, enough for 370 years at the 1980 consumption level of 703 million tons.

Coal is noteworthy in many respects: it has the most future reserves, the longest past history, the most fatalities, the worst labor problems, the greatest air pollution impact, and the lowest cost on a $/Btu basis.

A cautionary note: the RFF reserve estimates differ from those given previously for oil in Chapter 5 and natural gas in Chapter 8. This variation is itself caused by variation in specific assumptions relating to heat content (i.e., Btu per barrel, ton, or cubic foot), definition of types of reserves, grouping of fuel categories (e.g., NGL with oil), dates of estimates, and recovery rates. It should by now be evident that this inconsistency is a recurring characteristic of resource studies.

Defining Coal

In Chapter 3, coal was simply described as being primarily carbon in forms of peat, lignite, bituminous, or anthracite, and having an average Btu content of 22.14 MBtu/ton and 11,070 Btu/lb. In fact, there is considerable variation in the energy content of types of coal, and variation in the characteristics of coal from different locations.

[1]Sam H. Schurr, et al., Resources for the Future, *Energy in America's Future: The Choices before Us* (Baltimore: The Johns Hopkins University Press, 1979), p. 226.

Table 10-1. The Resources for the Future estimates of energy reserves in the United States

Resource	Units	Proven reserves	Possible additional reserves	Total ultimate	Quadrillion Btu	Percent Q
Coal	billion tons	260	1,543	1,803	37,863	93%
Oil & NGL	billion barrels	40	127	167	921	2%
Natural gas	trillion ft³	209	686	895	917	2%
Uranium	thousand tons	890	2,910	3,800	1,140	3%
					40,841	100%

Notes: The uranium resources are assumed to be used in light water reactors. If fast breeders are assumed, the same amount of uranium provides 68,400 Q instead of 1,140 Q. (Chapter 11 describes this aspect of the nuclear fuel cycle.) For reference, the 1981 values for consumption are 16 Q for coal, 32 Q for oil and natural gas liquids, 20 Q for dry natural gas, and 3 Q for nuclear power. Major imports were 11 Q of crude oil and petroleum products, and 1 Q of natural gas.
Sources: Schurr, et al., *Energy in America's Future*, p. 226, and U.S. DoE, *Monthly Energy Review*, February 1982.

Table 10-2. Representative types of coal

Type of coal rank	Lignite	Subbituminous	Bituminous	Anthracite
Power Plant Data				
Representative location	McLean, N. Dakota	Sheridan, Wyoming	Muhlenberg, Kentucky	Lackawanna, Pennsylvania
Physical composition				
moisture	37 %	22 %	9 %	4 %
volatiles	28 %	33 %	36 %	5 %
fixed carbon	30 %	40 %	44 %	81 %
ash	6 %	4 %	11 %	10 %
total	101 %	99 %	100 %	100 %
Chemical composition				
sulfur	0.9%	0.5%	2.8%	0.8%
carbon	41 %	54 %	65 %	80 %
other	58 %	45 %	32 %	19 %
total	100 %	100 %	100 %	100 %
Energy, Btu/lb	7,000	9,610	11,680	12,880
1979 National Quantities				
Sulfur content	0.7%	0.5%	2.2%	0.6%
Billion tons estimated reserves	42.9	182.4	241.9	7.3
Million tons consumed by utilities	37.9	110.7	406.8	1.1
Per MBtu price paid by utilities	$0.56	$1.00	$1.31	$0.80

Sources: U.S. DoE, *Coal Data; Cost and Quality of Fuels: 1979;* and *Demonstrated Reserve Base of Coal in the United States on January 1, 1979,* May 1981.

The most common classification scheme for coal is according to *rank,* and is based upon the geological process of metamorphism from plant material to high-carbon anthracite. Peat is not generally considered coal, but is a stage of development that precedes coal. There are thirteen standard ranks for U.S. coal; four major types are lignite, subbituminous, bituminous, and anthracite. The two most important characteristics in judging rank are carbon content and energy value, but volatility and moisture content may also be relevant.[2]

Several important characteristics of four representative coals are described in detail in Table 10-2. Moisture content declines from lignite through anthracite, and carbon content increases. The energy content ranking is anthracite, bituminous, subbituminous, and lignite, and, excepting anthracite, the prices paid per ton are closely linked in value.

Anthracite is distinct from the first three grades and, although Table 10-2 shows a high heat value, it frequently has less energy per pound than high-grade bituminous coals. In addition, the hardness of anthracite greatly reduces its usefulness for continuous utility boiler operation. But it was precisely this

[2]A good general description of coal rank is given in Douglas M. Considine, ed., *Energy Technologies Handbook* (New York: McGraw-Hill, 1977), pp. I-15–I-27.

solid quality which once created its popularity as a coal fuel in home and business furnaces. Pennsylvania by itself has 97 percent of anthracite reserves.

Lignite is the lowest quality of commercial coal, high in moisture and low in carbon and heat content. It crumbles and is not easily transported or stored. Consequently, it is frequently burned at the mine for power generation. Lignite deposits are concentrated in the plains and mountains states; Montana, North Dakota, Colorado, and Texas have 97 percent of American lignite reserves.

Subbituminous coal is intermediate in quality between lignite and bituminous coal. It generally has more moisture and less carbon and energy content than bituminous coal. There is, as with lignite, a tendency for this coal to be used in *mine-mouth power plants*. Montana and Wyoming have 93 percent of the subbituminous coal reserves in the United States.

It should be evident that *bituminous* coal is the most important class of coal being used. It is high in carbon and energy content, lowest in moisture, and physically usable in continuously operating boilers. Three-fourths of the coal used by utilities is bituminous, and the subbituminous is another one-fifth. Lignite provides 7 percent, and anthracite only two-tenths of 1 percent.

Unlike the three other classes, bituminous coal can be found throughout much of the country. Demonstrated reserves are found in 27 states. The four states with greatest reserves—Illinois, Kentucky, Pennsylvania, and West Virginia—have 71 percent of the country's demonstrated reserves.

But bituminous coal has one other important characteristic: it has the highest sulfur content, and this is one of the most significant problems impeding its use. A major portion of Chapter 13 will consider the coal-sulfur-air pollution problem, which is of considerable importance.

Coal is mined by several methods. The various technologies (long wall underground, contour stripping) are described by Earl Cook[3] and in other publications. The significant divisions are *surface mining* and *underground mining*. Surface mining produced only one-fifth of the coal in the late 1940s, but now accounts for two-thirds of U.S. coal production.[4]

The end of the Growth Era for oil and natural gas caused an increase in coal use. However, almost all of the growth in coal production has been in surface-mined coal and Western coal, rather than deep-mined Eastern coal.

When coal leaves the mines, it is usually cleaned in some mechanical manner, or washed to reduce the proportion of noncoal impurities. It is sometimes shipped by barge or truck, but two-thirds is shipped by rail. Because of transportation costs, there is more regional variability in coal cost

---

[3]Earl Cook, *Man, Energy, Society* (San Francisco: W. H. Freeman, 1976), pp. 83–100.
[4]From data in U.S. Department of Energy, Energy Information Administration, *Cost and Quality of Fuels for Electric Utility Plants,* annual and monthly, various issues, and *Coal Data,* September 1978; U.S. General Accounting Office, *Low Productivity in American Coal Mining: Causes and Cures,* March 3, 1981.

than in other major energy forms. In early 1981, New England Power in Massachusetts paid $66 per ton ($2.53/M̄Btu) for 1.2 percent sulfur coal, which probably was bituminous coal shipped from West Virginia from an underground mine. At the other extreme, Montana Power Company paid $7 per ton (42 ¢/M̄Btu) for 0.7 percent—seven-tenths of 1 percent—sulfur coal, which was probably subbituminous or lignite surface-mined in Montana. The great importance of transportation cost can be illustrated with Western coal. In 1979, Wyoming coal with 0.5 percent sulfur cost Wyoming utilities $8 per ton. Wyoming coal with the same sulfur content and higher energy content which was shipped to Illinois cost utilities in that state $35 per ton. As a national average, the delivered cost of about $30 per ton for early 1981 probably included an average transportation cost of about $10 per ton.

The geographic distribution of low-sulfur coal has much to do with air pollution politics and policy, as will be seen in Chapter 13. Coal produced in the mountains and plains states is between 0.5 and 0.67 percent sulfur, but coal from Ohio and western Kentucky averages more than 3 percent sulfur. Appalachian coal in West Virginia, Virginia, Pennsylvania, and the South has medium sulfur content: these districts average 1.5 percent sulfur. Although the western coal has lower energy content, sulfur per M̄Btu is nevertheless much lower in Western coal.

These variations in coal type, heat value, sulfur content, and location interact to create quite different economic conditions for the different coal types. Surface mining is usually employed for coal at depths less than 200 feet, and underground mining at greater depths. However, this is only a general rule: underground mining may be necessary at lesser depths, and surface mining may be deeper. A final anomaly: low-sulfur coal tends to be at lower depths. Consequently, about two-thirds of the known low-sulfur coal is believed accessible by underground mining, and most of this is in the West.

Bituminous and anthracite coal can be mined in seams of thickness of at least two feet.[5] At the other extreme, subbituminous and lignite coal are considered to be usable if they are in seams at least five feet thick. In Wyoming, coal beds may be 100 feet thick.

Recovery rates vary also. Pennsylvania anthracite from surface mines achieves a 90 percent recovery. For other coals, surface mining may achieve 80 percent recovery, and underground mining 50 percent.

Social Cost and Surface Mining

In the past 40 years, the term surface mining has come to be used more frequently than the older phrase, strip mining. In part, this is because the older

[5]This information is from U.S. DoE, Energy Information Administration, *Demonstrated Reserve Base of Coal in the United States on January 1, 1979*, May 1981.

term acquired a pejorative connotation during the period in which the process originated on a large scale in the coal states of Appalachia.

Harry Caudhill, a former Kentucky legislator and author, gave this unfavorable description of one stripped area in Kentucky thirty years ago:

> The community of Upper Beefhide Creek in Letcher County was stripped between 1950 and 1954. Highwalls ten miles long and forty feet in height were created. Earth and rock were tumbled down the mountainsides and raced through fields of grass and corn and blanketed lawns and vegetable gardens. Stones big as bushels were blasted high into the air like giant mortar shells, to rain down on residences, roads and cornland. One huge rock crashed through the roof of a house, struck a bed dead center and carried the spring and mattress into the earth under the foundation. Another fell on a cemetery, smashing a tombstone and crushing the coffin six feet underground . . . seven years after the stripping ceased three quarters of the population had moved away.[6]

In the 1980s, such events are and should be unusual and perhaps nonexistent. Throughout the 1960s and 1970s, state and federal legislation created new reclamation requirements for surface mining.[7] Ohio, for example, requires each layer of soil on good farmland to be stored, and replaced after mining. The crop yield is supposed to equal or exceed the yield before mining. Original hill or mountain contours must be restored in Kentucky and Pennsylvania. West Virginia prohibits a "highwall" that rises more than 30 feet above the floor of a surface mine or descends more than 60 feet below level ground.

By 1975, 23 states had reclamation laws. In 1977, Congress enacted the Surface Mining Control and Reclamation Act. Federal law is similar to those in the states. It requires restoration of original use capability, original land contours, and restoration of prime farmlands. Federal standards supercede weaker state regulations, but more stringent state standards are acceptable. One consequence of this federal minimum standard is that it significantly reduces the possibility of individual states or companies attempting to compete by operating surface mines without reclamation.

It is surprising to see that per-acre cost of reclamation shows little variation for standardized requirements. Walter Misiolek and Thomas Noser estimate that, if one foot of topsoil is required in each state, the per-acre cost ranges from $6,500 in Oklahoma to $8,000 in Ohio. However, the dramatic variations in coal seam thickness cause very high variations in per-ton cost:

---

[6]Harry M. Caudhill, *Night Comes to the Cumberlands* (Boston: Little, Brown, 1962), pp. 316, 317. Caudhill's later *My Land Is Dying* (New York: Dutton, 1971) contains photographic material with the same perspective as his cited quotation.

[7]The state reclamation programs are summarized in U.S. GAO, *Low Productivity*, pp. 68–70. The federal act is explained in U.S. Department of Agriculture, Rural Development Research Report, Number 19, *Costs of Strip Mine Reclamation in the West*, February 1980, pp. 7–10.

15¢/20¢ in Wyoming and Montana, but more than $2 in Missouri and Oklahoma. This is generally 5 to 10 percent of total mine coal cost.[8]

In one respect, the economic effect of surface mine regulation is comparable to the effect of underground safety and health regulation. The cost increase caused by labor and equipment allocated to topsoil storage, contour grading, and revegetation is effort that otherwise would produce coal. For major surface-mining states, the General Accounting Office study cited above reported that productivity in tons per miner generally declined about 43 percent from peaks in the late 1960s and early 1970s to lower levels at the end of the decade.[9]

In general, surface mine regulation should be judged a success as of 1982. In the abstract, the external cost has been reduced by regulation, and the market cost has been increased. It seems likely that total social cost—the sum of these two costs, as well as subsidies—has been reduced.

Labor

The generally available literature on energy and coal takes a managerial perspective in the sense that the focus is upon maximum productivity and minimum cost. The same perspective is taken here. Yet, coal production more than any other energy form is affected by labor relations. The labor history of coal includes men and women of epic proportions. Two of them have become part of national history as well. "Mother Jones" worked for decades as a labor organizer in the coal fields and faced armed confrontations, jail cells, and governors with equal militancy. John L. Lewis, the legendary leader of the United Mine Workers (UMW) earlier in the century, led the nearly complete unionization of the industry and oversaw the creation of hospitals, health care, and pension systems for union miners.

However, UMW history also has its dishonorable facets. After Lewis, A. C. "Tony" Boyle was UMW president from 1963 to 1972.[10] Boyle was convicted in 1974 of ordering the assassination of his rival for the union presidency, "Jock" Yablonski. Yablonski's campaign for union office was based upon health and safety issues as well as claims of Boyle's corruption,

---

[8]See Walter Misiolek and Thomas Noser, "Coal Surface Mine Land Reclamation Costs," *Land Economics*, 58:1 (February 1982), 77–79. The earlier Department of Agriculture study, *Costs of Strip Mine Reclamation*, estimates reclamation cost to be only 1 percent of total cost.

[9]U.S. GAO, *Low Productivity*, p. 72.

[10]Lewis was UMW president from 1920 until his retirement in 1960. He was succeeded by Thomas Kennedy, whose official term of office was 1960–1963. However, during this period Kennedy was in poor health, and it was Tony Boyle, then vice president, who actually controlled the union. Upon Kennedy's death in 1963, Boyle became president. He campaigned for his first elected term as president in 1964. See Elizabeth Levy and Tad Richards, *Struggle and Lose, Struggle and Win: The United Mine Workers* (New York: Four Winds Press, 1977), pp. 69–77.

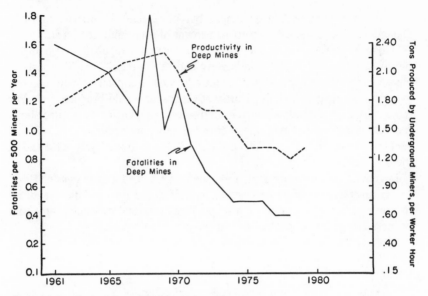

Figure 10-1. Productivity and fatalities in deep mines.
Source: U.S. GAO, *Low Productivity in American Coal Mining,* March 3, 1981, pp. 3, 40.

and these health and safety problems continue to be major issues in the coal industry.

Figure 10-1 shows underground mine productivity and fatality rates. It is clear that there is a correlation between the Coal Mine Health and Safety Act of 1969 (CMHSA) and the subsequent decline in both fatalities and productivity, and this relationship will be taken up below.

First, some perspective on the hazards in underground mining is necessary. In the early part of this century, the fatalities per year were always in the range of 2,000–3,000. As unionization increased, the fatality rate (measured per miner or per miner-hour) declined. In 1931, fatalities finally were less than 2,000. In 1948, fatalities declined to the 1,000 mark, and went below the 500 mark in 1953. However, it was not because underground mining became safer that the number of deaths declined after World War II. The fatality rate (deaths per 500 miners each year) was higher in 1968 than in any other year in the 1946–68 period. The decline in total fatalities immediately after World War II was caused by declining production. The data for the 20 years before 1968 show stable fatality rates and rising productivity.

Underground coal mining is one of the most hazardous occupations. The 1969 legislation created new standards and assigned their enforcement to the Department of Labor's Mine Safety and Health Administration.

The interaction of safety and productivity can be made easier to understand with an illustration of regulations adapted from the U.S. General Accounting

Office study, an excellent source for empirical and objective information.[11] The CMHSA prohibits any miner from working under an unsupported roof. The rock above the seam being mined is supported by "roof bolts" which, literally, bolt the looser rock at the ceiling to the more solid rock above. As a consequence, roof falls were only one-third as frequent in 1977 as they had been in 1969. The General Accounting Office study concludes that the roof-support regulation accelerated better safety in terms of reduced fatalities from roof falls.[12]

But there is a productivity loss. Mining machinery works at a coal face, withdraws for roof bolting, returns to work at the face, withdraws again for roof bolting, all in a continuous process. The regulation that requires better roof bolting requires more miner time at this task, and reduces actual coal-mining time. There is here a literal tradeoff between safety and productivity.

A similar relationship exists with respect to coal dust in underground mines. At the end of the 1960s, 30 percent of miners working at that time had a disease known as *coal worker's pneumoconiosis*, commonly known as black lung disease. It results from the retention by the lungs of very small particles.[13] It is possible that a full 40-year working life in an underground coal mine may create a better than 50-50 probability of experiencing this disease. In its most advanced form, it is similar to acute emphysema, and the simplest physical activity becomes painful.

The CMHSA requires considerable improvement in dust control, which may reduce pneumoconiosis to less than a tenth of the earlier incidence.[14] All the means by which dust is removed involve increased effort and reduced production. Companies must install stronger fans, ductlike air control systems, separate shafts for coal conveyors, water sprays, and increased dust sample monitoring. More time is required to implement these efforts, and the coal seam cannot be worked if the dust control system is being moved into place or is not working.

Analogous practices have been required to meet higher standards for methane monitoring and control. Methane seepage can accumulate in mine areas and be ignited by operating equipment or explosives. The positive results of methane reduction are a dramatic drop in fatality and injury rates from explosions. The tradeoff, again, means a negative impact on productivity.

In summary, the Coal Mine Health and Safety Act required better working conditions with respect to roof and wall collapses, dust and methane control, electrical equipment and machinery safety, more monitoring and sampling, more frequent inspections by MSHA personnel, and more severe enforcement

[11]U.S. GAO, *Low Productivity.*
[12]Ibid., p. 46.
[13]National Academy of Sciences, *Energy in Transition 1985–2010,* Final Report of the Committee on Nuclear and Alternative Energy Systems, 1980, p. 431.
[14]Ibid.

of regulations. One result is clearly a reduction in the number of deaths per ton in mining underground coal, and a probable reduction in pneumoconiosis. The other result is a reduction in productivity. Although surface mining productivity is considerably higher and fatality rates much lower, the CMHSA had the same impact. Both fatality rates per miner and productivity per miner were reduced by the act.

International comparisons support the same conclusions. British and German productivity rates in underground mines are both lower than American levels. But fatality rates per miner are lower, and this is in spite of mines which are on the average said to be deeper, have more stress on roofs, walls, and floors, produce more methane, and have thinner and steeper seams. The General Accounting Office report concludes that European mines invest more heavily in health and safety protection. (It should be noted that the lower productivity is apparently caused by less favorable coal geology as well as by greater safety emphasis.[15])

It would be logical to suppose that the problem of the productivity/safety and health tradeoff would lead to increased surface mining. This has in fact happened. Unfortunately, as the preceding section indicated, two-thirds of the low-sulfur coal reserve base is accessible through underground mines. The highest-grade coals are also underground: 80 percent of the bituminous coal and 98 percent of the anthracite.

Just as underground mining has serious labor, health, and safety problems, surface mining has serious environmental problems in terms of land and water impact. As explained in the preceding section, new legislation in the 1970s had an effect on surface mining comparable to the impact of the CMHSA on underground mining. Environmental degradation was reduced, but so was productivity.

Before we leave this subject, it might be worthwhile to note the present economic dimensions of some of these problems. In the 1980s, utilities purchase much more surface-mined coal than underground coal, and they pay considerably less for it. In 1981, the United Mine Workers began a nationwide strike which the union saw as linked to health and safety issues. The union's impact on nationwide coal production is less than it formerly was. Coal prices actually declined after the strike began. This, presumably, is the result of prestrike stockpiling and high levels of production of non-union coal. Coal mined by UMW members is less than half the national total; it formerly approached 80 percent. Although the UMW represents almost 90 percent of the underground miners, only one-half of the surface miners are members of any union.[16]

---

[15]U.S. GAO, *Low Productivity*, chap. 4.

[16]*United Mine Workers Journal*, March 1981, April–May 1981, discusses the strike from the union's viewpoint. During the strike, the average cost of coal actually declined from $31.68 per ton in March 1981 to $29.25 per ton in April 1981. These figures are from U.S. DoE, *Cost and*

In late 1981, electric utilities paid $1.60/M̄Btu for coal, $3.01/M̄Btu for gas, and $5.12/M̄Btu for oil. Not only has the cost advantage held by coal relative to other fossil fuels increased, but, in 1982, coal-generated electricity from new plants can be less costly than nuclear power.[17]

Chapters 11 and 13 analyze the comparative costs of generation with coal and nuclear power. The present cost advantage held by coal relative to both nuclear and oil generation may change if the Clean Air Act is modified. If the sulfur removal requirement for new plants is deleted, then low-sulfur coal will be a feasible and less costly option for many new plants. Conversely, uniform reductions in permissible sulfur emissions would increase the cost of coal generation.

In late 1981, the price premium for low-sulfur coal was modest: coal with a 3-percent-plus sulfur content cost $1.44, and coal with 1–1.5 percent sulfur cost $1.74. This is only 30¢/M̄Btu difference. Very low sulfur coal (less than one-half of a percent) was actually less costly than medium-sulfur coal; it cost $1.52/M̄Btu compared to the $1.74/M̄Btu for medium-sulfur coal. This suggests that, in general, present state limits on sulfur emissions are sufficiently high that large quantities of acceptable coal are available.[18]

For the residual oil used in power plants, the low-sulfur premium is significant. Oil with 2–3 percent sulfur cost $4.14/M̄Btu, but oil with less than 0.3 percent sulfur cost $2 more at $6.13/M̄Btu.

With all its problems, there is no doubt that coal consumption will continue to increase in the near future.

Coal Use

In Figure 10-2, coal consumption is seen to have gone through four stages from 1800 to the 1980s. It grew slowly in the early 1800s, then rapidly throughout the last of the nineteenth century and up to the end of World War I. Following the Depression decline and the World War II resurgence, consumption declined again. In 1943, consumption was 651 million tons; it did not reach this same level again until the late 1970s.

Each stage is marked by different technologies as they affected coal con-

---

Quality of Fuels, April 1981, p. 5. Michael Yarrow (in his "How Good Strong Union Men Line It Out: Explorations of the Structure and Dynamics of Coal Miners' Class Consciousness," Ph.D. dissertation, Rutgers University, 1982) estimates the strength of the UMW over several decades. The figure on current UMW membership is from The President's Commission on Coal, *Coal Data Book,* February 1980, p. 147.

[17]In "U.S. Average Electrical Generating Costs and Capacity Factors in 1979" (unpublished data), the Atomic Industrial Forum, Inc., reported that nuclear plants which began operation in each of the years 1975–79 were more costly than coal plants beginning operation in those years.

[18]The data on fuel costs (and on sulfur premiums) are from U.S. DoE, *Monthly Energy Review,* February 1982, and *Cost and Quality of Fuels,* September 1981.

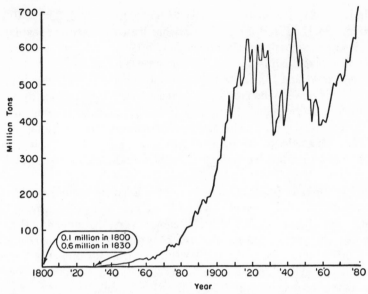

Figure 10-2. Coal: domestic consumption, bituminous and anthracite.
Sources: Historical Statistics of the United States; U.S. DoE, EIA, *Coal Data* and *Monthly Energy Review; Survey of Current Business;* and *Statistical Abstract of the United States, 1980.*

sumption.[19] Before 1865, anthracite consumption exceeded bituminous coal use. Anthracite was used for space heating in homes and businesses, and bituminous coal and anthracite were both burned for iron, steel, and metal smelting. Coal use in total grew from a small 108 thousand tons in 1800 to 25 million tons in 1865. In this period, coal was replacing woodfuel as well as being the preferred fuel for new technologies. The annual growth rate was a phenomenal 8.7 percent.

After the Civil War, coal emerged as the dominant energy source in steam engines as well as heating and smelting. In this second stage, from 1865 to 1917, total coal consumption grew 6.4 percent annually. Much of the increase was a result of growing rail traffic with steam locomotives, steel production, coal-fired steam engines on ships, and other steam uses. Bituminous coal use grew much more than anthracite consumption.

The increased use of petroleum and natural gas terminated this long period of coal growth. The kinds of substitutability outlined in Chapters 3 and 4 were dominant in the third stage. After World War I, rail travel declined. Not only

[19]This brief economic history is much assisted by Sam H. Schurr and Bruce C. Netschert, Resources for the Future, *Energy in the American Economy, 1850–1975* (Baltimore: The Johns Hopkins University Press, 1960), pp. 57–85; and by statistical data in U.S. Department of Commerce, Bureau of the Census, *Historical Statistics of the United States: Colonial Times to 1957*, 1960; and U.S. DoE, *Monthly Energy Review* and *Bituminous and Subbituminous Coal and Lignite Distribution*, 1979.

were railroads losing passengers and freight to petroleum-using cars, trucks, planes, and buses, but the railroads replaced coal-burning steam locomotives with diesel locomotives. Rail use of coal, at a peak of 35 million tons in 1920, is almost nonexistent in the 1980s. Similarly, ships used oil instead of coal, replacing steam with internal combustion. Homes, businesses, and industry replaced coal with oil, natural gas, and electricity. Coal use in the war years of 1917–18 and 1943–44 was not surpassed until 1979.

The decline in coal consumption was reversed by the fourth major historical phase. Electric utilities have, since the end of the Growth Era in oil and natural gas, made increasing use of coal. Utilities now use 80–85 percent of the coal consumed in the United States. Coal consumption began growing again in the 1960s, and—a dramatic change from the 1970s—coal is now usually preferred to oil or nuclear power as the energy source for new plants. Coal is America's only major energy export, and exports averaged about 10 percent of production throughout the 1970s.

The cost advantage cited above by which coal is less expensive than oil or natural gas on a $/MBtu basis guarantees continued growth in domestic consumption and exports for the forseeable future. The limitations which bound coal use are labor and environmental problems. Social cost as defined in Chapter 2 is the sum of producers' market cost, subsidies, and external nonmarket cost. The future of coal use will be determined in part by the political definition of these concepts in economic theory.

The Organization of the Industry: Monopoly or Competition?

Unlike the oil and gas industries, the coal industry cannot be considered to be independent. All the major petroleum companies have had oil and gas as their major economic activities. In contrast, most leading coal companies are subsidiaries or affiliates of oil, utility, or steel companies. For example: Consolidation Coal is owned by Conoco/Du Pont, Island Creek Coal is owned by Occidental, and Old Ben is owned by Sohio/BP. Since BP is controlled by the British government, Old Ben may be America's only nationalized coal company!

Table 10-3 defines the owners and affiliates of America's leading coal companies.

It would be incorrect, however, to assume that this structure restricts production and consumption. Since the resumption of growth in coal use in 1973, the coal producers which have noncoal interests have increased their production much more rapidly than other segments of the industry. Coal production by electric utilities has increased by 169 percent, in comparison to an overall growth in production of 39 percent. However, production increases by oil companies are more difficult to identify because the expansion in the 1960s

Table 10-3. Coal industry structure, 1980

Coal company or group	Parent or affiliate company or independent status	1980 production million tons
A. Largest 15 Producers of Bituminous Coal		
Peabody Coal Co.	Peabody Holding Co. (Newmont Mining, Williams Co., Boeing, Fluor, Bechtel, Equitable)	59.1
Consolidation Coal	Conoco/Du Pont	49.0
AMAX	Standard/Calif. owns 21%	40.5
Texas Utilities	Texas Utilities	27.6
Island Creek	Occidental Oil	20.0
Pittston	independent	17.8
NERCO	Pacific Power and Light	16.9
Arch Mineral	Ashland Oil	15.8
U.S. Steel	U.S. Steel/Marathon Oil	14.2
Amer. Elec. Power	Amer. Elec. Power	14.1
Peter Kiewit	independent	13.5
North American	independent	12.7
Westmoreland	independent	12.7
Total production of 15 largest producers		313.9
B. Petroleum Companies and Affiliated Companies as Bituminous Producers		
Oil companies in "15 largest" group above: Conoco/Du Pont, Occidental, Ashland, U.S. Steel/Marathon Oil		99.0
Monterey Coal, etc.	Exxon	11.4
Atlantic Richfield	Atlantic Richfield	11.4
Sun	Sun	11.2
Old Ben	Sohio/BP	11.1
Pittsburgh & Midway	Gulf	9.7
Empire	Standard/Indiana	0.8
Plateau	Getty	0.4
Large coal companies affiliated with petroleum industries above: Peabody Coal (Newmont, Williams et al.) and AMAX (Standard/California)		105.0
Total production of petroleum and affiliated companies		263.2

C. Composition of Bituminous Coal Industry

	1980 production	
	Amount	Percent
Petroleum companies, subsidiaries, and affiliates, above	263.2	32%
Electric utilities, including Texas Utilities, Pacific Power and Light, & American Electric Power, above	89.8	11%
Largest independents: Pittston, Peter Kiewit, North American, Westmoreland above	56.7	7%
Steel companies, excluding U.S. Steel/Marathon Oil	41.8	5%
Other independent and subsidiary coal companies	380.5	46%
Total production of bituminous coal	832.0	101%

Note: The bituminous coal data here include lignite and subbituminous coal.
Sources: Annual Reports of the listed petroleum and coal companies and the *1981 Keystone Coal Industry Manual.*

and 1970s was linked to acquisition of previously independent coal companies. It appears that these companies increased their physical production in the 1970s on about the same basis as the overall increase. Oil companies neither owned nor were affiliated with any significant production in 1960: apparent production was, literally, zero in this category. This is according to the American Petroleum Institute as well as independent analysis.[20] Part C of Table 10-3 shows that twenty years later, oil companies have 32 percent of bituminous coal production.

The increased participation of petroleum companies in coal leads to consideration of the influence this participation may have on coal mine health and safety. In this context, it should be recalled that Chapters 1 and 2 showed that each of the four theories studied gives distinctly different predictions about the magnitudes of production and external social cost. It might be argued that growing participation of oil companies in coal provides a monopoly shelter to provide greater safety expenditures than would be possible in a competitive industry. Alternatively, horizontal integration might be expected to provide sufficient political power to keep effective mine safety at low levels.

The empirical evidence is sparse but unambiguous. In a recent National Academy of Sciences study of underground mine safety, the 5 safest companies were all subsidiaries of other companies whose major activities are outside coal. This is shown by the Table 10-4 data on disabling injuries per 100 miners. The safest company is Old Ben, the subsidiary of Sohio/BP/ British government. The six independent coal companies as a group had an average safety rate of 15.3 compared to the 11.9 figure for the subsidiaries.[21]

The data on reserve ownership confirms the dominance of noncoal corporations in the coal industry. Approximately 400 companies report owning at least one million tons of reserves. The 30 smallest companies, holding 30 million tons, appear to be overwhelmingly concerned with coal as their major economic activity. The 30 largest companies report 120 billion tons of reserves, 75 percent of the amount for which ownership has been identified.[22] Fourteen of the major oil companies (including U.S. Steel/Marathon Oil) are in this group, and two other large owners of reserves are affiliated with

[20]American Petroleum Institute, "Concentration Levels in the Production and Reserve Holdings of Crude Oil, Natural Gas, Coal, and Uranium in the U.S., 1955–1976," Discussion Paper #004R (Washington, December 1977), p. 12. My own review of coal production data also concludes there was no measurable coal production by oil companies in 1960.

[21]The data in Table 10-4 are from National Academy of Sciences, *Towards Safer Underground Coal Mines,* 1982, cited in *United Mine Workers Journal,* September 1–15, 1982, p. 6. The *UMW Journal* also reports that, since 1980, Westmoreland has improved its injury rate to 12.3 per 100 miners. The large companies in Table 10-4 were identified in Table 10-3. Additional parent/coal operator relationships are Bethlehem, Jones and Laughlin, and Republic (steel), Zeigler (Houston Natural Gas Corporation), Freeman United (General Dynamics Corporation), and Valley Camp (Quaker State Oil).

[22]These data are from *Keystone Coal Industry Manual* (New York: McGraw-Hill Mining Publications, 1981), pp. 736–739.

Table 10-4. Annual disabling injuries per 100 miners, 1978–80

Coal company or group	Disabling injury rate	Coal company or group	Disabling injury rate
Old Ben	4.5	Rochester & Pittsburgh*	13.0
Island Creek	6.1	Eastern Associated*	15.0
Bethlehem	7.6	Zeigler	15.2
U.S. Steel	8.0	Freeman United	15.4
Consolidation	8.3	American Electric Power	15.7
Alabama By-Product*	9.1	Republic	18.7
Jones and Laughlin	11.8	Valley Camp	18.9
Pittston*	11.9	Westmoreland*	21.1
Peabody	12.3	North American*	21.8

*Independent coal companies.
Source: See footnote 21.

petroleum. Only four of the largest reserve owners are primarily coal companies. The remaining 10 coal companies in this group of 30 are generally owned by railroads, steel, and utilities. The early importance of coal as a locomotive fuel leaves an economic legacy: the largest owner of coal reserves in the United States is the Burlington Northern Railroad.

The analysis of the petroleum industry in preceding chapters leads to the expectation of similarities in the coal industry with respect to cooperative activities, regional versus national concentration, stability in leadership, and the prevailing frequency of private ownership among U.S. coal companies.

In the coal industry, large producers frequently purchase or market coal from small producers in a system analogous to the cooperative marketing relationships of major oil companies and retail dealers.[23]

National concentration in coal is limited. The four largest companies in Table 10-3 had 21 percent of national bituminous coal production in 1980. However, just as in the petroleum and natural gas industries, regional concentration is usually higher. In the middle 1970s, the average regional concentration of the four largest producers was 38 percent.[24] Again, as with the petroleum industry, it should be emphasized that national concentration in coal is lower than in almost all other industries surveyed in Table 6-9. Overall, concentration appeared to increase until 1973, but it has not done so since.

The stability in ranking of groups of major oil companies is not found in the coal industry. While Peabody and Consolidation have been the leaders for many years, there has been considerable change in other rankings as independent companies have been acquired by petroleum corporations, and utility production has accelerated.

[23]These relationships are described in Richard A. Schmidt, *Coal in America: An Encyclopedia of Reserves, Production, and Use* (New York: McGraw-Hill, 1979), p. 153.
[24]U.S. Federal Trade Commission, *The Structure of the Nation's Coal Industry*, November 1978.

Of the major energy industries, coal appears to be the closest to the definition of a competitive industry. Paradoxically, this is in part attributable to the high degree to which it is dominated by other industries seeking minimum-cost coal supplies. Coal production could be operated as a profit monopoly or growth monopoly only if its owners (oil, utilities, steel, etc.) joined with the large and small independent coal companies to pursue monopolistic policies. There is no evidence that this has ever happened. Indeed, the economic competition between coal and nuclear power as described in Chapter 11 is a major factor influencing both of these sectors.

Both coal and nuclear power are experiencing serious problems with respect to external social cost. The aspects of these problems that relate to the health and safety of mining have been reviewed in this chapter. As following chapters will indicate, there are many public health and environmental issues to be resolved.

11

The Economics of Nuclear Power

Coverage by the Western press of the accident at the nuclear reactor in Harrisburg, in which some basically minor unpleasant consequences were described in an extremely exaggerated manner, was an extension of the campaign against atomic power. The campaign was supported by fuel companies whose profits would be endangered by a shift to nuclear power.

—Anatoly P. Aleksandrov,
president of the Soviet Academy of Sciences

This quotation must be paraphrased: its meaning may not be understood by someone outside the Soviet Union because it is, literally, incredible. Aleksandrov makes three points. First, the Three Mile Island accident was minor and unimportant. Second, the American press campaigns against nuclear power. Third, petroleum corporations support this campaign.[1]

On the first point, the weight of the evidence is that the Three Mile Island accident was indeed serious. The second point is debatable, with legitimate arguments to be made on each side. However, the third point is simply incorrect. Petroleum corporations have become major participants in nuclear fuel development, a relationship to be investigated in a later section of this chapter. There is no evidence that the oil companies—investing in nuclear fuel and anticipating oil resource decline—work against nuclear power.

The Soviet Union differs markedly from the United States in its approach to nuclear power. It has lagged considerably behind America in nuclear power development, having less than one-third of the nuclear generating capacity of the United States in 1980.

Russia has given much less attention to the implications of nuclear plant safety. During the 1970s, no Russian nuclear plant had containment shells or emergency core cooling systems. Yet both safety features are believed to be essential in the United States, and the containment shell prevented widespread contamination in the Three Mile Island event. Russia has also experienced an

---

[1]*New York Times*, April 11, 1979.

explosion or other accident with military nuclear waste which contaminated several hundred square miles and forced the evacuation of 30 villages.[2]

Russia has operating breeder reactors. The United States, in contrast, has postponed an ultimate decision on breeder reactors because of the terrorism/ proliferation problem.

Overall, nuclear safety has been more important in the United States. It seems a simple matter to determine the elements of illogic in Aleksandrov's statement of Soviet views. It is less certain whether logic and illogic can be as easily differentiated in American perspectives on nuclear power. Neither committed advocates nor opponents of nuclear power will be wholly satisfied with the chapter that follows. It will, however, give an empirical introduction to nuclear power: its economics, its historical competitive advantage, its future problems, and the structure of the industry.

The Nuclear Fuel Cycle

Nuclear fuel has an important economic advantage over fossil fuels in cost. To demonstrate this key aspect of nuclear economics, compare the actual reported fuel costs for specific power plants in Table 11-1. The plants were selected because they possess typical characteristics of size, fuel cost, generation, and sulfur content in fuel. All are large plants, capable of generating billions of kilowatt hours per year. In 1978, these four plants alone produced 2 percent of the country's electricity.

The usual method of showing a plant's generating capability is with a simple engineering equation:

$$(1) \quad G \quad = \quad C \quad * \quad 8{,}760 \quad * \quad cf \quad \div \quad 10^6$$

| (generation in billion kWh) | (capacity in megawatts) | (hours per year) | (capacity factor; a percent) | (a conversion factor) |

For the Gavin coal plant, this would be:

$$(2) \quad 13.7 \text{ billion kWh} = 2{,}600 * 8{,}760 * 60\% \div 10^6.$$

Recall from Chapter 3 that a megawatt is 1,000 kilowatts. Dividing by 10^6 converts the answer for G into billion kilowatt hours. If this division were omitted, the answer would be $G = 13{,}700{,}000$ megawatt hours. These are large plants.

The significant aspect of Table 11-1 is the fuel cost advantage held by Commonwealth Edison's nuclear plant. It reported only 24¢/M̄Btu, far below

[2]*Science*, 18 July 1980, pp. 345–353, and 16 April 1982, p. 274.

Table 11-1. Four power plants: Reported fuel costs for 1978

	Nuclear	Coal	Oil	Gas
Plant data				
Plant name	Zion	Gavin	Brayton Pt.	Cedar Bayou
Location, and company	Illinois, Commonwealth Ed.	Ohio Ed.- Am. Elec. Power	New Eng. Power, Mass.	Houston Light & Power, Texas
Capacity rating	2,196 MW	2,600 MW	1,600 MW	2,295 MW
Capacity operation	70%	60%	60%	60%
Fuel Data				
Use	1.3 M grams U	8.1 M tons	14 M barrels	123 bill. ft^3
Cost	$3.6 million	$237 million	$160 million	$146 million
Price	$28/gram	$28/ton	$11/barrel	$1.18/mcf
Price per MBtu	24¢	$1.57	$1.78	$1.14
Btu per pound	53 billion	9,150	20,600	10,000 (approx.)
Sulfur content	0%	2.7%	2.1%	0%
National averages				
Price, average 1978	45¢/MBtu	$1.12/MBtu	$2.19/MBtu	$1.43/MBtu
Price, early 1981	56¢/MBtu	$1.42/MBtu	$5.40/MBtu	$2.54/MBtu
Sulfur content	0%	1.6%	1.1%	0%

Note: All sulfur content data are for 1980. The 1978 average nuclear fuel cost is the average for 42 plants without abnormally high cost, as reported in U.S. Department of Energy, Energy Information Administration, *Steam-Electric Plant Construction Cost and Annual Production Expenses: 1978*, December 1980. The 1981 nuclear fuel cost statistic is based upon the analysis in Gordon R. Corey, "An Economic Comparison of Nuclear and Coal-Fired Generation," Commonwealth Edison Company, Atomic Industrial Forum, Inc., December 1, 1980. Other data are from U.S. DoE, *Cost and Quality of Fuels for Electric Utility Plants: 1980 Annual*, June 1981.

the other three plants, and considerably less than the national averages for coal, oil, and natural gas. For the three fossil fuel plants, the fuel bill in 1978 averaged around $200 million. Surprisingly, the Gavin plant paid almost as much for coal in dollars per energy unit in Ohio as the Brayton plant in Massachusetts did for oil.

The national data in Table 11-1 confirm the point: nuclear fuel has for many years provided a Btu of energy at a considerably lower cost than fossil fuels. Note the energy content of a pound of each fuel. Coal is 9,150 Btu per pound, oil is 20,600, and natural gas at standard pressure is about 10,000 Btu per pound. But, nuclear fuel is *orders of magnitude* higher: 53 billion Btu per pound. It is the high-energy density of nuclear fuel that gives much of its economic advantage, and creates its potential problems in safe use.

Nuclear fuel is the product of a complex technological process. The technological difficulties at certain stages have significant economic implications, and therefore it is relevant to begin an economic discussion with a summary of the process. That process can be described through its various stages with reference to the amount of material which needs to be handled to provide fuel for a typical reactor for a year. The complete process through each of its stages is the *nuclear fuel cycle*.

The basic fuel requirement for a nuclear plant can be summarized with the relationship shown in Equation (3).

$$(3) \qquad F = \frac{d * cf * C/eff}{B}$$

Two of these terms were used in Equation (1). The *capacity factor* (*cf*) indicates the percentage of capacity that is—on the average—used over a year's time. Capacity in megawatts is again C. Days in the typical year are 365 (*d*). Efficiency is represented by *eff* and, as in the analysis in Chapter 3, defines the ratio of energy utilized (electricity) to energy supplied (energy in the uranium). The remaining factor is crucial: B, burn-up, the ratio of megawatt days of production per metric ton of uranium fuel. The F term represents the uranium fuel requirement. In previous work, I have adapted a model by K. B. Cady and A. C. Hui,[3] and use this numerical example:

[3]See K. B. Cady and A. C. Hui, "NUFUEL—A Computer Code for Calculating the Nuclear Fuel Cycle Cost of a Light Water Reactor," Cornell University, Ward Laboratory of Nuclear Engineering, CURL-51, August 1977. See also three publications by Duane Chapman: "The 1981 Tax Act and the Economics of Coal and Nuclear Power," U.S. House Interior Committee, Subcommittee on Oversight and Investigations, Hearing, *Nuclear Policy*, October 1981, pp. 95–104, 630–651, printed as Cornell University Department of Agricultural Economics Staff Paper No. 81-26; *Nuclear Economics: Taxation, Fuel Cost, and Decommissioning*, California Energy Commission, November 1980; and "Federal Tax Incentives Affecting Coal and Nuclear Power Economics," *Natural Resources Journal*, 22:2 (April 1982), 361–378. General discussions of the fuel cycle appear in: Spurgeon Keeny, Jr., et al., Ford Foundation/Mitre Corporation, Nuclear Energy Policy Study Group, *Nuclear Power Issues and Choices*, 1977; Paul Ehrlich, A.

$$(4)\ F = \frac{365 * .8 * 1,000/.33}{32,600} = 27.143 \text{ metric tons of uranium per year}$$

In this example, the burn-up is 32,600 megawatt days of power for each metric ton of uranium, the efficiency of conversion in the reactor is .33, the plant's capacity is 1,000 MW, and it operates on the average at 80 percent of capacity in a 365-day year. Consequently, it requires 27.1 metric tons of uranium each year.

It must be strongly emphasized that Equation (4) is an example to illustrate the relationships between conversion efficiency, fuel use, and capacity in a hypothetical reactor. Specific values appropriate to each reactor depend upon its design and the richness of the fuel it uses.

A major factor influencing technology and economics at each stage is the richness of the fuel which will enter the reactor. In its natural state, uranium occurs in three isotopes that have almost identical weights: U-234, U-235, and U-238. It is U-235 which provides the atomic reactions that are the basis for nuclear plant operations. Commonly, American reactors are designed to use uranium fuel which is about 3 percent U-235. In its natural state, however, the frequency of U-235 in uranium is only .711 of 1 percent. Uranium 238 constitutes 99.284 percent of natural uranium, and U-234 is only .005 of 1 percent of natural uranium. A major goal of the nuclear fuel cycle is to prepare usable uranium fuel that has about 3 percent U-235 instead of the naturally occurring .711 of 1 percent. This illustration will assume the uranium fuel used in the reactor is 3.2 percent U-235.

To provide the 27.1 metric tons of fuel, the fuel cycle goes through at least six stages: *mining* and milling, conversion, enrichment, fabrication, transportation, and spent fuel handling.

A typical sandstone ore being mined in the 1980s will contain about .1 of 1 percent uranium oxide, U_3O_8. The average grade of ore has declined slowly throughout the 1970s, and will probably continue to do so.[4] If the recovery rate[5] is 95 percent for a .1 of 1 percent grade ore, then 253,000 tons of sandstone ore will be mined which will contain 240 tons of uranium oxide. Note the different units: uranium ore is usually measured in American tons of 2,000 pounds each. When the fuel is prepared for reactor use, it is measured in metric tons (MT) of 1,000 kilograms each.

Uranium *milling* uses physical crushing and chemical processes to separate and concentrate the uranium oxide. The resulting material is commonly called

---

H. Ehrlich, and J. P. Holdren, *Ecoscience: Population, Resources, Environment* (San Francisco: W. H. Freeman, 1977); and June H. Taylor and M. D. Yokell, *Yellowcake: The International Uranium Cartel* (New York: Pergamon, 1979).

[4]U.S. Department of the Interior, Bureau of Mines, "Uranium," preprint from 1976 *Minerals Yearbook;* and U.S. Department of Energy, *An Assessment Report on Uranium in the United States of America,* October 1980.

[5]Recovery rate has the same meaning here as in Chapter 5.

"yellowcake." Suppose a 5 percent loss of U_3O_8 in milling. The yellowcake will contain 228 tons of U_3O_8. (Chemical leaching of uranium from the ground follows different processes, but is much less common than mining and milling.)

In the *conversion* stage, the solid yellowcake uranium oxide (U_3O_8) is transformed into another compound, uranium hexafluoride (UF_6). This UF_6 is a solid at ordinary temperature, but is a gas at the higher temperatures at which it is used. Recall that 228 tons of U_3O_8 entered the conversion process. The basic chemistry indicates that 84.8 percent of the U_3O_8 will be uranium.[6] So 228 tons of U_3O_8 contained 193.3 tons of natural uranium. Supposing a 1 percent loss in the conversion process, the uranium hexafluoride contains 191.4 tons of uranium.

At this point, it is conventional to change to metric units. So, 191.4 American tons of uranium will be expressed as 173.7 metric tons of uranium.

The UF_6 gas is shipped to *enrichment* plants. The uranium in UF_6 still has the same composition it exhibited in the ore. The objective of enrichment is to increase the proportion of fissionable U-235 to 3 percent or more. This is accomplished by a technology known as "gaseous diffusion." The gaseous UF_6 molecules have slightly different weights according to whether the U is U-238 or the slightly lighter U-235. Accordingly, gaseous UF_6 which has the lighter U-235 will be somewhat more likely to pass through a very fine membrane. With the UF_6 originally coming from uranium of which U-235 was .711 of 1 percent, passing the gas through more than 1,000 diffusion barriers can create UF_6 gas in which U-235 constitutes more than 3 percent of the U. In the hypothetical fuel being prepared here, the enriched fuel has uranium in the UF_6 containing 3.2 percent of the fissionable isotope U-235. The waste UF_6 in this example has uranium in which U-235 has declined to .25 of 1 percent from its original .711 of 1 percent. This enrichment waste, incidentally, is used primarily in manufacturing military ordnance. It is valued because of its heaviness, and its readiness to ignite.[7]

Gaseous diffusion is now the sole method used in the three operating enrichment plants, although other methods such as centrifuge and laser separation are being developed in the United States and in other countries. The gaseous diffusion process requires considerable electric energy, a fact that in part motivates the search for other enrichment methods.

To return to the numerical example: 173.7 metric tons of natural uranium (with .711 of 1 percent as U-235) entered the enrichment process. The result is 27.1 metric tons of uranium at 3.2 percent U-235, and 146.3 metric tons of depleted uranium at .25 of 1 percent U-235.

---

[6]The atomic weight of three uranium atoms is 3 * 238, and eight oxygen atoms are 128. U_3 is 84.8% of U_3O_8. Remember that natural uranium is overwhelmingly U-238; U-235 and U-234 are present in very small proportions.

[7]See U.S. Department of the Interior, Bureau of Mines, "Depleted Uranium," *Mineral Facts and Problems,* 1980 Edition, preprint from Bulletin 671.

Table 11-2. Nuclear fuel cycle costs, early 1980s, 1,000 MW nuclear plant, 7 billion kWh annually

Stage	Quantity in example	Typical price	Annual cost $ million	Cost, mills per kWh
Mining, milling			$18.2	2.6
mining	253,000 tons ore	$80,000/ton		
milling	228 tons U_3O_8	of U_3O_8		
Conversion	173.7 MT-U	$4.85/kg	0.8	0.1
Enrichment	27.1 MT-U at 3.2% U-235	——	11.3	1.6
Fabrication	27.1 MT-U	$150/kg	4.1	0.6
Spent fuel transport	27.1 MT-U 250 kg-Pu	$25/kg-U	0.7	0.1
Spent fuel disposal	27.1 MT-U 250 kg-Pu	$275/kg-U	7.5	1.1
Total			$42.6 million	6.1 mills/kWh

Sources: quantity information from text. Price data are from Corey and from Gibbs & Hill, Inc., "Economic Comparison of Coal and Nuclear Electric Power," January 1980. Spot market prices in 1982 for yellowcake have fallen below $80,000/ton U_3O_8, probably because of the considerable reduction in nuclear fuel requirements caused by the reduction in planned nuclear plants.

After enrichment, the gaseous UF_6 is shipped to a *fabrication* facility. Here, the enriched uranium in the UF_6 is transformed into a solid uranium oxide which is formed into pellets. These pellets are in turn placed in cylindrical fuel rods. Assuming no loss of uranium in fabricating the fuel pellets and rods, we now have the required 27.1 metric tons of fuel available for use in the reactor.

The nuclear power industry is justifiably proud of its achievement in processing nuclear fuel for use by utilities at a cost which—as seen in Table 11-1—is considerably less than that of coal, oil, or natural gas fuel. Table 11-2 collects the various quantities calculated in the example, and indicates their individual costs in early 1980. The example here is just that: the grade of enriched fuel may be on the high side, and the losses in the fuel cycle may be too low, but the result is representative of nuclear fuel economics in the early 1980s.

The 6.1 mills/kWh total cost in Table 11-2 is equivalent to about 56¢/ M̄Btu,[8] the figure for the estimated national average cost of nuclear fuel in Table 11-1.

Note that three stages have costs of less than one mill per kWh. The three

[8]Assume the nuclear plant uses 10,800 Btu to produce one kWh of electricity. Therefore, 56¢/ M̄Btu = 6.1 mills/kWh * .1¢/mill * 1,000,000 Btu/M̄Btu * kWh/10,800 Btu.

other stages (mining and milling, enrichment, and spent fuel disposal) each add more than one mill/kWh to nuclear power costs.

Detailed estimates of American uranium reserves indicate sufficient proven reserves to operate the 60,000 MW of existing nuclear capacity for about 75 years, at uranium ore prices within 200–300 percent of the present cost. The relationships are simple. Assume 228 tons of uranium oxide per year (as in Table 11-2) for a typical 1,000 MW reactor. The 60,000 MW capacity requires about 13,700 tons of ore each year. Proven reserves with an estimated cost of about $75/pound are estimated to be about 594,000 tons.[9] The result: 43 years.

However, if capacity grew to the 148,000 MW planned in the early 1980s, the present proven reserve supply at the $75/pound cost would be used much sooner.

If the focus is moved to lower cost ore of about $45/pound, there are an estimated 205,000 tons of proven reserves of U_3O_8. This is 15 years' supply for 60,000 MW, or 6 years' for 148,000 MW.

A paradox emerges. There are sufficient proven reserves of low-cost uranium in the United States to last into the next century—if nuclear power ceases to grow. An accelerating nuclear power industry will more quickly use today's proven reserves, and increase the economic demand for breeder reactors and for imported uranium.

Incidentally, it is interesting to contrast the first-stage requirements with the tonnage needed for coal plants. The Table 11-2 example needs 253,000 tons of ore each year. A similar 1,000 MW coal plant would need about 3.4 million tons of coal.

Reactor Types

The first four stages in Table 11-2 are called the *front-end* of the nuclear fuel cycle. They are the stages from the mine to the reactor core. The last two stages are the *back-end:* they are intended to resolve the spent fuel problem. The front-end of the cycle is (more or less) identical in concept for all of the present commercial reactors in the United States.

Originally, *spent fuel* was to have been *reprocessed.* Remaining fissionable uranium and newly created plutonium were to be removed. The reprocessed uranium and the plutonium would themselves be made into nuclear fuel. The plutonium would be the major input fuel for breeder reactors, and these

[9]The concept "forward cost" is used in uranium resource studies, and includes 50–67% of market cost. So, $75/lb market cost as discussed in the text refers to $50/lb forward cost in the uranium resource literature, and $45/lb market cost means $30/lb forward cost. Although the physical estimates of uranium in the ground have been stable in recent years, rising production costs have a significant impact on the cost-availability estimates. The highest-cost ore ($150/lb market cost, about $100/lb forward cost) adds an additional 300,000 tons of U_3O_8 to the 594,000 tons of $75/lb reserves. See U.S. Department of Energy, Grand Junction Area Office, *Statistical Data of the Uranium Industry,* 1982, and Stephen Sullivan, *Uranium Resources: A Review of Estimation Methodologies,* California Energy Commission, 1978.

breeder reactors would create new fissionable uranium for conventional reactors. The breeders would also, of course, generate electricity. Spent fuel now, however, is accumulating at each nuclear plant pending a resolution of the spent fuel problem, which will be considered in Chapter 13.

The discussion of the fuel cycle has made little reference to the differences among types of reactors. As an introduction, this is appropriate. However, broad classes of commercial power reactors can be distinguished as relevant to economics and resource use in the United States: converter reactors and breeder reactors. A *converter reactor* consumes more nuclear fuel than it creates; a *breeder reactor* creates more fissionable material than it consumes.

Among the converter reactors, the *light water reactor* (LWR) dominates commercial nuclear power plants in the United States today. "Light water" refers to the use of regular water in the reactor's cooling and steam systems. Heavy water reactors, on the other hand, make use of deuterium, an isotope of hydrogen that must be concentrated in electro-chemical processing plants because of its relatively low occurrence in ordinary hydrogen. The deuterium combines with oxygen to form heavier water molecules. While light water reactors require enriched uranium fuel as described above, heavy water reactors can use uranium that is only slightly enriched, or even natural unenriched uranium. However, the greater complexity and capital cost of the heavy water reactors have made the light water reactors less costly overall in the United States.

Light water reactors can be further subdivided into two major types. The pressurized water reactor uses fuel enriched to 3–3.2 percent U-235 and has a separate steam-turbine system which does not pass through the reactor. Because of its high pressure, this kind of reactor prevents boiling of the reactor water even though it is heated to about 600°F. In contrast, the boiling water reactor uses slightly less enriched fuel, has only one steam system common to turbine and reactor, and allows water in the reactor to boil into steam. The boiling water reactor has a small economic advantage in terms of responsiveness to variations in capacity requirements and need for fuel enrichment; the pressurized water reactor gains slightly more energy from its fuel. Overall, the two types of light water reactors are very competitive.

In breeder reactors, the fuel cycle becomes considerably more complex, and reprocessing of spent fuel to remove uranium and plutonium is necessary. If breeder reactors function effectively, the amount of energy obtainable from uranium resources may be increased 60 to 70 times.[10] Breeder reactors are not presently competitive with LWR's because of their higher initial capital cost and the present low cost of uranium ore. Difficulties in resolving problems in

---

[10]According to Sam H. Schurr, et al., Resources for the Future, *Energy in America's Future: The Choices before Us* (Baltimore: The Johns Hopkins University Press, 1979), p. 249, and Keeny, et al., p. 335.

reprocessing and terrorist access to plutonium add to breeder problems in the United States, although both France and Russia have initial breeder facilities.

This is an elementary introduction to a complicated topic. Several good sources exist on the subject, particularly publications by Keeny, Nero, and the National Academy of Sciences.[11] Henceforth, in discussing nuclear power plants I will generally mean LWR's in the United States.

Nuclear Economics and Coal Generation

The major point of Table 11-1 is that nuclear fuel has been, is, and will be less expensive than coal, oil, or natural gas fuel. Table 11-2 summarizes the separate stages of the fuel cycle. The item of largest cost is uranium ore. As noted, there is a considerable supply of domestic uranium reserves obtainable at costs less than or equal to $75/lb market cost. The front-end of the cycle is in hand, in the sense that nuclear fuel can be supplied at low cost.

In order to compare the overall costs of generation to a utility, it is necessary to spread the capital investment over the expected generating life of a plant. Since a nuclear plant has higher initial cost but lower fuel cost, the method of capital cost allocation allows the utility to compare different kinds of power plants on a minimum cost basis.

The typical method used is *levelized cost:* the capital investment is amortized over the life of the plant, and charged to each kilowatt hour. The method is identical in logic to the mortgage factor used in comparing different home heating systems in Table 3-8.

The cost data in Table 11-3 can illustrate the method. Assume, from the TVA data, that two large nuclear units of 2,400 MW total capacity will cost $12 billion. The basic relationship uses these three equations:

(5) $$LCC = K * FCR/G$$

(6) $$FCR = CRF + PD$$

(7) $$CRF = \frac{i(1+i)^n}{(1+i)^n - 1}$$

In Equation (5), *LCC* is the levelized capital cost. This is the amount in $/kWh which, if charged for every kilowatt hour over the life of the project, would wholly pay a loan and interest for the $12 billion capital cost. In addition, the *LCC* includes an allowance for small capital-linked costs such as

[11]Keeny, et al.; Anthony V. Nero, *A Guidebook to Nuclear Reactors* (Berkeley: University of California Press, 1979); National Academy of Sciences, *Energy in Transition 1985–2010*, Final Report of the Committee on Nuclear and Alternative Energy Systems, 1980.

property-tax-type payments or a decommissioning allowance. K represents capital cost, here $12 billion. Generation is defined as in Equation (1) at the beginning of the chapter. Applying Equation (1) with an assumed capacity factor of 67.6% for the 2,400 MW, we have

$$(8) \qquad G = 2{,}400 \text{ MW} * 8{,}760 * .676 \div 10^6 = 14.2 * 10^6 \text{ kWh}$$

For illustration, assume, then, that the nuclear plants would generate 14.2 million kWh in a typical year.

The fixed charge rate, FCR in Equation (5), is defined in Equation (6). In that equation, CRF is the capital recovery factor and PD is the allowance for property-tax-type payments and a decommissioning allowance. The capital recovery factor is the key to levelized cost: it is the term that represents the charge per kWh necessary to amortize a capital investment.

In Equation (7), CRF depends on the interest rate i and the expected operating life n. Suppose i is 15 percent and n is 35 years. CRF, then, is .151. Suppose the PD allowance is .004. Then, in Equation (6), the fixed charge rate FCR is .155.

The result, then, for Equation (5) is

$$(9) \qquad LCC = \frac{\$12 * 10^6 * .155}{14.2 * 10^6 \text{ kWh}} = \$0.131/\text{kWh}$$

It so happens that this 13.1¢/kWh is the TVA allowance for investment for nuclear power, shown in Table 11-3. This is intended coincidence: the numerical values in Equations (5)–(9) were selected in order to illustrate the method of levelized cost calculation.

Table 11-3 summarizes total cost estimates for power generation by means of coal and nuclear power from two utilities with both types of plant. Some liberties have been taken in estimating the relative amounts of individual components, but the total costs are precisely quoted. The two utilities are the Tennessee Valley Authority (publicly owned) and the Commonwealth Edison Company in Chicago.

Total costs for future coal plants are identical: 18¢/kWh for both companies. The estimated nuclear costs differ markedly. Commonwealth Edison expects nuclear construction cost to be less than one-half the TVA estimates that assume similar capacities and inflation rates. TVA expects a longer construction period. TVA also expects unplanned forced outages to be an economic consideration, while Commonwealth Edison does not. The nuclear estimates for operations, maintenance, and insurance are similar, but the other elements are not. And—for assumptions from which TVA and Commonwealth Edison estimate identical coal generation cost—the Tennessee Valley Authority expects nuclear power to cost almost 50 percent more than does Commonwealth Edison.

[210]

Table 11-3. Utility analysis of nuclear economics, cost in ¢/kWh

| | Tennessee Valley Authority | | Commonwealth Edison Co., Illinois | |
	Nuclear	Coal	Nuclear	Coal
Assumptions				
Capacity	2,400 MW	2,400 MW	2,240 MW	2,228 MW
Constr. period	13 years	8 years	11 years	11 years
Constr. cost	$12 billion	$6 billion	$5 billion	$4 billion
Inflation per yr.	8.5%	8.5%	7.5%	7.5%
Forced outage	10%	10%	0	0
Cost per kWh				
Investment	13.1¢/kWh	6.3¢/kWh	7.6¢/kWh	5.7¢/kWh
Fuel	3.9	10.0	1.5	9.7
O, M, I	1.9	2.1	1.8	1.7
Spent fuel	—	—	0.9	—
Decommissioning	—	—	0.2	—
Regulations	—	—	1.3	1.1
Total cost	18.9¢/kWh	18.3¢/kWh	13.3¢/kWh	18.2¢/kWh

Notes: O, M, I = operations, maintenance, and insurance. For Commonwealth Edison the 4.2¢/kWh nuclear fuel cost in their report was divided among spent fuel; O, M, I; and fuel according to their reported current relationships.
Sources: Tennessee Valley Authority, "Comparison of Central Station Generator Costs" (staff report), July 9, 1981, Tables 4 and 5; and Corey, p. 22.

Not surprisingly, Commonwealth Edison continues to build additional new nuclear plants in the early 1980s, and TVA does not.

Taxation and Nuclear Economics

In the last 30 years, the corporate income tax has been revised to encourage increased private investment as a stimulus to economic growth. One finding of recent research is that nuclear power has been a disproportionate beneficiary of these tax incentives. This is because nuclear power is perhaps the most capital-intensive technology in widespread use in the American economy. A new plant may have $20 million in assets per employee; large industrial companies have $60,000, and retail establishments $34,000 per employee.[12] The specific provisions that have a significant effect on utility costs are the investment tax credit, accelerated depreciation, short tax lives, and interest deductions. In Chapter 12, these provisions will be examined as they affect customer rates and utility economics. Here we are interested in the impact of tax provisions upon the economic competition between coal and nuclear power in electric utility operations.

One important economic consequence of these tax provisions is that they reduce nuclear power cost more than coal generation cost. Table 11-4 is based upon analyses undertaken at Cornell University, and indicates the effect of taxation on coal and nuclear power cost to a utility.[13] The underlying economic assumptions are similar to those in Table 11-2, and the results are comparable to the utility company estimates in Table 11-3.

Table 11-4 (column 1) shows nuclear cost in Pennsylvania to be 1¢/kWh less than coal cost. This appears significant since Pennsylvania is a coal-producing state. If nuclear power is less costly in Pennsylvania, then it will probably look attractive anywhere in the country. The second column indicates what costs to a utility would have been without any tax subsidies. Note that, in the absence of subsidies, coal generation would be less costly. The third column defines the subsidy as the difference between the first two columns. It is much greater for nuclear power. The tax subsidy to nuclear power is about 45 percent of utility cost, and the tax subsidy to coal generation is about 20 percent of after-tax cost to the utility.

The magnitude of the tax effect is significant. A 1,000 MW nuclear plant with 5.3 billion kWh generation attains a subsidy of $365 million per year.

---

[12]Assume the nuclear plant will cost $4 billion and employ 200 persons. *Fortune Magazine*, "The Fortune Double 500 Directory," 1982, gives the other two statistics for 1981.

[13]From work by D. Chapman and Kathleen Cole in Chapman's "The 1981 Tax Act and the Economics of Coal and Nuclear Power."

Table 11-4. The effect of corporate income taxation on nuclear and coal generating cost

	After-tax cost with 1981 Tax Act provisions	Total cost, no subsidies	Subsidy with 1981 Tax Act
Nuclear generation	15.6 ¢/kWh	22.5 ¢/kWh	6.9 ¢/kWh
Coal generation	16.5 ¢/kWh	19.8 ¢/kWh	3.3 ¢/kWh

Note: Both plants are assumed to be located in Pennsylvania, planned and built in the 1980s, and operated for power generation in the 1990–2019 period.
Source: Chapman, "The 1981 Tax Act."

Structure of the Nuclear Industry

Petroleum companies participate in the nuclear fuel cycle to a significant extent. On the basis of partial detailed company reserve data, I estimate that 12 oil companies own one-third to one-half of U.S. proved reserves of uranium oxide.[14]

More complete public information exists for mill ownership in the late 1970s, and that is shown in Table 11-5.[15] Petroleum companies owned 52 percent of the total capacity. Management affiliation in Table 11-5 means affiliated boards of directors as described previously in Chapter 7. Fourteen percent of the U.S. uranium milling capacity is so affiliated with oil companies. The remaining 7 companies all had subsidiary oil operations. As a result, 100 percent of American milling capacity was owned by oil companies, by companies affiliated with oil companies, or by companies with subsidiary oil production. There was an increase in direct ownership over the decade.

There are two UF_6 conversion facilities.[16] One is operated by the Kerr-McGee oil company, and the other by the Allied Chemical Corporation. This second corporation has its own oil company, the Union Texas Petroleum Corporation.

The country's three existing enrichment plants are owned by the federal government and operated by two chemical companies, Goodyear and Union Carbide.

---

[14]The American Petroleum Institute reports that oil companies owned 267,000 tons of reserves at the end of 1980. At that time, $45/lb market cost ore was estimated to have 470,000 tons of reserves, and the $75/lb market cost ore had a 787,000 ton estimate. At the end of 1980 the spot market price was $26/lb, and it declined throughout 1981 and 1982. Since oil companies generally view reserves as being economically recoverable under existing prices, the 470,000 national figure is probably more comparable to the 267,000 company figure. Sources: Footnote 9 above; American Petroleum Institute, *Market Shares and Individual Company Data for U.S. Energy Markets: 1950–1980* (Washington, 1981), p. 147; and *Wall Street Journal,* November 3, 1981.

[15]Chapman, "Horizontal Integration," Table 4.

[16]Taylor and Yokell summarize the organization of much of the fuel cycle in their chap. 2.

[213]

Table 11-5. Uranium milling capacity and affiliation with oil companies, 1973–78, tons of ore per day

Industry segment	1973 capacity	% of total	1978 capacity	% of total
Oil Company ownership				
Conoco-Pioneer	1,750		2,900	
Exxon	2,000		3,000	
Kerr-McGee	7,000		7,000	
Getty-Kerr-McGee	1,500		—	
Anaconda-ARCO	—		6,000	
Sohio-Reserve	—		1,660	
Total	12,250	40%	20,560	52%
Milling capacity owned by companies having *management affiliation* with oil companies:				
Union Carbide	2,300		2,500	
Anaconda	3,000		—	
Cotter (Commonwealth Ed.)	450		450	
Western Nuclear	1,200		1,700	
Federal Resources–TVA	—		950	
Total	6,950	23%	5,600	14%
Milling capacity owned by companies with *subsidiary oil production:*				
Susquehanna-Western	2,000		—	
Utah International	2,400		3,450	
Atlas	1,500		1,100	
Dawn Mining (Newmont)	500		400	
Homestake–UNC Partners	3,500		3,400	
United Nuclear	—		3,000	
Rio–Tinto Zinc	—		700	
Rocky Mountain Energy (Union Pacific–S.C.E.)	—		1,000	
Total	9,900	32%	13,050	33%
Milling capacity owned by companies with coal operations:				
Rio–Tinto Zinc	500		—	
Federal Resources–TVA	950		—	
Total	1,450	5%	—	
Total U.S. milling capacity:	30,550	100%	39,210	100%

Note: Several companies changed capacity, ownership, or affiliation between 1973 and 1978. Some companies could be placed in two or more categories.
Sources: D. Chapman, "Horizontal Integration," and *The Structure of the U.S. Petroleum Industry.*

Fuel fabrication plants for fuel rod preparation are operated by two oil companies (Exxon and the Gulf section of General Atomic) as well as the four major nuclear reactor manufacturers.

In late 1980, Westinghouse was the leading reactor manufacturer, with 29 reactor units. General Electric had 26, and two other large fabricators (Babcock & Wilcox, and Combustion Engineering) had each built 17 units.[17]

The three attempts at waste fuel processing have all been undertaken by companies otherwise active in petroleum or nuclear power, or both. The Nuclear Fuel Service site near Buffalo, New York, was managed by Getty Oil. After severe problems with employee contamination and earthquake potential, the facility was closed. The Morris, Illinois, facility was operated by General Electric on an experimental basis before being closed because of potentially severe operational problems.[18] (Considering the role of petroleum companies in nuclear fuel, it is interesting to note a reverse kind of relationship here. General Electric, a major power plant supplier, has acquired Utah International, thereby becoming active in a small way in oil production.)

The Barnwell, South Carolina, waste fuel processing plant was developed by a partnership of the Allied Chemical Corporation and the General Atomic Corporation. This latter corporation (mentioned above as a fuel fabricator) is a partnership between Gulf Oil and Shell Oil. It will be recalled that Allied Chemical operates its own Union Texas Petroleum. Barnwell has never been operated, in part because of the moratorium on plutonium extraction through waste fuel processing. This, in turn, came into being because of the terrorist problem and the ease with which nuclear bombs can be manufactured from such plutonium. However, although the reprocessing ban was revoked in 1981, Barnwell has no plans for operation. Whether it could perform better than General Electric or Getty Oil is therefore not known.

The nuclear power industry is obviously highly concentrated. Fewer than five corporations have participated in several stages of the fuel cycle. Conversion, enrichment, fabrication, reactor manufacturing, and attempted waste fuel plants have involved so few firms that conventional profit monopoly behavior would seem clearly possible. But there is no evidence to date of any such profit monopoly efforts in these five particular stages of the nuclear fuel cycle.

Apparently, however, an effective cartel did operate in mining and milling during the middle 1970s. There were 18 corporate groups involved in the Uranium Market Research Organization, with membership covering Great Britain, France, South Africa, Australia, Canada, and the United States.[19]

[17]According to the Atomic Industrial Forum, Inc., Public Affairs and Information Program, "Nuclear Power Facts and Figures," September 1980.
[18]*Science*, 30 August 1974, pp. 770–771.
[19]Taylor and Yokell, Fig. 7.1.

Geoffrey Rothwell summarizes its activities in this way: "Throughout its meetings from early 1972 to mid-1974, the Club attempted to allocate quotas among its participants, to discipline cartel cheaters, to determine the prices and uniform contracts to be offered, and to eliminate competition. After many of these goals were accomplished and the price had risen by a factor of seven, the cartel dissolved in 1975."[20]

Uranium oxide prices did rise from $6/lb in February 1973 to $41/lb in September 1976. The participants held meetings in Paris, Johannesburg, London, and the Canary Islands. Prices and quotas were established for cartel producers and their customers.

Rothwell concludes that the governments of Australia, Canada, France, and South Africa were active in promoting the cartel. In fact, government ownership of all or part of 8 of the 18 corporate groups was reported by Taylor and Yokell. This is a serious problem for economic theory, as shall be suggested in the concluding section.

The major American participant was the Gulf Oil Corporation. Gulf apparently thought that if its active participation was exercised through its Canadian subsidiary, it would be exempt from American antitrust prosecution. Gulf has been found guilty, pleaded no contest, made settlements, or been in litigation with 23 utilities, two oil company uranium suppliers (Exxon and Standard of Ohio), Westinghouse, and the Tennessee Valley Authority. Even its former spouse, the United Nuclear Corporation, was in court with Gulf.[21]

The general complaint of uranium customers is that Gulf and the cartel arranged contracts to buy low and sell high. In other words, the uranium suppliers thought their prices were too low, and that Gulf's low-price contracts were made with knowledge of the cartel's plans to force prices upward. The utilities felt their contracts set prices excessively high because they were unaware of the cartel's efforts to keep prices up.

Economic Theory and Nuclear Economics

The applicability of existing economic theory to the operation of the nuclear power industry is uncertain. If profit monopoly were clearly practical, it would seem to be in the heavily concentrated stages of the cycle: conversion, enrichment, fuel fabrication, nuclear power plant manufacture, and waste fuel reprocessing (if such plants should be operable). No evidence exists for production controls or excessive prices in these sectors.

Yet uranium seems rather clearly to have been the basis for a major cartel in the 1970s.

[20]See in Geoffrey Rothwell, "Market Coordination in the Uranium Oxide Industry," *The Antitrust Bulletin*, 25:1 (Spring 1980), 244.
[21]Gulf Oil Corporation, *1980 Annual Report*, Notes 13 and 14.

Petroleum companies take a dominant position in the nuclear fuel industry. This relationship implies that Academician Aleksandrov is rather more ideological than empirical: there is most definitely no basis to support his belief that petroleum companies are working against nuclear power.

The active involvement of governments and government corporations in the uranium cartel raises a difficult problem for advocates of public ownership of energy resources. How can it be argued that governments will not engage in profit monopoly activities when the uranium cartel, as well as OPEC, has had major government participation?

The competitive theory is not easily adapted to the structure of the industry. In four stages, four or fewer corporations hold 100 percent of the industry. The uranium cartel, of course, is at least historically relevant in indicating that uranium supply was briefly operated as a profit monopoly.

The growth monopoly theory would be significant if it could be established that nuclear power carries considerable external social cost. If the nuclear power industry is viewed as less than normally profitable at its present size, it can be argued that the high profit levels of the petroleum companies, their pursuit of growth, and their strong position in nuclear fuel interact to create a potential for an unprofitable, overexpanded industry requiring considerable public tax subsidy, and creating major external social costs.

Having put this position strongly, I must note that in terms of known public health and environmental damage, nuclear power's record is clearly superior to those of the other conventional energy forms. In Chapter 13, some balance will be sought between nonmarket social costs of nuclear and coal power generation. In the next chapter, the economics of the industry that operates these plants is explored.

Summary

Nuclear fuel is clearly the least expensive fuel available for steam power generation. Total generating costs include operating and capital costs as well as fuel cost. For those nuclear plants which operate successfully without problems, total after-tax generating cost is also less than the generating costs of coal or other types of steam plants.

The nuclear fuel cycle is complex, and consists of several separate stages: mining, milling, conversion, enrichment, fabrication, power generation, and spent fuel disposal. A typical 1,000 MW reactor might require 25–30 metric tons of ore at the beginning of the fuel cycle. A coal plant of similar capacity would require a much greater amount of material to be mined: nearly 3.4 million tons of coal each year.

There is sufficient high-grade ore in the United States to supply the nuclear industry into the next century.

The corporate income tax, because of its investment incentives, provides a major stimulus to nuclear power. Since nuclear power is more capital-intensive than coal power, nuclear power receives a greater tax benefit.

Formerly, utilities were in uniform agreement about the economic edge held by nuclear power. In the early 1980s, some utilities have come to believe that coal generation is less costly, while many continue to see nuclear power as less costly.

The nuclear industry is the most heavily concentrated at the national level of all the major energy industries. Conversion, enrichment, fabrication, reactor manufacturing, and spent fuel processing have each had five or fewer

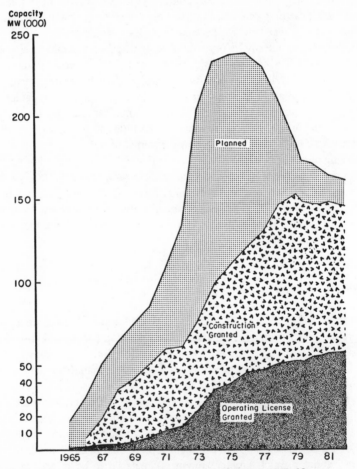

Figure 11-1. U.S. nuclear reactor capacity, operating and planned, 1965–82.
Sources: Monthly Energy Review (various issues); U.S. Nuclear Regulatory Commission, *Annual Reports,* 1975–78; U.S. Atomic Energy Commission, *Annual Reports,* 1965–74. Prepared by Dooley Kiefer.

firms involved, and the major oil companies are active participants and affiliates at every stage of the fuel cycle.

The uranium cartel was an attempt at conventional profit monopoly in uranium supply in the early 1970s. It included private corporations and government agencies. There is no evidence of profit monopoly behavior at other stages of the fuel cycle.

The significant theoretical question arises from the applicability of the growth monopoly theory to nuclear power. In the abstract, it is possible to propose that profitability in the petroleum industry permitted major oil companies to pursue the development of nuclear power at a more rapid rate than would otherwise be the case.

Although nuclear power has to date a clearly superior environmental record compared to coal generation, the small possibility of very large external social costs has a considerable impact on the future of nuclear power. In Chapter 13 these external social costs will be compared for coal and nuclear power.

The future of nuclear power cannot be known with confidence. Total planned American capacity, as shown in Figure 11-1, has declined from its peak in the late 1970s. Operating capacity continues to increase slowly. This is in contrast to nuclear generation in some Western European countries, Japan, and Russia. Nuclear generation in these countries has continued to increase at an annual rate above 20 percent since 1979, the year of the Three Mile Island accident. In the United States, nuclear generation in the early 1980s was only slightly above the 1978 level.[22]

[22]U.S. DoE, Energy Information Administration, *Monthly Energy Review*, November 1980, p. 70, and September 1982, p. 76.

12

Electric Utilities

> Insull knew little about how to manage a central station company, but he was sure of one thing: that the only sensible way to sell electricity was to have no competition. . . . Insull's thinking . . . was based on the premise that electricity should be considered as for everybody, "even the smallest consumer," and that it should be sold at the lowest possible price.
> —Forrest McDonald

With insight, a historian has perceived what motivated the behavior of a major figure in the history of electric utilities in the United States. Samuel Insull emigrated from England to become Thomas Edison's private secretary in 1881 before an electric utility industry existed. Fifty years later, he led a corporate grouping with operating electric utilities in more than 30 states. But in 1932, in the Depression, he was forced out of the management of these utilities. Whether by the design of opponents or by circumstance, he was unable to secure debt financing for the Insull companies, and he was replaced. Insull fled to Europe to avoid prosecution for fraud and embezzlement. The Turkish government seized him while his yacht was in Istanbul and surrendered him to the American government. He was tried for using the U.S. mail to defraud investors by sending them what was alleged to be fictitious information about assets and income.[1]

Insull was acquitted, basically because his business practices were no worse than normal for that period. The Chicago *Times* concluded in 1934: "Insull and fellow defendants—not guilty; the old order—guilty. That was the Insull defense, and the jury agreed with it."

There are obvious parallel paths of development in petroleum and electric utilities in the United States. In the last century, both were generally monopolistic in the sense that production and pricing decisions were made on a noncompetitive basis by small groups of companies. Both were dominated by the personality of an individual leader: Rockefeller in oil, Insull in electricity.

---

[1]Forrest McDonald, *Insull* (Chicago: University of Chicago Press, 1962); the first quotation from p. 57, the following quotation from pp. 332, 333.

In this century, each was the object of political controversy, and each became reorganized. Chapter 6 explained the antitrust dissolution of the Standard Trust in 1911. A following section of this chapter will note the growth of electric utility regulation.

Just as with petroleum products, real electricity prices were shown in Figure 1-3 to have declined throughout the full recorded history of electric utilities in the Growth Era. As with petroleum, this era of declining real prices was accompanied by exponential growth in electricity sales, equivalent to 8.4 percent each year from 1902 to 1.9 trillion kWh in 1973. Just as with petroleum and natural gas, the last year of the Growth Era marked the lowest real price for electricity, and the last year of exponential growth. Since 1973, electricity's average real price has experienced less change than prices of oil products and natural gas. In Figure 4-2, the 1980 real price is shown to be slightly below the 1975 price. Since 1980, real electricity prices have been increasing. However, as indicated in Figure 4-3, electricity demand growth is no longer exponential. One problem—to be addressed in a later section—is that utilities in the middle 1970s continued to order new plants on the assumption of a need for 3.2 trillion kWh in 1980, a figure that as a national average is 39% above the actual 2.3 trillion kWh needed in that year.

The basic economic characteristics underlying the Growth Era and the succeeding decade can be summarized succinctly for electric utilities.

First, technological innovation in production and consumption contributed to continuous improvement in the generation and usage of electricity. The past 100 years have seen the development of the steam turbine generator, alternating current, long distance transmission, complete residential electrification, electric light bulbs, refrigerators, air conditioning, household appliances, computers, electric furnaces in industry, and so on. But in the years since the Growth Era ended we have not experienced such innovations.

Second, economies of scale have been of crucial significance in power generation. For approximately 80 years, larger size meant lower cost of generation. In 1893, General Electric built two 0.8 megawatt generators. In the 1980s, 1,000 megawatt facilities are common. By integrating separate rural communities in the early 1900s, utilities increased *load factors*, enabling expensive capital investment to be used more hours in the day. In the 1980s, rural and urban areas are linked through interstate *power pools* that distribute electricity over large regions. As with innovation, however, economies of scale are less important in a period when the scale of the industry is no longer increasing. In addition, there may be a practical limit for large generating plants, in that *outages* for very large plants may be due to insurmountable scale factors which create *diseconomies of scale*.

Third, fuel cost for power generation is no longer declining in real terms. Large inexpensive hydropower sites have been developed, but in the future new hydro generation above about 300 billion kWh will be comparable in cost to coal or nuclear power. Until 1973, utility fuel prices declined in real terms. Since then, all have increased in a manner similar to the increased residential costs for oil and natural gas. As a broad consequence, future expansion of generating levels will cause average real costs to increase.

Fourth, the environmental impact of electricity use is very significant, as the next chapter will discuss. This is particularly important in regard to air pollution (much of which is caused by coal-burning power plants) and nuclear power (all of which is used for power generation).

Finally, the electric utility industry is generally believed to be, and is regulated as, a *natural monopoly*. Economies of scale in development to the 2 trillion kWh level have meant that competition within any geographic area is so costly as to be impossible. Duplication of transmission and distribution lines would create much higher cost without any improvement in service. Similarly, duplicated generating facilities would mean lower load factors and higher cost. In general, enforced competition and duplication would probably about double the cost of electricity to utility customers.

The points summarized above have important implications for several other aspects of the utility industry, particularly the regulation of rates and of environmental impact. Before proceeding to those subjects in this and the next chapter, we must investigate three other economic aspects of the industry: ownership, integration (horizontal, vertical, and circular), and taxation.

Ownership

As with oil, gas, and coal resources, the United States is unique among industrial nations of the world in the degree to which electric utilities are owned by private corporations. Table 12-1 portrays the sources of America's electricity generation. Three-fourths is produced by investor-owned utilities, one-fifth by federal, state, municipal and cooperative organizations, and very small amounts by industrial generation and importation.

The municipal *public utilities* range in size from the very large Los Angeles Department of Water and Power (1.2 million customers, 18 billion kWh sales) to Coldwater, Michigan (5 thousand customers, 115 million kWh sales). The biggest municipal or state utility is the Power Authority of the State of New York, which generates more than 30 billion kWh each year.

In contrast, some federal utilities are much larger. The three biggest are the Bureau of Reclamation (45 billion kWh), the U.S. Army Corps of Engineers (80 billion kWh), and the Tennessee Valley Authority (120 billion kWh). These federal agencies generally sell their power to the private or public utilities, which deliver the electricity to the final customer.

[222]

Table 12-1. Power generation by electric utilities in the United States, 1980, trillion kWh

Type of ownership	Amount	% of U.S. total
Investor-owned utilities	1.783	75%
Federal projects, authorities	0.235	10%
State projects & power districts	0.118	5%
Municipal utilities	0.087	4%
Cooperatives	0.064	3%
Total generation, utility industry	2.286	96%
Industrial generation	0.068	3%
Imports: Canada & Mexico	0.030	1%
Total generation and imports	2.385	100%

Source: Edison Electric Institute, *Statistical Yearbook of the Electric Utility Industry,* 1980, pp. 16, 20. Rounding error responsible for discrepancies in sums and totals.

Although public utilities exist throughout the country, they sell more than one-half of the electricity in only four states: Nebraska (100 percent of the state's electricity is sold by publicly owned utilities), Tennessee (98 percent), Alaska (93 percent), and Washington (73 percent).

The largest private utility corporations are described in Table 12-2. These six companies provide 26 percent of the sales of all U.S. *investor owned* private utilities.[2] Four of the six are operating companies without major subsidiaries. Two are *utility holding companies,* meaning that they own other companies in the actual business of producing and distributing power. Holding companies earlier in this century were a common vehicle of financial abuse, in that they sold paper assets and fictitious services to their subsidiaries. In the 1980s, the combination of state and federal regulation with a much higher standard of business ethics means that earlier abuses are probably nonexistent.

Electric utilities work in a quite different financial environment from that of petroleum corporations. The revenues of the largest utility in Table 12-2 (Pacific Gas and Electric) are less than the revenues of the smallest major oil company (Cities Service; see Table 6-1). Profitability is considerably less as well: the 10 percent average rate of return on shareholders' equity is much less than the 22 percent earned by major oil companies. Generally, utilities have been earning less profit per dollar of investment than do other industries.

The six utilities have an average $436,000 assets per employee, making them more capital-intensive than major oil companies ($279,000 assets per employee).

The leading shareholders of these utilities, shown in Table 12-3, are similar to those of the petroleum companies, in that they are banks, insurance, invest-

[2]In 1979.

Table 12-2. Largest electric utilities, economics data, 1980

Company	Pacific Gas and Electric	Consolidated Edison	American Electric Power	Southern Co.	Southern California Edison	Commonwealth Edison	Total or average, six largest utilities
Headquarters	San Francisco, California	New York City	Columbus, Ohio	Atlanta, Georgia	Los Angeles, California	Chicago	
Holding or operating company	operating	operating	holding	holding	operating	operating	
States in which utility has retail sales	California	New York	Indiana, Kentucky, Michigan, Ohio, Tennessee, Virginia, West Virginia	Alabama, Florida, Georgia, Mississippi	California	Illinois	
Revenue	$5.3 billion	$3.9 billion	$3.8 billion	$3.8 billion	$3.7 billion	$3.3 billion	$23.1 billion
Net income	$525 million	$335 million	$348 million	$344 million	$318 million	$382 million	$2,252 million
Return on equity	10.6%	9.5%	9.5%	9.5%	10.6%	11.2%	10.2%
Federal income tax charged to expenses	$72 million	$64 million	$35 million	$73 million	$39 million	$103 million	$381 million
Sales	58 bill kWh	37 bill kWh	112 bill kWh	93 bill kWh	60 bill kWh	62 bill kWh	422 bill kWh
Assets	$11.3 billion	$7.5 billion	$11.0 billion	$11.5 billion	$7.7 billion	$10.2 billion	$59.2 billion
Employees	27,582	23,156	26,659	27,940	14,157	16,400	135,894

Sources: Annual Reports and U.S. Department of Energy, Energy Information Administration, Statistics of Privately Owned Electric Utilities in the United States, annual. Federal income tax amounts charged to expenses may include taxes not actually paid because they are offset by credits.

ment, and pension companies. But, in general, the leading shareholders in large utilities are not the same financial institutions that are the leading shareholders in major oil corporations.

A pattern of management affiliation similar to that of the oil companies exists for electric utilities. Table 12-4 summarizes these affiliations for shareholding, affiliated directors, and accountants. Each of the six largest utilities is associated with each of the other five. Also shown in Table 12-4 are affiliations between these utilities and General Electric and Westinghouse. They are extensive: only the American Electric Power/Westinghouse pair is empty.

The economic motivation for this pattern is like that described in Chapter 7 in the context of petroleum companies. For electric utilities as for petroleum companies, major shareholders are almost exclusively banking, insurance,

Table 12-3. Leading shareholders, largest private utility corporations, percent of common stock voting shares owned or managed, 1980

Company	% of stock	Company	% of stock
PACIFIC GAS & ELECTRIC		CONSOLIDATED EDISON	
Bancal Tri-State Corp.	10.4%	Connecticut General	1.5%
First Nat'l Boston Co.	0.9	FMR Corporation	1.1
Prudential Insur.	0.7	Ford Foundation	1.0
Wells Fargo	0.7	Miller-Anderson	0.8
TIAA/CREF	0.6	Travelers Corp.	0.8
J P Morgan	0.6	First Union Bancorp	0.7
	13.9%		5.9%
AMERICAN ELECTRIC POWER		SOUTHERN COMPANY	
Michael Pescatello	0.8%	Harris Bankcorp	0.8%
Prudential Insur.	0.5	Wells Fargo	0.7
Wells Fargo	0.3	U.S. Steel, Carnegie Pension	0.6
Bankers Trust	0.3	Bankers Trust	0.5
Walter E. Heller	0.3	Lord Abbett & Co.	0.3
Citicorp	0.2	Walter E. Heller	0.3
	2.4%		3.2%
SOUTHERN CALIFORNIA EDISON		COMMONWEALTH EDISON	
Western Bancorp	6.1%	Northern Trust	1.1%
Capital Group	1.6	Prudential Insur.	0.9
Crocker National	1.5	Bancal Tri-state Corp.	0.8
TIAA/CREF	1.4	University of Calif. Pension	0.8
Loews Corp.	1.2	Sears Roebuck	0.8
Wells Fargo	1.1	Loews Corp.	0.7
	12.9%		5.1%

Source: U.S. Senate, Committee on Governmental Affairs, *Structure of Corporate Concentration,* Committee Print, December 1980.

Table 12-4. Largest shareholders and management affiliation: Electric utilities and electrical equipment manufacturers, 1980

Utility Companies	(2) Comm. Ed.	(3) Consol. Ed.	(4) PG & E	(5) Southern Co.	(6) So. Cal. Ed.	(7) General Elec.	(8) Westing- house
(1) Amer. Elec. Power	S2,D	S	S2,A	S3,D	S	S2,D	*
(2) Commonwealth Ed.		S,D	S4	S,A	S3,A	D	D2
(3) Consolidated Ed.			S	S	S	D2	D2,A
(4) Pacific Gas & El.				S	S2,D	S3,D2	S,D
(5) Southern Co.					S,A	S,D	S
(6) So. Calif. Ed.						S,D2	S,D
(7) General Elec.							S,D5
(8) Westinghouse							

S = shared institutional shareholders among largest 6 in each company.
D = the same firm is represented on the boards of directors of a pair of electric utilities.
A = shared accountants.
* = no affiliations.
Source: U.S. Senate, *Structure of Corporate Concentration,* and Annual Reports.

pension, and investment companies. As with petroleum, the utilities' high level of capital intensity leads directly to major investments by the large investment companies. In the same way, given the complexity of the modern utility industry, multiple holdings and directorates enable the investment company to attain a more accurate and confident understanding of the industry than is possible with a single large holding. Similarly, economy of scale in accounting leads to multiple affiliations of accounting companies with utilities, and concern about conflict of interest leads to the avoidance of complete accounting responsibility by a single firm.

If Chapter 7's benchmark is adopted, ownership exceeding 10 percent is necessary for a single owner to have significant influence or control. The absence of an ownership share exceeding 10 percent defines management control. On this criterion, five of the six utilities in Table 12-3 are management controlled. It is not known if employee stock and pension plans are large shareholders in utilities as they are in petroleum companies.

In Chapter 11, General Electric and Westinghouse were seen to be the two most important nuclear reactor suppliers. The affiliations between General Electric, Westinghouse, and major utilities may, on the one hand, be viewed as promoting efficient planning between related industries. Alternatively, if the planning data on construction cost, spent fuel disposal, and decommissioning are erroneously low in comparison to actual costs, then these affiliations between reactor suppliers and utilities can inhibit efficient planning by utilities.

Integration: Vertical, Horizontal, and Circular

The electric utility industry always owns and operates its generation, transmission, and distribution facilities. Vertical integration (as defined in Chapter 6) is limited since most utilities do not own fuel resources. However, there are exceptions to this generalization. Many of the combination natural gas–electric utilities may have developed gas resources, and electric utilities in oil-producing states may have oil properties. Most electric utilities, however, do not own oil and natural gas reserves.

The situation is different with respect to coal and nuclear fuel. In Chapter 10, it was pointed out that utilities mined at least 90 million tons of bituminous coal in 1980, 11 percent of the national total. Nineteen utilities were involved in this production. Utility involvement in coal production is likely to increase because at least 52 electric utilities now own coal reserves.

Utility participation in nuclear fuel is much less common than in coal. No utility owns or operates conversion, enrichment, fabrication, or disposal facilities. Only twelve utilities are known to own uranium reserves.[3]

The economic significance of vertical integration can depend upon the nature of cost-of-service regulation. If a utility is so regulated that investment in coal facilities must earn the same rate of return as investment in electric utility plant, then coal investment may have little influence on fuel cost, utility profit, or the price charged to customers. However, if the utility can charge itself a price for coal, then the possibility exists for raising profits and rates by overcharging for coal. A survey by the Federal Energy Regulatory Commission examined this question.[4] Fifteen utilities were studied: 11 charged themselves more for their own coal than they paid for outside contract purchases. For example, one of the holding corporations listed in Table 12-2 owns coal companies, and these coal companies sell coal to the subsidiary operating companies owned by the holding company. In 1978, the holding company charged itself $36 per ton for its own coal while it paid $25 per ton for contract-purchased coal. (Incidentally, the commission's survey included two publicly owned electric utilities, and both charged themselves more for their own coal than for the coal they purchased.)

The term *combination utility* refers to a company that sells both electricity and natural gas. This form of horizontal integration is common. The economic question is similar to that posed by electric utility participation in coal and nuclear fuel: is it efficient? Alfred Kahn's summary of empirical evidence

[3]See Edison Electric Institute, "Investor-Owned Electric Utility New Business Ventures: A Survey of Diversification Activities," October 1981, and the Tennessee Valley Authority *Annual Report 1980*, Vol. II, p. 26, showing investment in uranium and coal properties.

[4]U.S. Federal Power Commission, "Electric Utilities Captive Coal Operations," June 1977, and U.S. Federal Energy Regulatory Commission, "Updated Tables," June 1979.

had two somewhat contradictory findings: (1) combination utilities have higher costs and charge more for electricity than do electric-only utilities, and (2) combination utilities may have lower costs and charge less for gas than do gas-only utilities.[5]

Cooperative activities similar to those found in production, refining, and marketing in petroleum and natural gas are typical in the electric utility industry. Joint ownership of large new generating facilities is common, although most plants and transmission lines have a single owner.

Of greater economic significance is the system of power pools that exist throughout the country. These power pools are cooperatives of utilities which trade, buy, and sell electricity between member companies. Generally, they attempt to provide minimum-cost electricity by selecting the most efficient power plants available, and dispatching excess power from utilities with surpluses to those utilities with different periods of peak demand. For example, New York City utilities have their maximum load in the summer, but the upstate utilities have theirs in the winter. When the two kinds of needs are marked up in a single integrated system, the cost is made lower for all the customers of all the utilities. Each utility may be trading electricity with as many as 20 others over a year's time. In addition, interconnections among utilities permit them broad access to electricity from the country's large inexpensive public hydropower projects. As a result of this cooperative system, a little more than half of the country's electricity is sold by utilities that did not generate it. In the absence of such cooperation, each company would need to build sufficient capacity to supply all of its customers' needs. A speculative guess is that the system of power pools and interconnections reduces costs by about 25 percent. These networks of cooperating utilities might be termed circular integration, since the power pool networks permit each utility to purchase, sell, *wheel*, or interchange power with adjacent and distant utilities.[6]

A power pool works effectively when the companies in it have capacity in conformance with the demand in the region. If surplus capacity exists, then each utility must decide whether certain costly plants should be kept in service

[5]Alfred E. Kahn, *The Economics of Regulation: Principles and Institutions*, 2 vols. (New York: John Wiley, 1970 and 1971), 2: 276–280. Rodney E. Stevenson, in his "X—Inefficiency and Interfirm Rivalry: Evidence from the Electric Utility Industry," *Land Economics*, 58:1 (February 1982), 52–66, concludes that combination gas-electric utilities are less efficient producers of electricity than are electricity-only utilities.

[6]The estimate of the proportion of traded electricity is based upon data in two U.S. Department of Energy, Energy Information Administration annual publications: *Statistics of Privately Owned Electric Utilities in the United States* and *Statistics of Publicly Owned Electric Utilities in the United States*. Further integration would not yield the magnitude of savings already attained; see Gordon R. Corey, "Effect of Power Grid Size on Economics," presented to the International Fuel Cycle Evaluation/Non-Proliferation Alternative System Assessment Program, June 1979, and U.S. Federal Energy Regulatory Commission, *Power Pooling in the United States*, December 1981.

in order to qualify for continuation in the rate base from which revenues are allowed.

Cost-of-Service Regulation

The general public opposition to private monopoly in the last century led to the formation of state regulatory commissions. This began in Massachusetts in 1887 when private electric light companies were made subject to commission regulation. All privately owned electric utilities throughout the country are now regulated by state commissions which determine the aggregate revenues as well as the individual rate schedules that can be used by utilities in their states. The basic relationship can be shown with an equation:

(1) $REV = r * RB + DEP + FUEL + PURCH + AOMT + TAX$

This says that the allowed revenue for a utility is basically the sum of six separate terms: return on rate base ($r * RB$), depreciation, fuel cost, the cost of purchased electricity (if any), operating and maintenance expense, and a tax allowance.

The *rate base* is the company's investment in facilities, less normal depreciation. This normal depreciation is taken as *straight line depreciation* in which the amount of depreciation is the same each year for a plant, transmission line, or other item.

The *rate of return* is determined by a regulatory commission as yielding a fair percentage profit for shareholders. This rate of return is typically an average of interest rates on debt and profit to shareholders. The rate of return (r) is multiplied by the rate base to give an allowed return on capital investment.

The second term, *DEP,* represents the same straight line depreciation. The third term is equally direct: fuel cost. There is some complexity, however, because a *fuel adjustment clause* may be in force which allows the utility to pass along higher fuel costs automatically to its customers without a regulatory proceeding and approval. It is conceivable that a utility which overcharges itself for its own captive coal could abuse this automatic adjustment mechanism.

PURCH represents the cost of purchased power. (For some utilities, of course, that sell more to other utilities than they purchase, *PURCH* would be replaced with *SELL*.)

Annual operations, maintenance, property taxes, and similar expenses (*AOMT*) are, as with fuel, collected dollar for dollar from customers.

The relationship of the revenue allowance to income tax payment is sufficiently complex (as suggested in Chapter 11) to require special consideration.

[229]

Coal, Nuclear Power, and Utilities

Table 12-5. Rate regulation: Cost-of-service determination for hypothetical General Edison

		$ million
Return allowed on rate base:	12.7% * $1,200 =	$152
Depreciation, straight line		40
Fuel cost		127
Purchased power, state hydro		20
Annual operation, maintenance and property taxes		140
Federal corporate income tax allowance		78
Total allowed cost of service		$557

For ease of understanding, assume for the moment that *TAX* is the federal 46 percent rate on taxable income. Imagine that taxable income is simply profit or net income before income taxes.

With this simplified view of cost-of-service regulation, an example is easily developed. Suppose General Edison has four 500 MW coal plants which it built over the last 35 years. The cost was $1.6 billion dollars, and this includes transmission lines, distribution systems, and so on. The company estimates each plant will last 40 years, so straight line depreciation is $1.6 billion divided by 40 years, or $40 million per year.

Accumulated depreciation is $400 million. So, in Table 12-5, the rate base is currently $1.2 billion. The regulatory commission approved a 12.7 percent rate of return on rate base, so this return was $152 million.

Fuel cost is $127 million for the coal plants.

General Edison purchases 2 billion kWh from its state hydro authority at 1¢ per kWh, or $20 million.

The company's operation and maintenance costs are $100 million and its property and other state and local taxes are $40 million, so AOMT is $140 million.

In this simple example, the federal corporate income tax will be applied only to the return on rate base, and only to that portion of return on rate base which exceeds interest payments to bond holders. Suppose the interest payments are $60 million. So the tax allowance must be large enough to permit a 46 percent levy on taxable income. The precise amount is $78 million.[7]

Given all these data, the company and the commission can proceed to estimate the average cost per kWh that needs to be collected from—on the average—each customer.

Let us assume a 70 percent load factor for General Edison's four plants and 2,000 MW total capacity. This develops 12.3 billion kWh, as in Equation (2):

(2)   Generation = Capacity * load factor * hours in the year
        12.3 billion kWh = 2000 MW * 70% * (365 * 24)

This is simply another application of Equations (1) and (2) from Chapter 11.

[7]It can be determined by $TAX = ((z/(1 - z))(r * RB - \text{interest})$ with z representing the corporate income tax rate. So: $TAX = (.46/.54)(152 - 60)$, or $78 million.

Table 12-6. Generation, cost of service, and average price, hypothetical General Edison

Generation by four plants, 2,000 MW, from Eq. (2)	12.3 bill kWh
Purchases from state hydro authority	2.0 bill kWh
Total supply	14.3 bill kWh
Transmission loss (10%)	1.4 bill kWh
Available for sales to customers	12.9 bill kWh
Cost of service, from Table 12-5	$557.0 million
Average cost and allowed price per kWh	4.3¢/kWh

Remember that General Edison—as do many private utilities—purchased 2 billion kWh from the state hydro authority. It has a 14.3 billion kWh supply. Suppose the transmission loss is 10 percent, so 12.9 billion kWh remain for sale to customers. The allowed cost and price will be 4.3¢/kWh. All this is shown in Table 12-6.

Observe three other points about the example. First, the company is much smaller than the very large utilities listed in Tables 12-2–12-3. In this, General Edison is representative of the average utility. Second, buying from the state hydro power at 1¢/kWh is much less costly than coal generation. This is typical also, and was the basis for the numerical illustration in Chapters 1 and 2. Third, there is no excess capacity or outage problem, and the load factor is an efficient 70 percent.

Tax Factors

In the period before 1954, this simple summary would have been an adequate guideline to regulatory policies on revenue determination. From that year up through the enactment of the Economic Recovery Tax Act of 1981, the corporate income tax has been continuously revised to encourage general industrial development. As we saw in Chapter 11, those tax incentives are now of considerable importance for utilities. Without attempting a full discussion, we can summarize the net effect as indicating that a typical utility building new capacity will be exempt from income tax payments for a lengthy period. The new tax act prohibits state regulatory commissions from attempting to capture these *tax benefits* for customers in the years they are earned. Instead, commissions and companies now use *normalization,* a term meaning that the company collects the tax benefits, then distributes them to customers almost on a normal straight line basis over the expected 30-to-40-year operating life of the facility.

It was shown in the previous chapter that the effect of tax policy on coal and nuclear generating cost can be considerable. The large utilities described in Table 12-2 have a typical pattern. In 1979, two paid actual federal income taxes: Pacific Gas and Electric, and Consolidated Edison. Both of these are operating companies. Two other operating companies, Commonwealth Edison and Southern California Edison, were negative tax payers: they received monies from the Internal Revenue Service through the provisions of the income tax. Both these companies earned net income of $300 to $400 million. The two large holding companies (Southern and American Electric Power) each had subsidiaries that were negative payers although the total for each holding company was positive.[8]

The impact of normalization on rates can be heavy.

Formerly, under *flow-through* regulation, the tax benefits were passed along to customers immediately. In this approach, the *TAX* term for regulation would be the actual current tax payment. Suppose General Edison is building a new 1,000 MW nuclear plant and has, in fact, a zero or negative tax liability. With flow-through regulation, the corporate income tax allowance in Table 12-5 would be the actual amount: zero. Consequently, cost of service would be a lower $480 million, and the average kWh would now cost 3.7¢/ kWh.

If General Edison's average residential customer uses 8,400 kWh per year, that average customer is being charged $50 in this hypothetical year for income taxes which are not being paid.

But a very important reminder is necessary: taxes deferred now become due in the future. The utility company is not earning additional profit: it is sacrificing future profit for present profit. The illustration is realistic, because this typical utility will be using the tax benefits from its new nuclear plant to finance that plant. If investigation of tax liability and its absence has created an interest in these provisions themselves, they may be examined in summary form in Appendix 12-B.[9] Remember that the 1981 tax act prohibits flow-through regulation.

CWIP, AFUDC, and Paper Money

The financial strain placed on utilities by major construction programs during a period of less-than-expected growth is creating new concern about other aspects of the economic regulation of utilities.

Traditionally, a new plant is not placed in the rate base until it begins generating electricity. There are two basic reasons for this practice. One is the principle that actual users of a facility should pay for it, so customers should

[8]The data on tax payments are from U.S. DoE, *Statistics of Privately Owned Electric Utilities: 1979,* and from Carl Goldfield, "More IOU Tax Favors?" *Public Power,* 39:3 (May-June 1981), 43–45.

[9]Other discussions of the subject are Duane Chapman, "Federal Tax Incentives Affecting Coal and Nuclear Power Economics," *Natural Resources Journal,* 22:2 (April 1982), 361–378; Donald W. Kiefer, "Accelerated Depreciation and the Investment Tax Credit in the Public Utility Industry," National Regulatory Research Institute, Ohio State University, April 1979; and Christopher P. Davis, "Federal Tax Subsidies for Electric Utilities," *Harvard Environmental Law Review,* 4:2 (1980), 311–358.

not be required to pay for a plant until it is in use. A corollary is that capital on new plants should be raised from shareholders and lenders, and not collected from customers without their consent. A second reason is that the exclusion of construction expenditures from the rate base may be an incentive to utility management to avoid needless extension of construction periods, and unnecessary plants. The formal accounting term for this category of expenditures is *construction work in progress:* cumulative expenditures on construction of new facilities which are not yet placed in service. It is abbreviated CWIP. As 1980 began, the average electric utility had about 40¢ in CWIP for each dollar in its rate base. It is understandable that utility managements would like to see CWIP placed immediately in their rate bases.

However, an older economic concept is generally incompatible with placing CWIP in the rate base. This complicated concept is AFUDC: *allowance for funds used during construction.* It is intended to allow utilities to enlarge rate bases, on the assumption that additional value is merited for companies that build new capacity. AFUDC is calculated as if the company were earning interest on its construction, as if the construction expenditures were in a savings account, and collecting interest. Table 12-7 shows AFUDC being collected for General Edison's hypothetical nuclear plant. The AFUDC rate is

Table 12-7. Correcting income for funny money: General Edison analyzed

	$ million
A. Reported conventional net income data	
Sales: 11.0 billion kWh	
Revenue (+)	475
Operating Expenses (−)	365
fuel and purchased power 175	
operations, maintenance, property taxes 150	
normal straight-line depreciation 40	
Reported income tax expense (−)	23
Operating income (=)	87
AFUDC allowance (+)	50
Interest payments (−)	60
Net income (=)	77
B. Corrected income statement	
Revenue (+)	475
Operating expenses (−)	365
Actual income tax payment (−)	0
Operating income (=)	110
Interest payments (−)	60
Corrected net income (=)	50
C. Return on shareholders' equity	
Shareholders' equity: $600 of 1,200 rate base	
Net income, conventional method: $77 (from Part A)	
Return on shareholders' equity 12.8%	
Corrected net income: $50	
Return on shareholders' equity 8.3%	

determined by the regulatory commission on the basis of the company's interest expenses on actual debt, its allowed returns on preferred and common stock, and the impact of the corporate income tax on these amounts. Current AFUDC rates are about 10 percent. AFUDC is counted as income (or as an offset to interest expense) in the year it is calculated. However, this is not real money: AFUDC has nothing to do with revenue collected from customers. It is an accounting entry which records an allowed increase in rate base for some future year when a facility enters useful service. AFUDC is recorded as income now because it recognizes the creation of the basis for future earnings.

There are two major problems, then, with interpreting the financial reports of electric utilities which involve the present-year treatment of future events. AFUDC counts earnings that have yet to be received, but the income tax account records an expense that is not yet paid. It should be kept in mind that both AFUDC and tax benefits arise from construction programs. Surprisingly the two "funny money" items almost cancel each other.

Table 12-7 is developed from the income statement of the hypothetical General Edison. It shows the effects of some common problems facing utilities. First, the company has seriously overestimated demand growth. Instead of the 12.9 billion kWh it expected to sell, customers actually purchased 11.0 billion kWh. Second, the company underestimated operating expenses. The actual amount is $365 million rather than the $327 expected. The reported tax expense is $23 million, 46 percent of the excess of revenue over operating expense and interest payments.

The AFUDC allowance is $50 million, and actual interest payments are $60 million. Net income will be revenue plus the AFUDC allowance less operating expenses, less reported income tax expense, less interest payments. The result is the $77 million net income in Part A of the table.

The correction for funny money is in Part B. The actual tax payment is zero, and the AFUDC allowance will not be counted. Corrected net income, then, is revenue, less operating expenses, less actual income payment, less interest payments. Corrected net income is thus $50 million in this example.

The difference is considerable. In terms of conventional economics, General Edison's $77 million in net income appears to earn 12.8 percent on the $600 million shareholder investment in rate base. When the funny money correction is made to exclude future taxes or future value from AFUDC, the corrected net income is $50 million, and an 8.3 percent return. If general inflation is 10 percent, shareholders are losing money by maintaining their investment in General Edison.

There is a good deal of variation in customer prices which follow from variations in regulatory treatment of the same investment. One method of showing this is to determine what the average price of electricity would be in each year according to flow-through, normalization, or no-subsidy regulation. Kathleen Cole has prepared such an analysis for this chapter, using the representative coal and nuclear plant cost assumptions from Chapter 11.

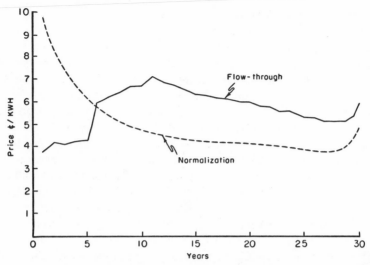

Figure 12-1. Average prices with flow-through and normalization regulation (hypothetical nuclear plant, 1980 dollars)

In Figure 12-1, the normalizing utility is charging more than twice as much as the flow-through utility for the same nuclear plant in the first two years. The flow-through utility uses its investment tax credit evenly in the first five years of operation, and this and other tax benefits are wholly captured for customers. This causes the major discontinuity in the figure. As expected, the normalizing utility is returning tax benefits to customers over the life of the facility rather than when those tax benefits were gained, so after the fifth year the normalized price is always below the flow-through price. The no-subsidy price, incidentally, would be off the top of the graph in the first few years.

Cole prepared a similar analysis for the coal plant. Although it is not shown, its characteristics are identical.

The 1981 tax act will reduce the frequency of this kind of tax-related variation because of the prohibition of flow-through capture of any tax benefits for customers for new facilities. Again, it should be remembered that normalization does not necessarily benefit management and shareholders at the expense of customers. The profit earned in the two methods is comparable. Also note the years of lower prices resulting from normalization in Figure 12-1.

Rates and PURPA

Given an estimated cost of service and a derived cost per kWh, the utility and regulatory commission determine acceptable *rate schedules* which will bring the utility the amount estimated. Typically, there are several rate schedules for different types of industrial, commercial, and residential customers. These are differentiated by the costs of serving different customers as well as by the ability of very large customers to negotiate rates with utilities.

The average industrial customer will use nearly two million kWh per year, compared to nine thousand for the average residential customer. This large industrial consumption reduces the cost of distribution lines, meters, and billing for the large customer, justifying a lower price. In addition, an industrial average annual electric bill exceeding $50,000 creates a negotiating incentive. For very large industrial customers such as aluminum companies, nuclear fuel enrichment plants, refineries, and automobile companies, both cost and negotiating incentives are strengthened. The net result of both influences is that industrial customers pay less per kWh than commercial and residential customers. In 1980, the difference meant that, nationwide, the average residential customer paid 5.4¢ for the average kilowatt hour, 1.7¢ more than the average industrial kilowatt hour.

The simplest rate schedules apply to residential customers, and one is given in Table 12-8. The customer charge is a flat rate of $3 per month, and represents the fixed monthly cost of serving a typical customer. The first 500 kWh are each billed at 5.25¢/kWh. All kWh above 500 are billed at 4¢ each. A fuel adjustment factor adds 0.73¢ per kWh, and a state hydro authority credit deducts a very small 0.04¢ for each kWh. In addition, General Edison collects a 3 percent state sales tax.

The application is not difficult. General Edison's typical household uses 700 kWh per month. First, there is the $3 charge. Second, the 500 kWh portion at 5.25¢/kWh adds $26.25. Third, the remaining 200 kWh (above 500) are billed at 4¢/kWh, adding $8.00. Fourth, the fuel adjustment factor is 0.73¢/kWh for all 700 kWh: $5.11. Finally, the state hydro credit is .04¢/ kWh on all 700 kWh, and subtracts a modest 28 cents from the bill. Summing all five components, the result is a monthly bill of $42.08. The sales tax adds another 3%: $1.26, so the total bill is $43.34. This will be the amount the residential customer pays.

There are many kinds of complicated amendments to the rate schedule just described. Billing may in fact be bimonthly. There may be a late charge.

Table 12-8. Typical residential rate schedule

	Monthly charge
Customer charge	$3.00
Electricity charge, first 500 kWh	5.25¢/kWh
Electricity charge, use over 500 kWh	4.00¢/kWh
Fuel adjustment factor	0.73¢/kWh
State hydro authority credit	−0.04¢/kWh
Sales tax: Add 3%	

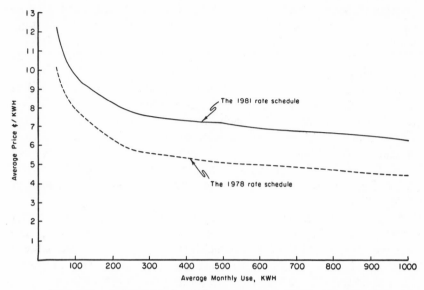

Figure 12-2. General Edison's residential rates: Average price and level of use

Major discounts may be available for electrically heated homes, for off-season use, or for nighttime use. There may be other variations for climate zones, or for the customer's willingness to allow the utility to control certain appliances. But the simple procedure described is the general basis for residential rates.

Observe that use over 500 kWh has a lower charge than smaller levels of use, and that $3 each month is charged to the customer regardless of use levels. The effect of the rate schedule in Table 12-8 is to lower the average price per kWh as use increases, and raise the average price as use declines. For this specific illustration, increasing consumption by 10 percent (70 kWh) raises the bill by 7 percent. Conserving 10 percent (70 kWh again) reduces the bill by 7 percent. Rate schedules continue to have a modest promotional effect for increased use, and a modest penalty for conservation.

Figure 12-2 represents the relationship of average price to level of use, and shows how it declines with greater use, and rises with reduced use.

During the Growth Era, promotional rate schedules were economically justified. They encouraged increased sales, which allowed larger plants to generate electricity at lower average cost. This is not true in the 1980s, and typical residential rate schedules are giving consumers the wrong incentives.

Commercial and industrial rate schedules are considerably more complex than residential schedules. They are more likely to include variations. And they typically differentiate a capacity investment kW charge from a kWh generation charge. The theory here is that a kW charge is linked to capacity investment cost, and a kWh charge to fuel cost for power generation. Thus an industrial customer with a modest kW requirement will, logically, cost the utility less to serve than a second industrial customer with the same kWh usage but periodically a very high kW requirement. The first customer needs less capacity for the same kWh.

With this background, it is possible to integrate the cost-of-service revenue allowance, the rate schedule, and expected use levels. Table 12-9 requires careful attention. The first line shows residential class data. It includes the 700 kWh average monthly use, the average yearly use, and the number of customers. These factors define the expected 3 billion kWh of residential sales. From the Table 12-8 rate schedule, the average price of 6.01¢/kWh is defined, giving the utilities expected revenue from residential customers of $180 million.

Similar procedures define the commercial, industrial, and public sector figures. (The public sector means street lights, city subways, buildings, etc.) Observe in Table 12-9 the very large differences in average use between classes, and the related difference in average prices. If all goes as planned, General Edison sells 12.9 billion kWh for its allowed $557 million.

It should be remembered that preceding sections explained some of the ways in which these matters do not go as planned. Demand growth is less than expected, and fuel and construction cost and interest rates are higher than anticipated.

The 1978 Natural Gas Policy Act (discussed in Chapter 8) was accompanied by the *Public Utility Regulatory Policies Act,* known familiarly by its acronym PURPA. The act was intended to promote energy conservation. In this it differs markedly from the government-utility consensus of the Growth Era, in which both parties favored continued exponential growth in an economic environment of declining costs and declining real prices to customers.

PURPA has several provisions, and it is simplest to list them:[10]

(1) Rates charged to each customer class are to be based on the cost of service to provide electricity to that class. This means that one class is not supposed to subsidize another.

(2) Declining rates are discouraged—the kind of typical rate schedule shown in Table 12-8.

(3) Utilities are encouraged to base rates upon seasonal and daily variations in cost, and to offer load-management techniques.

(4) Master meters for apartment buildings are prohibited, in order to ensure that each customer knows his or her monthly bill and consumption.

[10]See the act itself ("Public Utility Regulatory Policies Act of 1978," Public Law 95–617, 95th Congress); Paul L. Joskow, "Public Utility Regulatory Policies Act of 1978: Electric Utility Rate Reform," *Natural Resources Journal,* 19:4 (October 1979), 787–810; and current editions of *Public Utilities Fortnightly* and the Environmental Action Foundation's *Powerline.*

Table 12-9. General Edison's revenue expectations by customer class (Goal: $557 million, cost-of-service allowance from Table 12-5)

Customer class	Expected average sales data					Revenue for each class, $ million
	Monthly use, kWh	Annual use, kWh	Number of customers in class	Annual sales, bill kWh	Average price, ¢/kWh	
Residential	700	8,400	357,143	3.00	6.01	180
Commercial	3,750	45,000	63,333	2.85	4.10	117
Industrial	200,000	2,400,000	2,563	6.15	3.66	225
Public	15,000	180,000	5,000	0.90	3.85	35
Total or average			428,039	12.90 bill kWh	4.32 ¢/kWh	$557 million

(5) Fuel adjustment clauses must be reviewed regularly.

(6) Rate schedules must be published and provided to customers.

(7) Electric service cannot be ended by the utility if that endangers the health of customers unable to pay electric bills.

(8) Political and promotional advertising cannot be charged to customers, but must be paid for by shareholders.

(9) Consumer intervention in rate proceedings is encouraged. In some circumstances the state or the utility may be required to pay the cost of lawyers, economists, and engineers who represent the consumer intervention.

(10) Utilities are required to purchase electricity from small hydropower producers and from cogeneration facilities. Other parts of the act define the legal environment for interlocking directorates, wheeling, pooling, gas utilities, and several other matters.

There is little evidence in 1982 that the act has had much influence upon electric utilities. There is no significant increase (yet) in power generation from non-utility hydro or cogeneration sources. Perhaps most important, rate schedules have not been much affected. The upper curve in Figure 12-2 showed an average price curve for an actual utility, used for the hypothetical General Edison. The lower curve for 1978 is similar in shape: both the 1978 (pre-PURPA) and the 1981 schedules promote usage and (slightly) penalize conservation. The only significant difference is the shift: in 1981, rates are higher at every level of consumption.

One reason for the lack of impact is the short time: three years is not long in the utility industry. A second reason is that most of the first nine points were already standard in some states, so the act formalized existing policy.

Theory, Monopoly, and Ownership

Given the average cost pricing mechanism inherent in cost-of-service regulation, it is logical that, during the Growth Era, the growth monopoly concept was for electric utilities the most applicable of the three behavioral theories examined in Chapters 1 and 2. Nothing pejorative is involved: electric utilities had been assumed to be natural monopolies, they had pursued and achieved sustained growth, real prices declined regularly, and profit levels in the post–World War II period were not above normal.

Since the end of the Growth Era, the situation is less easily described. Much of the industry has continued to build new capacity although sales growth has declined. As a consequence, excess installed capacity is considerable, exceeding 40 percent of the hour of peak demand in most regions. The

West and Southwest have continued to experience growth in sales, and have lesser excess capacity.[11]

Criticism of this apparently unneeded growth should be tempered for two reasons. First, the demand studies of Mount, Chapman, and others (shown, for example, in Figure 4-3) were generally disbelieved, and most observers expected continued growth in the 1970s. Second, as Chapter 11 indicates, nuclear power was expected to be far less costly than coal or oil generation. Nuclear capacity was expected to replace imported oil, reduce air pollution, lower costs, and contribute to the accelerating growth in demand. Hence the growth in nuclear capacity.

Finally, some observations about public and private utilities. In Insull's era, private utilities advocated their own regulation for two reasons. One reason was to forestall the growing public power movement. The second was to eliminate the political bribery which was endemic to the utility industry prior to regulation. Utilities wanted their operations reviewed by professionals rather than opposed by politicians who expected personal compensation for the performance of public responsibilities.

In the last several years the public/private issue has been dormant. Private utilities concede that the public utilities are efficiently run. In fact, the public utilities now claim superiority to the private utilities because they sell electricity at considerably lower average prices.[12] There is less here than meets the eye. Much of the lower public cost results from greater access to low-cost public hydropower, less tax responsibility, and access to low-interest tax exempt bonds. Neither public nor private utilities can make a clear economic case for the superiority of their form of ownership. In the 1970s the plants with the poorest air pollution control were operated by the Tennessee Valley Authority, a problem that will be explored in the final chapter, on public policy. In nuclear power growth, the public utilities match the private. So far as conservation is concerned, the average public residential customer increased consumption more in the 1970s than did the average private residential customer. This is also true for commercial and industrial sales, which have grown considerably more for the publics than for the privates.

The public utilities are equally well described as growth monopolies. The last chapter in this book reviews the economic arguments for and against public and private ownership of energy resources and energy corporations. The experience in utilities suggests the following problems for advocates of

[11]From data in U.S. DoE, Energy Information Administration, *Electric Power Monthly,* summer months of 1981. Installed capacity is greater than available capacity; the latter definition excludes installed capacity which is unneeded or unavailable.

[12]This argument is made in Richard Morgan, Tom Riesenberg, and Michael Troutman, *Taking Charge: A New Look at Public Power* (Washington: Environmental Action Foundation, 1976), and in Goldfield.

each view: Private ownership advocates must consider the equal efficiency of public utilities (and, of course, BP/Sohio), and the justification for public tax subsidies for the support of private corporations. Advocates of public ownership must explain the tax-exempt status of public utilities and their poor record in conservation and in pollution.

Future growth in coal and nuclear power for both kinds of utilities will be much determined by the economic significance of the fourth economic theory: social cost. In a phrase, can we afford safe nuclear power and clean air? That question leads directly to the following chapter.

Appendix 12-A. Federal Corporate Income Tax Provisions Affecting Electric Utilities

Note: Much of this material appeared previously in Duane Chapman, Kathleen Cole, and Michael Slott, "Energy Production and Residential Heating: Taxation, Subsidies, and Comparative Costs," U.S. Environmental Protection Agency, Ohio River Basin Energy Study, March 1980; and in Duane Chapman, *Nuclear Economics: Taxation, Fuel Cost, and Decommissioning,* California Energy Commission, November 1980. Information on the 1981 and 1982 tax acts is from the *Economic Recovery Tax Act of 1981,* Public Law 97-34, August 13, 1981; Price Waterhouse & Co., "Economic Recovery Tax Act of 1981 Provisions Affecting Electric Utilities," (1981?); and the *New York Times,* September 12, 1982.

AFUDC Income

The allowance for funds used during construction (AFUDC) has two components. One is an equity component which is added to operating income in arriving at total income. The other, the debt component, reduces actual interest expense in arriving at net interest charges. Net income, while being the difference between total income and interest charges, always includes AFUDC as a positive amount.

The significance of AFUDC, of course, arises from its inclusion in accumulated rate base, which is the basis for future rates.

AFUDC when earned is wholly excluded from federal income taxation. However, the Internal Revenue Service (IRS) does treat income derived from AFUDC rate base as normal income. The rationale is that AFUDC is an accounting entry rather than an actual income item, so no tax liability should be imposed until actual cash income is received.

By way of illustration, a nuclear plant with construction cost of $3.4 billion might have an 8% AFUDC rate applied to actual plant expenditures and to nuclear fuel inventory acquisition. For a representative $3.4 billion plant having a 10-year construction period from 1980 to 1990, AFUDC would add $900 million to the plant rate base and $50 million to the fuel rate base. None of this is taxed as earned, and all is defined as part of net income.

Interest Deductions

Interest expense payments are generally viewed in the United States as ordinary business expenses and therefore deductible from taxable income. However, stock and equity—the other form of capital contribution—have payments made to them subject to tax liability. Consequently, utilities in some circumstances prefer debt to new stock issues in part because a dollar of new debt reduces overall tax liability while a dollar of new equity does not.

Value-added taxation of corporate revenue is widely used in Europe. In this form of taxation, taxable value equals revenue less cost of goods, wages, and salaries. Therefore interest, as well as dividends, is subject to this form of corporate income tax.

During the period of plant operations, bond payments to amortize debt may have 90% to 100% of the payments going to interest in the early years.

Investment Tax Credit

The investment tax credit (ITC) is a direct reduction in tax liability. At the maximum rate, it is now equal to 10% of qualified investment. Qualified investment is essentially construction cost excluding land and structures. AFUDC is not included. Qualified investment is thus approximately 95% of construction cost. The maximum effective rate, then, is 9.5% of actual construction cost.

This is a significant tax subsidy, its sum for the hypothetical new plant mentioned above being about $325 million.

Accelerated Depreciation

For net income determination as well as rate making, depreciation expense is defined on the normal straight line basis. Depreciation expense is simply assumed to be spread equally over each year of the plant's life, and is each year equal to $3\frac{1}{3}\%$ of original cost for a plant with a 30-year life.

Accelerated depreciation literally speeds up depreciation for tax purposes. By placing larger deductions in earlier years, it shelters significant income in those years from tax liability. The 1981 and 1982 tax acts utilize the double-declining-balance depreciation concept. This technique is described in general accounting texts. As an illustration, the 200% double-declining-balance method doubles the normal depreciation rate from 3.33% to 6.67%. This percentage is applied to the undepreciated basis at the beginning of each year, and the result is current depreciation for tax purposes.

[243]

The 1982 tax act requires accelerated depreciation to begin from a basis which reflects a reduction equal to one-half of the investment tax credit. The original depreciable basis is lowered to $3.2375 billion—i.e., $3.4 billion less the 50% of the $325 million credit.

Tax Lives

The arbitrary tax lives assigned to electrical power equipment provide an additional tax subsidy. The IRS permits depreciation for nuclear plant to be based upon a 10-year period rather than the 30-year expected life. Consequently, the present method, as actually applied to a 10-year tax life, apparently gives a 34% depreciation expense in the first three years in which a plant is operated.

Similar arbitrarily short federal tax lives apply to other utility property: 15 years for fossil fuel generating systems and transmission and distribution equipment.

For our hypothetical $3.4 billion nuclear plant with its post-ITC basis of $3.2375 billion, federal depreciation tax deductions are $1.1 billion in the first three years. Normal plant depreciation for rate base investment is $340 million. The plant is wholly depreciated for federal tax purposes by 1999, and no further depreciation expense deductions can be applied to taxable income for the federal corporate income tax.

This discussion is simplified for expository purposes. In reality, a plant beginning operations on December 31 has considerably accelerated tax depreciation in comparison to one that begins operations on January 1, and other provisions define maximum credits and the transfer of credits and deductions between years and between companies.

Nontaxable Dividends

As effective tax management brings the utility into a position with no significant tax liability, the utility may exempt its dividend payments from income tax liability for the recipients of the dividends.

Suppose a company normally has a positive and significant net income and net cash receipts: it is in a position to make dividend payments if it elects to do so. Suppose it has, for tax purposes only, no taxable profits. Then, all its dividends would be tax-exempt for dividend recipients: it is essentially a fictional capital repayment.

If dividend payments total $X million, and taxable profit is a smaller $Y million, then Y/X% of each dividend is taxable for recipients.

In determining nontaxable dividends, taxable income is recalculated as "earnings and profits." Essentially, depreciation is recomputed on a straight line basis with tax lives rather than expected operating lives.

For the dividend recipient, these tax-exempt dividends remain exempt until they sum to the original purchase price of the stock. At that point, additional tax-exempt dividends become liable to capital gains tax.

It can be noted that this provision increases the value of tax subsidies pertaining to new construction by creating deductions which can be passed along to shareholders. One New York utility reported 85% of its dividend payments was tax-exempt in 1977.

The 1981 tax act permits individuals to reinvest $750 each in dividends in utilities on a tax-free basis.

Conflict of Interest

Under pre-1981 federal tax law, the last 1½% of the then-existing 11½% in the investment tax credit could be used directly to finance employee stock ownership plans. The maximum rate (11½%) required employees to match the final ½% contribution.

Put in its simplest terms, this portion of the investment tax credit used public funds to increase the compensation of utility managers who chose to construct a new plant. This interpretation has not been seen as invalid by Treasury Department personnel with whom I have discussed this problem.

This investment-related aspect of employee stock tax credit plans expires in 1983.

13

Air Pollution, Nuclear Power, and Social Cost

Chapter 2 treated the theory of social optimality in the same language as other chapters have described electric rate schedules, insulation, or Saudi crude production cost. The uncriticized presumption is that nonmarket factors may be measured and added to market economics.

Suppose it were to be reported that coal burning produces air pollution causing $400 damage per ton, and nuclear power causes only seven-tenths of a death per reactor year? Would this enable us to determine the proper amount of conservation and production, coal and nuclear generation? Would Figure 2-4 actually be used to determine the most desirable level of generation and consumption?

Perhaps at some future date this will be so. There is growing interest on the part of utilities in undertaking this kind of benefit-cost or risk-benefit analysis. At our present level of knowledge, it is possible only to define the boundaries of major problems. And that is the objective of this chapter: defining boundaries.

A large fraction of the many existing environmental problems arise from energy production and use. These include problems resulting from automobile production and accidents, surface mining, acid rain, and potential climate change. This chapter, however, focuses on the comparative external social cost of coal and nuclear power. The motivation for this emphasis arises from the serious problems associated with each, and the significant amount of research undertaken on those problems. (One interesting aspect of this subject is the comparison of the records of public and private utilities on environmental protection noted in the preceding chapter.)

It should be recognized at the outset that, measured by results, regulation of air pollution and nuclear safety has achieved its goals. As of this writing, no one has yet died in an accident involving radioactive materials at an operating nuclear power plant. And sulfate and particulate air pollution were at lower levels in 1980 than in 1970, this notwithstanding the sizable increase in coal use over the period. The Nuclear Regulatory Commission and the Environmental Protection Agency merit considerable commendation for their leading roles in achieving these records, and the electric utility industry also merits a generally high rating for the overall degree of cooperation to date.

Exposure and Effect

Radiological and air pollutant contamination share several characteristics. Both have significant potential negative effect on human health, both apparently act without a threshold of wholly safe exposure, and each has complicated pathways from a power-generating source to its actual or potential effect on health.

Figure 13-1 represents three of the possible forms of a dose-response relationship. The horizontal axis represents the concentration of the hazard, perhaps in air (measured in micrograms per cubic meter) or in water (measured in micrograms per liter).

Micrograms are common to both air and water measurements, and one microgram is one-millionth of a gram. It might be helpful to realize that 10 micrograms per cubic meter (10 $\mu g/m^3$) is a concentration equal to one hundred millionth of an ounce per cubic foot.

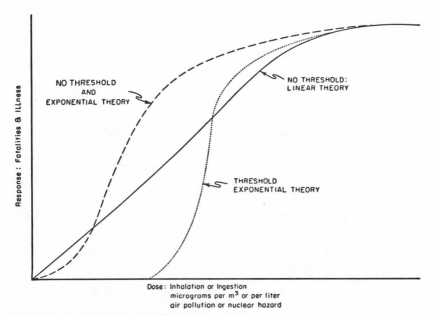

Figure 13-1. Dose-response theories.

The vertical axis in the figure represents the impact on an exposed population, perhaps in fatalities per million people. All three theories in Figure 13-1 turn concave at the end in recognition of the fact that, as toxic concentration approaches some fraction of the total air or water, all the exposed population dies or becomes ill.

The linear theory in Figure 13-1 says that any amount has some negative effect, and the health effect is always directly proportional to the concentration over low and moderate concentrations. For example, one finding suggests that, for one million persons, each μg of sulfate concentration might cause 33 deaths; two μg, 66 deaths; 8 μg, 264 deaths per million, etc.[1]

The threshold concept in Figure 13-1 implies that there is a safe level which has no negative effects. In radiation exposure, for example, it was thought in the 1960s that very low levels of anthropogenic radiation had no harmful effect. (*Anthropogenic,* incidentally, means created by human activities.)

An exponential theory hypothesizes that over some level of concentration the hazard increases more rapidly than the rate of increase in the population.

It is interesting to observe that many scientists reject threshold theories on empirical grounds as they relate to radiation hazard and air pollution.[2]

With this introduction, it is useful to define *dose-response relationship;* it is a quantitative expression of the effect of exposure to concentration levels of a toxic substance for a given time, and indicates the proportion of a population that experiences a specific health effect. Here, the concentration level is the "dose," and the health effect is the "response."

Another important concept is the *pathway:* the sequence of events over time and space in which a toxic substance is created, passes through the environment, has physical contact with humans, and causes a health effect.

Two pathways can be described as examples, one each for coal and nuclear power generation. In the first pathway, high-sulfur coal is burned in a power plant without a sulfur removal process. The sulfur leaves the plant in the form of sulfur dioxide (SO_2) and passes from a high stack into the atmosphere. In the atmosphere, it frequently interacts with sunlight, water, and air to form H_2SO_4, sulfuric acid. Carried by wind, the sulfuric acid passes over a town at the foot of a mountain range. The combination of particulate emissions in the town and the elevation effect of the mountains causes the particulate/sulfuric acid combination to fall into the lower atmosphere. It is inhaled, settles in the lungs, destroys lung tissue, and impacts upon the breathing capacity of the person inhaling it. In this illustration, 12 different parts of the pathway lie

---

[1]Richard Wilson, Steven D. Colome, John D. Spengler, and David Gordon Wilson, *Health Effects of Fossil Fuel Burning* (Cambridge, Mass.: Ballinger, 1980), pp. 197 and 219.

[2]See Edward P. Radford's discussion of the work of his National Academy of Sciences Committee on the Biological Effects of Ionizing Radiation in "Cancer Risks from Ionizing Radiation," *Technology Review,* 84:2 (November–December 1981), 66–78; also Lester B. Lave and Eugene P. Seskin, *Air Pollution and Human Health* (Baltimore: The Johns Hopkins University Press, 1977), p. 51.

between burning in the coal plant and the health effect. Most of these 12 segments have branching possibilities whereby the SO_2 may have been deposited on the ground near the plant, may have been carried elsewhere, or may even have had a positive environmental effect if it falls on an agricultural area deficient in sulfur. But the problem, of course, is that the 12-step pathway is sufficiently common to create serious health hazards.

For a *radionuclide,* a possible pathway begins in a reactor where the fission process creates more than 50 fission products of modest-to-large significance. Strontium-90 is one such spent fuel component. The spent fuel is shipped to a storage site such as West Valley, New York. It is liquefied and placed in a tank. If the tank leaks, strontium-90 (Sr-90) may pass into the water supply of a city. It is ingested/drunk and, since strontium-90 is a cousin to calcium, the Sr-90 particle lodges in a bone. The Sr-90 emits radiation into adjacent tissue which causes a cancerous mutation.

While the illustration of the air pollution pathway is real and affects many millions of people at a low but significant level, the Sr-90 pathway illustrates a dangerous possibility which has not yet actually occurred.

Readers interested in a better understanding of these environmental processes are encouraged to make use of the many excellent books and papers on the subject.[3]

Air Pollution

The success of electric utilities, other industry, and the EPA in controlling air pollution emissions is evident in Figure 13-2. This graph makes several points which merit attention. *Sulfur oxide* emissions peaked in 1973, and have declined in the succeeding seven years. This is equally true for three other major air pollutants: *particulates, volatile organic compounds,* and *carbon monoxide.* Only *nitrogen oxide* emissions were higher in 1979 than in 1970, this because of increased emissions from diesel trucks and electric utilities.[4]

Observe that SO_2 emissions from utilities and from coal-burning power generation are unchanged, although coal use increased two-thirds and oil use

[3]These include Lave and Seskin, Wilson et al., and Radford, cited above; National Academy of Sciences, *Energy in Transition 1985–2010,* Final Report of the Committee on Nuclear and Alternative Energy Systems, 1980; Paul Ehrlich, A. H. Ehrlich, and J. P. Holdren, *Ecoscience: Population, Resources, Environment* (San Francisco: W. H. Freeman, 1977); Ronnie D. Lipschutz, *Radioactive Waste: Politics, Technology, and Risk,* a report of the Union of Concerned Scientists (Cambridge, Mass.: Ballinger, 1980); Charles Hall, Cutler Cleveland, and Robert Kaufman, *Energy, Resource Quality, and Economic Systems* (New York: Wiley, in press); Bernard L. Cohen, "High-Level Radioactive Waste from Light-Water Reactors," *Reviews of Modern Physics,* 49:1 (January 1977), 1–20.

[4]This information has been published regularly in U.S. Environmental Protection Agency, Office of Air Quality, "National Air Pollutant Emission Estimates," Research Triangle Park, N.C. Each year's report summarizes current data and historical progress.

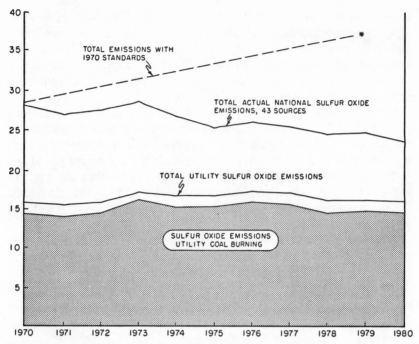

Figure 13-2. Sulfur oxide emissions, "Teragrams"

by one-half. This is because utilities have increasingly turned to low-sulfur coal and oil to reduce sulfur emissions. Sulfur removal processes have not yet been significant: low-sulfur coal and oil have made the difference. Utilities have nonetheless become the most significant source of sulfur emissions, now producing 67 percent of the national total. The total for all 42 other sources is obviously much less than emissions from utilities alone. In fact, the largest 200 power plant producers of sulfur pollutants are themselves responsible for more than half of all the sulfur emissions.[5]

The asterisk in the upper right of the figure indicates the level that sulfur emissions would have reached if 1970 standards had been continued: 150 percent higher than the actual 1979 emissions.

Declining emissions have generally been followed by reduced exposure levels for persons. In 1977, representative counties had an average of only 9 days per year in which sulfur oxide concentrations were above the health standard.[6] These figures show considerable improvement over the early 1970s.

[5]Data from E. H. Pechan & Associates, *Estimates of Sulfur Oxide Emissions from the Electric Utility Industry*, September 1981, prepared for the U.S. Environmental Protection Agency, p. 10, and U.S. EPA, "National Air Pollutant Emission Estimates."

[6]U.S. Council on Environmental Quality, *Environmental Quality*, Annual Report of the Council, 1979, p. 52.

The health standard referred to has been a concentration of 365 $\mu g/m^3$ over a 24-hour period. In Appendix 13-A, the best judgments of qualitative dose-response relationships are shown. Observe there that the 365 $\mu g/m^3$ exposure is the boundary judged to separate no effect from mild aggravation in susceptible persons. If the SO_2 concentration rises five and a half times (to 2,000 $\mu g/m^3$), then premature death of the ill and elderly is expected.[7]

As with SO_2, concentration levels for other air pollutants have also been declining.

The importance of avoiding the very high levels of pollution shown in Appendix 13-A is well known from studies of the major pollution episodes that occurred throughout the world in the 1950s, 1960s, and 1970s. A carefully studied incident in London of December 5–9, 1952, shows that both sulfur dioxide and soot reached average daily concentrations of 4,000 $\mu g/m^3$. The result of the four-day incident was the death of 3,900 more persons than expected from the normal 2,100 deaths per week. The air pollution victims died of heart and lung diseases. This description, incidentally, is from Wilson et al.[8]

Economic Benefits of Pollution Control

Given the interest in social optimality, and its political implications, many analysts working under contract for the utility industry and the EPA have attempted to quantify the economic benefits from air pollution control. The earliest EPA studies began in the late 1960s and considered damage to health, property loss, and vegetation. One conclusion reached then is still widely held: sulfur oxides and particulates are believed to be the two most significant causes of economic loss.[9]

One recent analysis by A. M. Freeman III, sponsored by the Council on Environmental Quality, concluded that the benefits from air pollution control that brought about a 20 percent improvement in air quality were most likely to be $21.4 billion per year. A review of this important study illustrates the difficulty of this kind of work as well as the importance of the subject.

In Table 13-1, the most important entry is the estimated benefit from improved health. This is so significant that it requires particular emphasis. It will be taken up again after reviewing Freeman's other estimates.

Home maintenance means that air pollution soot and corrosion make painting, window cleaning, car washing, and so on more necessary, and the

[7]U.S. Council on Environmental Quality, *Environmental Quality*, 1980, pp. 156–157.
[8]See Wilson, et al., pp. 164–166 and I-6.
[9]Thomas E. Waddell and Larry Barrett are the best known pioneers in this early work: see Waddell, *The Economic Damages of Air Pollution*, U.S. Environmental Protection Agency, May 1974.

Table 13.1 Freeman's CEQ analysis of air pollution control benefits, 1978, $ billion

Type of damage	Value of reduced air pollution		
	Utilities and other stationary sources	Automobiles, trucks, and other mobile sources	Total
Health: death and illnesses	$16.8	$0.2	$17.0
Home maintenance (total)	($2.0)		2.0
Vegetation	0	0.7	0.7
Materials	0.7	0.2	0.9
Property values	2.3	0.4	2.7
Totals	$19.8	$1.5	$23.3
Freeman's total, excluding double counting			$21.4

Note: Basis is a postulated reduction of 20% in average concentrations of sulfur dioxide and particulates nationwide, from 1970 to 1978. From A. Myrick Freeman III, *The Benefits of Air and Water Pollution Control,* prepared for the Council on Environmental Quality, December 1979, pp. 117–120.

benefit from improved air quality—according to several studies reviewed by Freeman—is $2 billion annually.

Vegetation in Table 13-1 includes agricultural crops throughout the country, forest products, and suburban hedges. The Freeman review attributes current benefits wholly to improved pollution control on mobile sources such as autos and planes. He estimates the value of improved air quality for vegetation to be $700 million annually.

Air pollution also causes corrosion and deterioration in a wide variety of materials including clothing, electrical equipment, bridges, and so on. The effect of improved air quality here was estimated to be $900 million, most being attributable to reduced stationary emissions.

The relationship of property value to air pollution has been studied by statistical analysis which defines a quantitative relationship between housing price, size and quality of house, location, and air pollution. The statistical relationship between housing cost and pollution is used to calculate the economic loss from pollution, or the benefit from pollution reduction. Freeman's judgment, from empirical studies, leads to the property value figures in Table 13-1.

Before we turn to the health question, three points should be made. First, Freeman believes that some air pollution effects have been counted twice. Paint corrosion and unhealthy air may both affect property value. So he estimates the benefit to be $21.4 billion rather than the $23.3 billion sum.

Second, Freeman calculated a range of values. He wanted to define a range which he thought would have a 9 in 10 probability of including the true value. The range is from $4.6 billion to $51.2 billion. The $21.4 billion total in Table 13-1 is his best estimate. (The upper limit of his range is more than 11 times the lower limit. Caution is advised in reliance upon particular numerical values.)

The third point is that Freeman studied the benefit from a 20 percent improvement in air quality, not the damage from its actual 1978 level. Examine Figure 13-1. If the linear theory is appropriate, then Freeman's work implies a 1978 level of damage of $85.6 billion. In the absence of the $21.4 billion benefit from pollution reduction, damage would have been $107 billion.

Health Effects and Criticism

The method of statistical analysis utilized to study health/pollution relationships is identical to the method used to study other aspects of pollution benefits. It is regression analysis, a method used to define the best possible quantitative descriptions of a relationship.

Figure 13-3 is an illustration.[10] The observations on sulfur concentration (the "dose") and mortality (the "response") are, literally, all over. The line shows the best fit: a tendency for mortality to rise in every city as sulfur concentration increases.

Lave and Seskin examined hundreds of possible statistical relationships. One actual relationship which they found of particular interest is given in Equation (1).[11]

(1) $\dfrac{\text{mortality rate per}}{\text{100,000 population}}$ = 343 (a constant)
$\qquad$ +0.67*(sulfur pollution)
$\qquad$ +0.48*(particulate pollution)
$\qquad$ +0.08*(population density)
$\qquad$ +6.88*(percentage persons over 65)
$\qquad$ +0.04*(percentage minorities)
$\qquad$ +0.04*(percentage in poverty)
$\qquad$ −0.28*(the tenth root of total population)

Obviously, an individual has a higher probability of not being dead if she or he avoids air pollution, concentrations of people over 65, dense populations, and is white and not living in poverty.

This relationship, incidentally, explained 83 percent of the city-by-city variation in mortality rates.[12] It is clear from the statistical record that there is

[10]The hypothetical illustration in Figure 13-3 might be compared to an actual regression in Lave and Seskin, p. 130

[11]Ibid., p. 31.

[12]This is a linear relationship. Lave and Seskin use minimum, mean, and maximum air pollution variables. Incidentally, the poverty and density variables were not significant in relationship to all of the other variables.

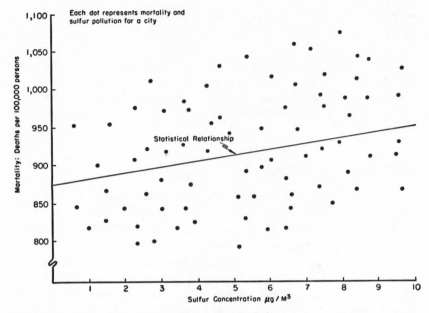

Figure 13-3. Representation of mortality and sulfate pollution

a cause-effect relationship between air pollution and mortality. Yet, as Figure 13-3 shows, it is difficult to be confident or precise about the exact numerical values.

Statistical studies of illness rather than mortality are similar in findings. The diseases affected by air pollution are known to be asthma, bronchitis, emphysema, cancer, heart disease, lung function, and other respiratory diseases.

It is valuable to pause for a summing-up of important points before proceeding to criticism. First, air pollution is known to have a real effect on illness and mortality. Second, in extreme cases such as the episodes which occurred 20–30 years ago, deaths increase significantly. Third, air pollution has been reduced considerably in the 1970s. Fourth, sulfur oxides and particulates are currently believed to be the most damaging pollutants. Fifth, EPA and CEQ research projects find the economic damage of the health effects of air pollution from sulfur oxides/particulates to be the single worst problem.

However, it is evident that several major obstacles prevent us from correctly gauging the relationships involved. In particular, there is uncertainty about the value of life in dollars; about the relationship of concentration levels to death and illness; and about the impact of emissions on concentration levels. These are severe obstacles to a confident understanding: remember Freeman's high benefit estimate was 11 times the low estimate; remember the

Table 13-2. Comparison of modified pathway analyses of economic benefits of improved air quality

	Freeman (Council on Environmental Quality)	Perl & Dunbar (National Economic Research Associates)	
	Sulfates & Particulates	Particulates	Sulfates
1. Improvement in air quality, 1978	20%	18%	27%
2. Percent change in mortality from 1% pollution reduction	5%	2.5%	0.5%
3. Normal deaths per million persons	8,710	8,710	8,710
4. Exposed population, millions Metropolitan	159		
Total U.S.		218	218
5. Value of life and health Value of life, each	$1 million	$548,000	$401,000
Value of less illness, %	21.2%		
Total value per life	$1,212,000	$548,000	$401,000
6. Total health benefits	$16.8 billion	$3.0 billion + $1.4 billion	
		Total = $4.4 billion	

Note: Multiply the entries for each column to arrive at the totals. E.g., for Freeman, $16.8 billion = 20% * 5% * 8,710 * 159 * $1,212,000.
Source: Perl and Dunbar, p. 6-38.

almost-but-not-quite randomness in Figure 13-3. And it is upon these problems that the utility industry bases its criticism of the CEQ/EPA work.

National Economic Research Associates Inc. (NERA) is an important consulting research corporation that works closely with many electric utilities. In 1980, Lewis Perl and Frederick Dunbar prepared a detailed review of Lave and Seskin and the EPA and CEQ work for The Business Roundtable, a corporate group.[13] The NERA conclusion: health benefits from air pollution control are much less than previously supposed. NERA finds $4.4 billion to be the economic benefit from better air quality compared to Freeman's $16.8 billion.

The studies share three assumptions. The most significant area of agreement is the assumption that pathways are sufficiently well known so that the multiplicative stages in Table 13-2 can be assumed to give valid results. Second, they assume the same mortality rate in 1978: 8,710 deaths per million persons. Third, they agree that sulfates and particulates are the most important air pollution health problems, at least in the early 1980s.

NERA and Freeman are in rough agreement on the magnitude of improved air quality: row 1 in Table 13-2.

[13]See Lewis J. Perl and Frederick C. Dunbar, *Cost-Effectiveness and Cost-Benefit Analysis of Air Quality Regulations*, prepared by National Economic Research Associates for The Business Roundtable, November 1980.

[255]

However, NERA's review concludes that mortality is less affected by pollution than did Freeman's CEQ study. The NERA finding is three-fifths as large. If sulfate and particulate concentrations are reduced by an average 20 percent, Freeman expects a 1 percent gain in reduced mortality (20% * 5%). NERA would expect 0.6 percent gain from the same reduction (20% * (2.5% + 0.5%)).

The studies also disagree on the size of the exposed population. Freeman considers urban areas, NERA all of the country.

Finally, they hold different positions on the value of saved life. Freeman chooses one million dollars, in part from statistical studies with higher and lower values, and in part, perhaps, because it is a large round number. NERA emphasizes a study by Robert Thaler and Sherwin Rosen which compared wages and deaths in various high-risk manual occupations. The average death counts as a $548,000 loss, and it is assumed that an average person would be willing to pay $548 to avoid a one-in-one-thousand risk of death.

Perl and Dunbar go on to note that elderly persons have fewer remaining years and less remaining income, hence a saved life for an aged person is worth less than a saved life of a younger person. In their words,

It seems reasonable to suppose that beyond some point these values are a decreasing function of age. Willingness to pay for decreases in mortality risk would be expected to decline as life expectancy falls. Since the mortality effects of air pollution are highly concentrated in the older segment of the population, the values assigned to these risks would in all likelihood be lower than those derived for the average person in the labor force.[14]

Perl and Dunbar believe that persons exposed to sulfate pollution are on the average older than those exposed to particulate pollution. Hence, in row 5 of the table, a saved life arising from sulfate control has less economic benefit than a (younger) saved life from particulate control.

It should be emphasized that acceptance of either the Freeman or NERA estimates requires acceptance of the basic methodology common to both. Economic benefit arises from protection of earnings of the exposed population. Freeman assumed implicitly what NERA describes. The logic, inexorably, is to value—for the purpose of estimating economic health benefits—at a lesser rate the elderly, minorities, poor, and women, and to value the highest income classes at a higher rate. It has been pointed out that this approach gives no weight to yesterday's contributions by today's elderly, thereby raising another problem with the method.

Another perspective on the value-of-life issue is given by reference to the economic value added to the economy by each employed person. This is, formally, a ratio of Gross National Product to civilian employed workers. It is

[14]Ibid., p. 6–26.

[256]

$29,000 per worker in 1981. The value over 45 years is $1.3 million per person if no discounting is used. But discounting strongly affects this total life-worth figure. If society assumes each person and each dollar is, in real terms, equally important regardless of the calendar, the $1.3 million value is correct. But if economic value added per worker is weighed by a 3 percent real discount rate, the total worth is $700,000. If a nominal 10 percent discount rate is applied, the value of life falls to $300,000.[15]

It should be recognized, then, that application of value-of-life estimates to benefit analysis has two severe logical problems. First, inexorably, the method leads to a lower value for elderly, women, minorities, and poor. Second, the method is very sensitive to variations in assumed discount rates: with the same GNP value added per worker factor, apparently logical but different discount rates lead to value-of-life estimates that vary from $0.3 to $1.3 million.

Perl and Dunbar raise an additional question about the air pollution/health studies. They argue that statistical relationships of the type in Equation (1) may seriously overestimate the health effects because they do not explicitly consider the impact of occupational hazards, smoking, or other personal and social habits.

An opposite implication arises from the absence of clear recognition of the time delay between exposure and illness in these health studies. Studies of smoking and radiation effects indicate that time lag is important, and it is logical to expect it to be significant in the air pollution/health relationship. If time delay is important, then the Lave and Seskin studies understate that relationship.

Table 13-3 reports the overall estimates of costs and benefits by Freeman and NERA. Both conclude—on the basis of the kind of material described here—that automobile/truck emission standards have not produced economic benefits equal to the cost. Both view this pollution problem as less serious than the utility and stationary-source problem. On sulfates and particulates, they agree the health problem is significant. But NERA finds the benefit/cost ratio to be only 0.9, while the CEQ concluded that a 2.3 ratio is more likely.

One final criticism of these studies: Both Freeman and NERA compare 1978 air pollution levels to 1970 levels to determine benefits, and use 1978 alone to evaluate control costs. Costs, then, are derived from considerably higher levels of coal, oil, and gasoline use than existed in 1970. It seems logical that 1978 benefits from pollution control should have been based upon protecting air quality and health from the effects of actual levels of energy use

---

[15]Discounting is the inverse of Chapter 11's levelized cost. Recall LC = CRF * IAMT; levelized cost equals the capital recovery factor times the initial amount. Discounted present value is DPV = LC/CRF. At 3% discount rate for 45 years, CRF = .041. The discounted present value of $29,000 per worker annually for 45 years is DPV = $29,000/.041 = $707,317, rounded to $700,000.

Table 13-3. Comparative summary cost and benefit estimates for air quality improvement, 1970–78, $ billion

	Freeman (CEQ)	Perl & Dunbar (NERA)
Utilities and stationary sources		
Benefits from improved air quality	$19.8	$7.7
Costs of pollution control	$ 8.6	$8.6
Benefit/cost ratio	2.3	0.9
Automobiles and mobile sources		
Benefits from improved air quality	$ 1.5	$0.8
Costs of pollution control	$ 8.0	$8.0
Benefit/cost ratio	0.2	0.1

Sources: Table 13-1 above, from Freeman, and Perl and Dunbar, pp. 6-35, 6-41.

in 1978. Recall Figure 13-2: Sulfur oxide emissions in 1979 would have been 50 percent higher if 1979 coal and oil use had taken place with 1970 standards. On particulates, the amount expected from actual 1979 coal and oil use would have been two and a half times greater in the absence of new pollution control standards and practices.[16] In combination, 1979 levels of coal and oil consumption would have created 62 million metric tons (68 million American tons) of sulfur oxide and particulate emissions if those fuels were burned with 1970 standards. The actual emissions were far less: 34 million metric tons (37 million American tons).

Basing estimates on avoided pollution would mean that both CEQ and NERA would almost double their benefit estimates for sulfur oxide and particulate control.

This problem can be given a sharper focus with a hypothetical illustration. First, use the actual 1970 emissions of 49 million metric tons of SO_x and particulates as a reference. Now assume that continued growth in coal and oil consumption leads to 1985 emissions of 51 million metric tons, even with the regulations now being followed. Assume that, in the absence of present standards, emissions in 1985 would be 100 million metric tons.

According to the approach used by CEQ and NERA, air pollution control would have negative benefits because the 1985 total would have risen above the 1970 level. But, if benefits were based upon avoided pollution, then we would conclude that emissions would be half the amount which would have been expected in 1985 without standards.

Stepping back from the economic perspective, we can ask whether the general public—informed about the techniques employed in benefit/cost analysis—may simply reject this approach to the determination of standards. The alternative approach, of course, is to use the political process to determine relevant standards.

[16]U.S. EPA, "National Air Pollutant Emission Estimates, 1970–1979," p. 28.

Nuclear Power

The range of uncertainty in determining the health effects and economic benefits of air pollution control is very great, as the preceding sections indicate. But however large this uncertainty, it is known that air pollution is causing premature deaths and illness. Freeman concluded that a 20 percent reduction in sulfates and particulates would save 14,000 lives each year; perhaps, then, air pollution in total causes 70,000 deaths each year. The Wilson group believed that a figure of 50,000 deaths per year arising from air pollution is a good approximation of the actual but unknown number.[17] In comparison to the total U.S. deaths each year, these figures are not large. In the 1980s, two million Americans die each year. But on the other hand, the known health effects to date from nuclear power use lead to an expectation of only 22 deaths per year. The calculation is simple: four-tenths of a cancer death per 1,000 MW per year, multiplied by 56,000 MW of operable nuclear capacity in 1981.

This kind of calculation is a dose-response analysis identical in logic to the air pollution/health analyses. The present view is that each large 1,000 MW nuclear plant contributes 2,000 *person-rems* of radiation exposure. It is also estimated that one cancer death arises from each 5,000 person-rems of exposure.[18] Hence the calculation above: 0.4 cancer deaths per 1,000 MW nuclear plant per year. This figure is believed to include all parts of the nuclear fuel cycle, and includes both occupational and general population exposure.

The average person in the United States has to date received 8,500 times as much radiation from natural background sources as from nuclear power plants. Obviously, radioactive strontium embedded in a bone can be a greater danger than the same quantity of radiation in the form of cosmic solar radiation. Appendix 13-B, comparable to Appendix 13-A, summarizes the anticipated health effects from various levels of exposure to radiation. In the early 1980s, however, nuclear power has given little exposure to the public, and presumably has had insignificant health effects. (On the average, of course. Some workers and miners probably have received fatal long-term doses.)

This is the dilemma in comparing the social costs of coal and nuclear power. To date, nuclear energy has been used with far less negative impact on public health. But there is a very small possibility of very large damage.

The major hazards now appear to be in three parts of the nuclear fuel cycle: (a) safe reactor operation, (b) spent fuel waste disposal, or reprocessing of spent fuel, and (c) reactor decommissioning.

---

[17]A. Myrick Freeman III, *The Benefits of Air and Water Pollution Control*, prepared for the Council on Environmental Quality, December 1979, p. 66. Wilson et al., pp. 219 and 5.

[18]National Academy of Sciences, *Energy in Transition*, p. 439.

To date, the actual operations of nuclear plants have been without a major impact on public health. The most serious accident has been at the Three Mile Island Unit 2 plant, and the radiation effect during the episode is expected to be less than one cancer death and less than one genetic defect.[19]

Considerable scientific effort has been applied to the problem of determining probabilities of accidents that have not yet been experienced.

The pioneering work of the Reactor Safety Study in 1975 continues to influence our understanding of reactor safety. The study was directed by Professor Norman Rasmussen of the Massachusetts Institute of Technology, and sponsored by the Nuclear Regulatory Commission (NRC). The report concluded that the worst plausible accident would be a core meltdown with major radioactive release, and adverse weather and population exposure. The probability was reported to be .000,000,001 per reactor year, and the consequences were said to be 3,000 immediate deaths, 45,000 delayed cancer deaths, and 26,000 genetic defects.[20] In summary: a very low probability of a serious accident.

However, a Ford Foundation/Mitre Corporation review of the Rasmussen study illustrates the difficulty in estimating probability for events which have not taken place. (And in this instance, of course, are not desired.) The Ford/Mitre 1977 study pointed out that the Rasmussen group collected data on average wind patterns and average populations, but separately. But for the Indian Point nuclear plants in New York, and for the Zion plants in Illinois, common wind patterns can lead directly to large populations in the New York and Chicago/Milwaukee areas. Correcting for this actual versus average weather/population relationship, Ford/Mitre finds the upper limit (their phrase) on an extremely serious accident to be fifty thousand times higher. Now multiply .00005 (extremely serious accidents per reactor year) by 5,000 (reactor years by the year 2000). The result is a 25 percent upper limit probability of a very serious accident by the year 2000.[21]

Risk analysis studies in the 1980s continue to conclude that risk is small, but the estimates may be increasing. In 1982, two NRC-sponsored studies reported higher risk and damage. The Accident Precursor Study was performed at the Oak Ridge National Laboratory, and analyzed 169 actual events which, in a theoretical sense, could have led to severe reactor core damage. The study's major finding is that such a severe accident can be anticipated with a frequency of between 1.7 and 4.5 times for each 1,000 reactor years of operation.[22]

[19]Ibid., p. 461.
[20]Ibid., pp. 463–464.
[21]See Spurgeon Keeny, Jr., et al., Ford Foundation/Mitre Corporation, Nuclear Energy Policy Study Group, *Nuclear Power Issues and Choices*, 1977, p. 231.
[22]J. W. Minarick and C. A. Kukielka, *Precursors to Potential Severe Core Damage Accidents: 1969–1979; A Status Report*, prepared for the U.S. Nuclear Regulatory Commission, June 1982, pp. 4–21, 4–22.

This Oak Ridge study analyzed data in the 1969–79 period before the Three Mile Island accident. The NRC's director of Risk Analysis suggested that improvements made in safety procedures and technology since 1979 may outdate the Oak Ridge study in the sense that the same methodology applied to current circumstances would give a lower frequency of severe core accidents, perhaps one in 10,000 reactor years.[23]

The second NRC study analyzed damage rather than probability. Performed by the Sandia National Laboratory, this study is similar to the Ford/Mitre analysis in that it examines consequences of the worst possible interactions of core meltdown, safety system failure, weather, and population exposure. On this basis, the Salem, New Jersey, plant, if it experienced this worst-case combination, would cause the deaths of 100,000 persons within a year of the accident's occurrence. In terms of economic loss, the greatest potential estimated damage exists for the Indian Point 3 reactor which, with a worst-case accident, would cause $300 billion in damage excluding health care cost or loss of productive economic activity.[24]

A final qualification: since 1977, the expected growth in nuclear reactors has declined (see Figure 11-1). It is unlikely that 5,000 American reactor years of experience will develop by 2000. Therefore, if other things are assumed constant, the probabilities of worst accidents were overestimated. As illustration, assume that the average number of reactors will be the average of present and planned reactors. In this case, there would be about 1600 reactor years experienced between 1985 and 2000.[25]

Ford/Mitre concludes that, overall, nuclear power risks compare favorably to other technologies. In the end, subjective opinion may be inevitable, and necessary.

The strong safety emphasis of the U.S. Nuclear Regulatory Commission and its predecessor, the U.S. Atomic Energy Commission, has led to the closure of eight utility-operated nuclear power plants as of early 1982. Table 13-4 lists these nonoperable facilities and, in brief, the most important reasons for the closings.

Spent Fuel and Decommissioning

In addition to the nonoperating power plants listed in Table 13-4, each of the three spent fuel processing plants is inoperable. Consequently, spent fuel

[23]*New York Times,* July 6, 1982.

[24]*Washington Post,* November 1, 1982.

[25]Reactor years are here expressed in 1,000 MW equivalents. With planned capacity of 148,000 MW and existing capacity of 60,000 MW, the midpoint value is 104,000 MW. This, at 15 years, gives (approximately) 1600 reactor years. Capacity data as of June 1982 are from *Monthly Energy Review,* September 1982, pp. 76, 78.

Table 13-4. Decommissioned and inoperable nuclear generating facilities, 1983[a]

Name and Location	Reactor type[b]	Power rating (MW)	Date of shutdown	Nature of shutdown	Expected action or action taken	Year of construction	Principal owner
Pathfinder Sioux Falls, S.D.	B	190[c]		Experimental use completed.	Mothballed 1972	?	Northern States Power Co.
Indian Point 1 Westchester Co., N.Y.	P	275	1974	Plant had no emergency core cooling system. Shutdown in compliance with new AEC regulations.	Simultaneous decommissioning planned—Indian Point 1 & 2. Plant will remain shut down until decommissioning of Indian Point 2 in 2003.	1962	Consolidated Edison Co.
Fermi 1 Monroe Co., Mich.	L	200[c]	11/27/72	Reactor accident	Mothballed 1975	1963	Power Reactor Development Co.
Humboldt Bay 3 Humboldt Co., Cal.	B	65	7/12/76	Plant located in earthquake fault. Closed to study possible modifications of plant.	Undetermined. Current study being completed of alternatives.	1963	Pacific Gas & Electric Co.
Peach Bottom 1 York Co., Pa.	H	40	?	?	Mothballed 1975	?	Philadelphia Electric Co.
Dresden 1 Grundy Co., Ill.	B	209	5/28/80	Radioactive corrosion products inside piping. Company must at some point clean and restart unit or decommission it.	Undetermined. Decision has been made to postpone decision to clean and restart unit.	1960	Commonwealth Edison Co.
Three Mile Island 1 Harrisburg, Pa.	P	871	3/28/79	Temporarily closed until completion of NRC hearings on reactor safety.	Scheduled for eventual reopening.	1974	General Public Utilities: Metropolitan Edison
Three Mile Island 2 Harrisburg, Pa.	P	961	3/28/79	Major reactor accident.	Clean-up. Tentatively scheduled reopening.	1978	General Public Utilities: Metropolitan Edison

[a]Commercial reactors with power ratings under 50 electrical megawatts have been omitted.
[b]Reactor type codes: B = Boiling water reactor; H = Gas cooled, graphite moderated reactor; L = Light water breeder reactor; P = Pressurized water reactor.
[c]Power rating is expressed in thermal megawatts, rather than in electrical megawatts. One electrical megawatt = approximately 3 thermal megawatts.

Source: Sally Hindman, "Estimates and Accounting Practices in Nuclear Power Plant Decommissioning," unpublished, December 1981, based upon Smith et al., U.S. DoE, Energy Information Administration, Inventory of Power Plants in the United States: 1980 Annual, June 1981, and personal communications with Pacific Gas and Electric Company, Consolidated Edison Company, and General Public Utilities Corporation.

is accumulating at each reactor site pending some solution to this problem. The scientific problem with spent fuel is succinctly summarized by a Presidential Review Group on Nuclear Waste Management:

The central scientific fact about radioactive material is that there is no method of altering the period of time in which a particular species [of radioactive waste] remains radioactive. . . . The pertinent decay times vary from hundreds of years for the bulk of the fission products to millions of years for certain of the actinide elements and long-lived fission products. Thus, if present and future generations are to be protected from potential biological damage, a way must be found to isolate waste from the biosphere for long periods of time, to remove it entirely from the earth, or to transform it into nonradioactive elements.[26]

This is not a metaphorical statement: it is literal. "Remove entirely from the earth" means space dumping by missiles. "Transform it into nonradioactive elements" means, literally, transmutation—a process by which radioactive waste elements might, through other fission or atomic processes, be transformed into some other element. Since the transmutation process might itself increase radioactive waste, it is not being considered as a solution.

The report is again literal in noting that some elements require isolation for millions of years. Plutonium-239, for example, has a *half-life* of 24,000 years. This means that, after 24,000 years, it has dissipated one-half of its radioactivity.

Suppose we have 100 kg of Pu-239. In Table 11-2, remember, the representative reactor produced 250 kg of all plutonium isotopes each year. Suppose also that inhalation and lung deposition of Pu-239 gives a 50 percent probability of lung cancer with one ten-thousandth of a gram.[27] It will take the 100 kg of Pu-239 on the order of 700,000 years to decay to the original radioactivity level of one ten-thousandth of a gram. This is illustration only, and modestly understates the relationship between time and the hazard of spent fuel.

In the Department of Energy's environmental impact analysis of spent fuel disposal, calculations were made on the amount of water needed to dilute all of the 58 major hazardous isotopes.[28] For the 27 metric tons of radioactive waste produced in the illustrative nuclear plant in Chapter 11, three quadrillion gallons of water are required for safe dilution after one year of storage. After one million years, the water needed for safe dilution is 168 billion gallons.

Hence the interest in safe storage. The California Energy Commission reviewed proposals for waste storage and arrived at this conclusion: no waste

[26]U.S. Interagency Review Group, *Report to the President on Nuclear Waste Management,* Draft, October 1978, pp. iii, iv.

[27]The dosage was estimated in Ehrlich et al., p. 454.

[28]U.S. Department of Energy, Office of Nuclear Waste Management, *Management of Commercially Generated Radioactive Waste,* Final Environmental Impact Statement, October 1980, p. 3.38 and Appendix A.

disposal concept has been confirmed as feasible.[29] In contrast, the U.S. Office of Nuclear Waste Management's detailed review concludes that the societal risk of underground spent fuel storage is small, and that the cost will be one-tenth of a cent per kilowatt hour.[30] This estimate is similar to the figures in Tables 11-2 and 11-3, above.

The specific geological and physical problems can be studied elsewhere.[31] It should be noted that several states have prohibited additional nuclear plants because of spent fuel and safety concerns. These states and their policies are summarized in Table 13-5. It is interesting to note that concern about waste fuel disposal motivated at least five of the moratoria on new plants. The following section of this chapter will examine the cost implication of different assumptions about spent fuel cost.

Decommissioning has been seen as important in two of the states with moratoria. If a reactor functions successfully without incident over 30 years of normal operations, it is expected that the reactor can be decommissioned without difficulty. This means the nuclear plant is shut down, and the public is assured protection from radioactive hazards. Three approaches to decommissioning are being considered. *Dismantlement* means that all hazardous radioactive material is removed from the site, and the property is usable for any other purpose. *Storage and dismantlement* means that the shutdown power plant is physically protected for some period up to 100 years. During this storage period, much of the radioactivity decays through normal fission processes. At the end of the storage period, the plant is made safe for unrestricted use as in dismantlement. This is also termed "mothballing," as in Table 13-4.

Entombment would have the reactor encased in concrete until all harmful radioactivity had decayed. As with spent fuel, the times involved are important. The basic NRC study says: "the induced niobium-94 and nickel-59 activities in the reactor vessel and its internal structures are well above unconditional release levels, and since nickel-59 has an 80,000 year half-life and niobium-94 has a 20,000 year half-life, the radioactivity will not decay to unconditional release levels within the foreseeable lifetime of any man-made surface structure."[32]

An addendum to the basic study pointed out that the reactor vessel parts containing these two long-lived elements could be removed, and the remaining plant entombed. Referring to the original full-entombment concept, the

[29]From Emilio E. Varanini III and Richard Maullin, California Energy Resources Conservation and Development Commission, Nuclear Fuel Cycle Committee, "Status of Nuclear Fuel Reprocessing, Spent Fuel Storage, and High Level Waste Disposal," January 1978, p. 34.
[30]U.S. DoE, *Management of Commercially Generated Radioactive Waste*, p. 1.32.
[31]Ibid., pp. 7.50, 7.51.
[32]R. I. Smith, G. J. Konzek, and W. E. Kennedy, Jr., *Technology, Safety, and Costs of Decommissioning a Reference Pressurized Water Reactor Power Station*, 2 vols., prepared by Battelle Pacific Northwest Laboratory, 1978, 1: p. 4–6.

Table 13-5. State prohibitions against new nuclear plant construction

State	Reasons for prohibition
California	The state's Energy Commission has concluded that there is at present no feasible means of waste disposal, and legislation therefore bars additional plants.
Connecticut	Legislation prohibits additional plants unless the state's Environmental Protection Department finds that the federal government has a satisfactory waste disposal method.
Montana	State-wide referendum in 1978 requires voter approval of new plants, full company liability for accidents, and a bond to insure decommissioning costs.
New York	The governor and the state Energy Office have since 1978 indicated their plans to disapprove any applications for new nuclear plants.
Maine	State legislation prohibits new plants unless the state Public Utilities Commission finds that the federal government has established a demonstrable waste disposal method.
Massachusetts	The House and Senate in the state passed 1979 resolutions opposing any additional plants.
Oregon	A 1980 referendum prohibits new nuclear plants unless the federal government licenses an adequate waste repository.
Wisconsin	The Public Service Commission in 1978 prohibited additional plants because of concern about spent fuel and reactor decommissioning.

Sources: Data prepared by Dooley Kiefer from information provided by the U.S. Nuclear Regulatory Commission, Environmental Action Foundation, and Nuclear Information and Resource Service. Generally these prohibitions permit the completion of new nuclear plants that were begun before the prohibition became effective.

addendum said: "However, in the case where the reactor vessel internals are also entombed on the site, there is no likelihood that the contained radioactivity will decay to unrestricted release levels within times conceivable to man, about 100,000 years, and the surveillance and maintenance programs will have to be continued in perpetuity."[33]

Although dismantlement is preferred by the NRC, it requires some specific location where a state allows the removed material to be deposited. As with spent fuel, no such disposal site has yet been located. None of the seven or eight nonoperable reactors in Table 13-4 has been dismantled. The question, then, is whether the dismantlement will be used, or whether mothballing will lead of necessity to entombment. Nuclear engineers generally hold an optimistic view. The Atomic Industrial Forum study concludes that the necessary technology is presently available, and that dismantlement (the most expensive form of decommissioning) will have an average cost of less than $100 million per reactor.[34]

[33]R. I. Smith and L. M. Polentz, Addendum to *Technology, Safety, and Costs of Decommissioning,* prepared by Battelle Pacific Northwest Laboratory, 1979, p. 4–10.
[34]See Atomic Industrial Forum, Inc., *Analysis of Nuclear Power Reactor Decommissioning Costs* (Washington, 1981).

Another uncertainty lies in the conditions under which the plant may become nonoperating. The preceding discussion assumes a plant works without difficulty for its normal operating life. A major accident or contamination has a strong effect on decommissioning costs even in the absence of employee or public exposure. The problem, then, is an economic one. What is the cost of decommissioning a normal reactor, and what is the cost of decommissioning one closed because of contamination problems?

With this question, the focus now moves to the cost of maintaining reactor and fuel safety and air pollution control. As noted in the beginning of the chapter, the regulatory practices of the EPA and NRC, in cooperation with the electric utility industry, have achieved a major reduction in air pollution and a good safety record to date in nuclear power operation. The cost, however, has been considerable.

The Economic Cost of Health and Safety and Emerging Problems

Table 13-6 gives a general indication of the high cost of environmental protection that has been achieved. The new plants referred to are all in New York and nearing completion. The Somerset coal plant is being built in western New York.

The cost increases experienced by the Nine Mile 2 and Shoreham nuclear plants have drawn a good deal of public attention, but there is no firm evidence that cost escalation has other causes in addition to the environmental requirements and general inflation. There seems to be a tendency to attribute higher cost to management rather than to management's response to rapidly changing regulations. New NRC bulletins and circulars on reactor safety and design have averaged one per week.[35]

Both coal and nuclear power cost have been increasing at 20–25 percent each year. Figure 13-2 indicated that—in spite of major increases in coal and oil use by utilities—emissions and air quality have both improved. The high standards for new plants, successfully implemented by utilities, will mean that coal use can continue to grow without increasing total emissions. But Table 13-6 shows these standards may constitute 68 percent of the cost of a new plant.

One problem introduced in Chapter 10 was the concentration of low-sulfur surface coal in the western United States. This geologic fact has interacted with political processes to create a dilemma in sulfur control. The present regulations for plants built after 1978 require sulfur removal from all coal, regardless of sulfur content. This manner of expressing regulations was

[35]Charles Komanoff, *Power Plant Cost Escalation*, Komanoff Energy Associates, 1981, p. 147.

Table 13-6. Costs of environmental regulation, 1969–81

Coal generation	
Estimated cost in 1969 for new plant	$ 183/kW
General inflation	$ 191/kW
Cost of new environmental protection	$ 795/kW
Total cost of new plant, 1981	$1,169/kW
(Somerset, New York)	
Change in air pollution emissions, 1969–81	
Sulfur dioxide	−95%
Particulates	−97%
Nitrogen oxides	−80%
Average reduction	−91%
Kinds of new equipment now required	
cooling tower to prevent heat discharge and fish kill	
improved particulate removal	
sulfur oxide scrubbing system	
protected sulfur sludge and ash	
taller stack	
treatment of in-plant waste	
continuing standards revisions: average one per week	

Nuclear generation	
Estimated cost in 1969 for new plant	$ 226/kW
General inflation	$ 236/kW
Cost of new environmental protection	$1,539/kW
Total cost of new plants, 1981	$2,001/kW
(Nine Mile 2 and Shoreham)	
Major features now required	
cooling tower	
emergency core-cooling system to reduce probability of core melt	
massive containment structure to hold radioactive releases	
increased training	
strict record and report requirements	
much higher earthquake standards	
protection against sabotage attacks	
emergency power system	

Sources: The 1969 estimates are reported in R. R. Bennett and D. J. Kettler, "Dramatic Changes in the Costs of Nuclear and Fossil-Fueled Plants," Ebasco Services Incorporated, September 1978. Analysis of change in coal emissions is from Komanoff. Discussion of new safety requirements in nuclear plants is in each of the above, and in E. P. O'Donnell and J. J. Mauro, "A Cost-Benefit Consideration of Nuclear and Nonnuclear Health and Safety Protective Measures and Regulations," *Nuclear Safety,* 20:5 (September-October 1979), 525–540. The Somerset, Shoreham, and Nine Mile 2 costs are from New York Power Pool, *Report of Member Electric Systems,* 1981. AFUDC figures, which are not included, would raise the values for all three plants.

favored by environmental groups because it means—if followed—that new plants will have perhaps only 10 percent of the sulfur emissions of old plants built in the 1960s. The policy was supported by the United Mine Workers, because it discriminates against non-UMW western coal, and it is also supported by many eastern coal companies, again because of the discrimination against western coal.

Appendix 13-C summarizes the very complex question of sulfur regulation as it applies to states in the Ohio River basin, the source of nearly half of all utility sulfur emissions.[36] In this appendix, it is estimated that the moderately stringent emission level of 1.2 lb SO_2/MBtu could be met with the use of coal having .5% S. But the present law prohibits the use of such coal by a new coal plant to obtain the same objective. Sulfur removal is required regardless of sulfur content. It should be recalled from Chapter 10 that low-sulfur coal resources exist in large quantity, and are obtainable at a cost equal to or only slightly greater than the cost of high-sulfur coal. It is difficult to determine the extra cost imposed on utilities by the mandatory sulfur removal requirement which excludes low-sulfur coal use. It probably adds about 5–10 percent to the total revenue requirements for the average all-coal utility. For a single new power plant, however, the additional imposed cost may be 10–20 percent. (This figure is generating cost at the power plant; the utility figure percentage for revenue requirement is smaller because transmission and distribution costs are unaffected, and old coal plants are unaffected.)

For nuclear power, Table 13–7 shows that safety standards may be 75 percent of the cost of a new plant.

In 1983, it is impossible to know if coal and nuclear power costs have stabilized. There are new problems emerging with each technology.

The acid rain problem arises from the impact of sulfates and nitrogen oxides on natural ecosystems and vegetation. The acids formed from these pollutants in high-altitude lakes are sufficient to initiate chemical processes which eliminate all fish life.[37] This problem is the subject of intensive research. Since coal burning by electric utilities is the major source of airborne sulfur oxide, new standards on sulfur and nitrogen oxide emissions beyond those noted in Table 13-6 will have significant effects on coal generation cost.

The carbon dioxide problem is also closely linked to coal-burning electric utilities.[38] Because coal is primarily carbon, and natural gas and oil both contain hydrogen, the burning of coal releases more CO_2 per Btu than does combustion of oil or natural gas. The present scientific opinion is that fossil fuel burning is increasing the earth's temperature. If this should come to be viewed as a major problem, it will have a significant impact on coal generation costs.

Nuclear power will almost certainly continue to be closely regulated. Its safety record is one that the industry finds a justifiable source of pride—but, as Table 13-6 shows, at a cost. Spent fuel disposal and reactor decommission-

[36]Pechan, p. 5.
[37]See U.S. EPA, "Acid Rain," 1980, and National Academy of Sciences, *Atmosphere-Biosphere Interactions*, Report of the Committee on the Atmosphere and the Biosphere, 1981, chap. 8.
[38]See J. Hansen et al., "Climate Impact of Increasing Atmospheric Carbon Dioxide," *Science*, 28 August 1981, pp. 857–966, and Don G. Scroggin, Robert H. Harris, and Lester B. Lave, "The Carbon Dioxide Problem," *Technology Review*, 84:2 (November–December 1981), 22–31.

Table 13-7. Overall health impact from coal and nuclear power generation, deaths per year per 1,000 MW

	Coal	Nuclear
Mining and milling	0.8	0.4
Transportation	0.8	*
Fuel processing	0.1	*
Power plant operation	0.1	0.2
Spent fuel disposal	*	*
Air pollution	10.0	*
	11.8	0.6

* Less than .05 deaths per year per 1,000 MW

Source: Leonard D. Hamilton, "Comparative Risk Assessment in the Energy Industry," in *Health Risk Analysis: Proceedings of the Third Life Sciences Symposium,* ed. by Chester R. Richmond, Phillip J. Walsh, and Emily D. Copenhaver (Philadelphia: The Franklin Institute Press, 1981), pp. 54, 55. Midpoint values are shown here, whereas Hamilton gives ranges.

ing are two emerging problems that may have a significant effect on nuclear costs. Table 11-3 presents an analysis of Commonwealth Edison costs in which decommissioning and spent fuel disposal are expected to contribute together only 1.1¢/kWh to nuclear cost. However, this author examined extreme cases with severe problems in spent fuel disposal and decommissioning, and found that—in the extreme—these problems could double nuclear power cost.[39] The Three Mile Island experience will continue to be important. In 1980, the General Public Utilities Corporation released cost estimates for decontamination and repair which were approximately $1.25 billion.[40] The actual cost may be greater.

Nuclear safety regulation has its own perverse nature. In at least three instances, the complexity of the safety systems themselves may have caused accidents. The Browns Ferry (Alabama) accident in 1975 was caused by an electrician using a candle in an attempt to locate air leaks. The resulting electrical fire disabled the plant's emergency core-cooling system. The Fermi partial meltdown was caused by a loose metal plate especially installed to reduce the probability of a meltdown.[41] And, most spectacularly, the TMI accident was caused by the interaction of faulty design and operating error in interrelated safety systems.[42]

[39]Duane Chapman, "The Economic Status of Nuclear Power in New York," New York State Assembly, Special Committee on Nuclear Power Safety, Hearing, *The Economics of Nuclear Power,* February 28, 1980, p. 7.; printed as Cornell University Department of Agricultural Economics Staff Paper No. 80-7.

[40]General Public Utilities Corporation, *TMI-2 Recovery Program Estimate,* August 1, 1980. This report gives an $855 million figure in 1980 dollars, (p. 7). An inflation adjustment of about 50% gives the $1.25 billion estimate.

[41]Keeny et al., p. 233, and *New York Times,* March 26, 1975.

[42]See John G. Kemeny et al., *Report of the President's Commission on the Accident at Three Mile Island,* October 1979, and Mitchell Rogovin et al., Nuclear Regulatory Commission Special Inquiry Group, *Three Mile Island: A Report to the Commissioners and to the Public,* January 1980.

It has been argued that nuclear power cost could be reduced with less regulation. Yet, General Public Utilities has sued the Nuclear Regulatory Commission, alleging insufficient regulation.[43] Relaxation of regulatory standards is improbable.

Summary

How serious are the health, safety, and environmental problems associated with coal and nuclear power? In terms of current economic impact, motor vehicle accidents are probably America's most serious external cost of energy use. In 1980, 53,000 people died, and five million were injured in 24 million reported accidents. The insurance industry estimates the economic loss involved in terms of morticians' fees, insurance payments, car repairs, and so forth at $57 billion.[44] If a life is valued at a million dollars apart from economic loss, the sum of the direct economic loss and the assumed value of loss of life is $110 billion.

Air pollution damage, according to the Freeman review, is of comparable significance. Deaths from air pollution may be about 50,000 annually. Air pollution levels have been significantly reduced since 1970. Three of the major policies responsible have been: (1) vehicle emission control systems, (2) electrostatic precipitators to remove utility soot and particulates, and (3) the use of low-sulfur coal to reduce sulfur emissions.

The Council on Environmental Quality (CEQ) estimates the value of improved air quality from utilities and stationary sources to be $20 billion. National Economic Research Associates (NERA), a utility research and consulting company, estimates this benefit to be a lower $7 billion. The CEQ finds benefits to be more than twice costs; NERA finds benefits to be less than costs.

Studies of air pollution damage and benefits of pollution control have several problems. These include: (1) estimating a dollar value of a saved life, (2) interaction of multiple pollutants, (3) lack of knowledge of the significance of personal factors such as smoking and occupational hazards in influencing pollution impact, and (4) the inability to represent time delays in health and environmental impact. A major defect in both the CEQ and NERA studies is that benefits are based upon the comparison of 1970 pollution levels with 1978 levels. Since pollution control costs are based upon actual levels of coal and oil use, benefits also should be based upon these levels of coal and oil use, which are much higher than in 1970.

[43]General Public Utilities Corporation v. United States of America, Complaint filed December 3, 1981, United States District Court, Eastern District of Pennsylvania.

[44]U.S. Department of Commerce, Bureau of the Census, *Statistical Abstract of the United States: 1981*, p. 622.

Both air pollution and radiation may have linear no-threshold dose-response curves, meaning that both cause increasing damage at any level. Both also are fatal to all persons in large doses.

Effective regulation by the EPA and NRC has reduced air pollution and allowed nuclear power plants to operate to date with no known fatalities caused in the general population. It is believed that the worst nuclear accident to date at Three Mile Island might cause one case of cancer.

The difficulty in comparing the environmental effect of coal and nuclear power is that the probabilities of consequences are almost incomparable. The impact of coal use on mine health and safety and on air pollution is known to be significant and predictable. In contrast, nuclear power has a very low possibility of catastrophic accidents. The historical record clearly favors nuclear power to date. Table 13-7 reviews estimated fatalities from all stages of coal and nuclear power generation, and shows coal generation to be more harmful.

Several large nuclear power facilities are now inoperable, as are all three spent fuel reprocessing plants. General policy is to decommission nuclear plants by immediate dismantlement or by mothballing for up to 100 years, followed by dismantlement. However, none of the inoperable large facilities has been dismantled. Entombment would require physical isolation and security for about 100,000 years for each power plant.

Decommissioning is expected to be inexpensive, on the order of $100 million per reactor in cost. However, decontamination after a major accident may cost as much or more than the $1 billion estimated cost at Three Mile Island.

No permanent site now exists as a depository for dismantled power plant components or for spent fuel disposal. Spent fuel generally accumulates at each operating nuclear plant. Eight states have prohibited the initiation of construction of new nuclear plants. Their reasons for doing so are usually perceived problems with spent fuel disposal, decommissioning, and/or reactor safety.

While the utility industry and regulatory agencies have been successful in reducing air pollution and avoiding major impact from nuclear power, the economic cost is high. Approximately 75 percent of the cost of both coal and nuclear plants is now spent for safety, health, and environmental protection.

The economic theory of social optimality is based upon the presumption that social cost consists of the market costs of production, subsidies, and external social costs. It is not possible to obtain confident quantitative estimates of these latter two terms. It is evident, however, that subsidies and external environmental cost are particularly important for coal and nuclear power generation.

In theory, it should be possible to determine the optimal levels of coal and nuclear generation by estimating the maximum positive differences between

[271]

social value and social cost. This was done hypothetically in Chapter 2. In reality, our inability to estimate the economic impact of external social cost creates a situation in which objective determination of benefits, costs, and social welfare is not possible. Consequently, it is logical to suppose that market decisions, subsidies, and external social cost will be viewed subjectively and politically by individuals, groups, and corporations. The political process transforms external cost into market cost by regulation and environmental protection, a shift that is making coal and nuclear power increasingly expensive.

For economic reasons as well as subjective evaluation and political decision making, America and the world are increasingly aware of another problem: the proper role for solar energy, renewable resources, and synthetic fuels. The book concludes with this subject and the basic question of public and private ownership and regulation in the following two chapters.

Appendix 13-A. Pollutant Levels and General Health Effects

Pollutant levels					Health	
Particulates (24-hr) µg/m³	SO_2 (24-hr) µg/m³	CO (8-hr) mg/m³	O_3 (1-hr) µg/m³	NO_3 (1-hr) µg/m³	Effects	Warning
0–74	0–79	0–4.9	0–119	NR		Good condition.
75–259	80–364	5.0–9.9	120–239	NR		Moderate condition.
260–374	365–799	10.5–16.9	240–479	NR	Mild aggravation of symptoms in susceptible persons, with irritation symptoms in the healthy population.	Unhealthful condition. Persons with existing heart or respiratory ailments should reduce physical exertion and outdoor activity.
375–624	800–1,599	17.5–33.9	480–899	1,130–2,259	Significant aggravation of symptoms and decreased exercise tolerance in persons with heart or lung disease, with widespread symptoms in the healthy population.	Very unhealthful condition. Elderly and persons with existing heart or lung disease should stay indoors and reduce physical activity.
625–874	1,600–2,099	34.0–45.9	900–1,099	2,260–2,999	Premature onset of certain diseases in addition to significant aggravation of symptoms and decreased exercise tolerance in healthy persons.	Hazardous conditions. Elderly and persons with existing diseases should stay indoors and avoid physical exertion. General population should avoid outdoor activity.
875 and above	2,000 and above	46.0 and above	1,000 and above	3,000 and above	Premature death of ill and elderly. Healthy people will experience adverse symptoms that affect their normal activity.	Hazardous conditions. All persons should remain indoors, keeping windows and doors closed. All persons should minimize physical exertion and avoid traffic.

NR = No index values reported at concentration levels below those specified by "alert level" criteria.
Source: U.S. CEQ, Environmental Quality 1980, pp. 156–157.

Appendix 13-B. Biological Effects of Ionizing Radiation

Exposure range	Chronic exposure over extended period	Acute exposure in 24 hours
Less than 1 rem	No observable effects; equivalent to exposure from background radiation for 5–10 years. Cancer risk: $1-2 \times 10^{-4}$ per rem for adult; greater than 4×10^{-4} per rem for fetus (may be as high as 6×10^{-4} to 6×10^{-3} per rem for children; 7×10^{-3} for adults).	No observable short-term effects.
1–50 rems	Chromosomal aberrations in blood. 0.3–30 leukemia cases per 10,000 person-rem observed in this exposure range; 0.5–1.2 thyroid cancers per 10,000 person-rem observed. Occupational exposure range.	Slight blood changes; decreased head circumference and increased leukemia risk in fetus.
50–100 rems	Observable increase in spontaneous mutations.	Mild symptoms of radiation sickness possible. Vomiting in 5% of those exposed to 100 rems within 3 hours.
100–200 rems	Observable increase in cancer.	Vomiting in 5% of those exposed to 100 rems to 50% of those exposed to 200 rems within 3 hours. Also, fatigue, loss of appetite, moderate blood changes that persist. Recovery within several weeks. Increased cancer risk; cataracts possible.

200–600 rems	Limited experience with regard to chronic exposure over 200 rems. Large increase in incidence of leukemia and other cancers. Uranium miners exposed to 700–1,000 rads, with maximum exposures estimated to be as great as 10,000 rads. Excess of lung cancer deaths may reach 600–1,100 in a population of 6,000 miners.
	Vomiting: 50% at 200 rems within 3 hours; 100% above 300 rems within 2 hours. Also, loss of hair, other symptoms of radiation sickness. Death in 0–80% of those exposed within 2 months from hemorrhage and infection; recovery for survivors in 1 to 12 months.
600–1,000 rems	Vomiting within 1 hour, severe blood changes, hemorrhage, infection, loss of hair, damage to bone marrow. Death in 80–100% of those exposed within 2 months from hemorrhage and infection. Long convalescence for survivors.
1,000–3,000 rems	Vomiting within 30 minutes; radiation sickness. Gastrointestinal syndrome within 5 to 14 days, including diarrhea, fever, severe blood changes, damage to bone marrow. Death in 90–100% of those exposed within 2 weeks due to circulatory collapse.
More than 3,000 rems	Vomiting within 30 minutes. Central nervous system syndrome within 2 days, including loss of muscular control. Death in 100% of those exposed within 1 to 48 hours due to respiratory failure and brain edema.

Source: Lipschutz, pp. 16–17, used by permission of the Union of Concerned Scientists.
Note: "Rem" here has the same meaning as "person-rem" in the text. "Rem" measures the biological effect of radiation; "rad" measures the physical energy. One rad indicates the absorption of 100 ergs per gram of tissue.

Appendix 13-C. Emission Limits for SO_2 on Coal Plants in the Ohio River Basin

There are four general categories of regulations applying to plants of different ages and locations:

(1) "SIP Urban": old coal-burning plants which are in or near cities and were operating before 1977. These SO_2 limits are set separately by a state for each plant in the state. "SIP" means State Implementation Plan.

(2) "SIP Rural": old coal-burning plants outside urban areas. These plants are also regulated by each state, and were operating before 1977.

(3) "NSPS": New Source Performance Standards. All plants in the United States which began operation between 1977 and 1983 are supposed to meet the 1.2 pounds SO_2/M̄Btu standard. The standard can be met either by low-sulfur coal or by sulfur removal.

(4) "RNSPS": Revised New Source Performance Standards, the regulations applying to new plants discussed in the text. All plants must remove sulfur from the coal; the percentage removal depends upon original sulfur content. High-sulfur coal must have 90% removal and meet a 1.2 lb SO_2/M̄Btu standard. Low-sulfur coal must have 70% removal and meet a 0.6 lb SO_2/M̄Btu standard. Every coal shipment, then, has a different standard. The values for RNSPS below were estimated averages.

Note: Consider a low-sulfur coal having .5% sulfur by weight and 19 M̄Btu/ton. This is 1.05 lb SO_2/M̄Btu, by this relationship:

$$lb \ SO_2/\bar{M}Btu = (2 * proportion \ sulfur * 2000 \ lb/ton) \div \bar{M}Btu/ton$$

SO_2 is twice the weight of S. This quality coal would meet the NSPS standard and the upper limit of the RNSPS.

Emission limits, SO_2 lb/M̄Btu

State	SIP Urban	SIP Rural	NSPS	RNSPS
Pennsylvania	0.60	1.80	1.20	0.41
Ohio	1.40	4.50	1.20	0.41
Indiana	1.21	5.00	1.20	0.55
Illinois	1.80	6.00	1.20	0.55
West Virginia	1.60	3.20	1.20	0.45
Kentucky	1.21	5.70	1.20	0.45

Source: Teknekron Research Inc., *Selected Impacts of Electric Utility Operations in the Ohio River Basin, (1976–2000): An Application of the Utility Simulation Model,* April 1980, prepared for the U.S. Environmental Protection Agency, p. 146.

PROBLEMS IN
PUBLIC POLICY

14

New Technologies and Microeconomics

The years 1970–72 ended the Growth Era for the United States. America had access to low-cost oil throughout the world. Imported crude oil was bought by American oil companies at an average price of 6¢ per gallon. American domestic resources of oil and natural gas had just passed their peak of production, and gasoline and natural gas prices had attained the minima of their full history of use. Because oil and natural gas were inexpensive, and because the cost of nuclear safety and pollution control had not yet been experienced, electric rates were also at their historical lows. Petroleum, natural gas, and electricity consumption were each growing at exponential rates. These three years provided the lowest energy prices to consumers for every one of the major energy forms, lower (real) prices than have been experienced before or since. Energy consumption now is declining (oil), stable (natural gas), or growing slowly (electricity).

These points have been developed in earlier chapters. Now the focus changes. What is the logical response? How should business and consumers and government address present and future possibilities? This chapter examines some of the technologies that may emerge in the remainder of the century: gasohol, synthetic fuels, rail transportation, solar energy, woodfuel, and energy-efficient building construction and use. As will be evident, not all these new technologies have an economic future on a large scale. They have been selected because they have been seriously investigated by industry, government, and university research. These technologies have been studied because they have been perceived as being the closest to the margin of economic feasibility. Other new technologies may be equally interesting, such as oil-shale production or liquefied natural gas or wind power, but they are not studied here in detail.

National Economics

While economists are in agreement about the veracity of data on energy prices and use, they are in considerable disagreement about the relationship of economic welfare to energy use, and about the potential contribution of new energy technologies.

The possibility of a strong relationship between economic welfare and energy use is suggested by the basic statistics. Real, deflated, disposable personal income per capita grew regularly at an annual rate of 2.1 percent from 1946 to 1973. Since 1973, to the middle of 1982, it has risen and fallen irregularly, averaging approximately a 1.2 percent annual increase. This difference in real income growth before and after 1973 implies that economic growth requires growth in energy consumption. It can be argued that increased production of energy from new technology in the United States would cause real income to grow more rapidly.

A logical consequence of this position is that energy supply should be encouraged, and new energy supply technologies should be promoted.

A contrasting position is that renewable energy resources and conservation are necessary to maintain economic standards.[1]

Consider Table 14-1: in 1972, at the end of the Growth Era, the United States ranked first in Gross National Product per capita. Canada, the other North American country, stood third. Eight years later, in 1980, America and Canada were eleventh and fourteenth. Eight countries in Western Europe now lead the United States.

Because GNP per capita can be statistically misleading about material living standards, Table 14-1 also shows the *physical quality of life index* for these countries. It is based upon life expectancy, infant mortality, and literacy. The United Arab Emirates and Kuwait descend in ranking to last among wealthy countries, but the positions of America and Western Europe are relatively unchanged.

The figures for energy consumption per capita show the United States and Canada to be the leading energy consumers. The question arises as to whether the differential economic growth in North America may be linked in some way to differential levels of energy consumption. It may be that reliance upon energy-intensive consumption patterns in North America has made economic growth difficult since 1972. Certainly, in terms of energy resources, Canada and the United States exceed Western Europe. The only European country approaching North America in energy use per person is Norway. And Norway

---

[1]The most detailed articulation of this conservation position is in the study by the Solar Energy Research Institute (SERI), *Building a Sustainable Future*, 2 vols., Committee Print, U.S. House Committee on Energy and Commerce, April 1981. For a detailed exposition of the need for increased energy use to support economic growth, see Michael Halbouty, "The Halbouty Report," *Energy Daily*, December 4, 1980, pp. 2–7.

Table 14-1. Living standards after the Growth Era

Country	GNP per capita, U.S. $				Physical quality of life index	1979 energy consumption per capita, MBtu
	1980	(rank)	1971	(rank)		
United Arab Emirates	26,850	(1)	3,220	(10)	34	111
Kuwait	19,830	(2)	4,090	(4)	74	158
Switzerland	16,440	(3)	3,940	(5)	95	128
West Germany	13,590	(4)	3,390	(8)	93	165
Sweden	13,520	(5)	4,480	(2)	97	212
Denmark	12,950	(6)	3,670	(6)	96	149
Norway	12,650	(7)	3,340	(9)	96	297
Belgium	12,180	(8)	3,210	(11)	93	168
France	11,730	(9)	3,620	(7)	94	124
Netherlands	11,470	(10)	2,840	(12)	96	168
United States	11,360	(11)	5,590	(1)	94	308
Saudi Arabia	11,260	(12)	550	(15)	29	39
Austria	10,230	(13)	2,410	(13)	93	130
Canada	10,130	(14)	4,440	(3)	95	335
Japan	9,890	(15)	2,320	(14)	96	106
World	2,510		981		65	57

Sources: World Bank, *World Development Report* (New York: Oxford University Press, 1981 and 1982), and *Atlas,* 1974; energy data reported there are converted here into MBtu by a constant 24.9 kBtu/kg coal equivalent. The physical-quality-of-life index is from Overseas Development Council, *Measuring the Condition of the World's Poor,* 1979, cited in "World Population Estimates 1981," Environmental Fund.

is unique among European countries in that it is a significant exporter of energy. Its major source of energy is renewable hydropower, equivalent to 75 percent of its aggregate consumption.

Another national problem is inflation. In 1981 and 1982 inflation slowed considerably. A combination of stable petroleum prices and recession reduced overall inflation to annual rates of 2–3 percent. However, inflation was a serious problem from 1973 to 1981, and it seems improbable that recession and unemployment will be a satisfactory long-run alternative. The desired economic policies that affect energy and the national economy should reduce the inflationary impact of resource scarcity and high energy prices, and do this in a manner which permits growth in employment. Figure 14-1 indicates that energy price increases between 1973 and 1981 were the fastest growing sources of inflation. Surprisingly, new-car cost at 7 percent growth is less than the overall 9.3 percent increase in the consumer price index (CPI). While public transportation prices are almost constant in real terms, gasoline prices grew 17 percent.

Another anomaly is housing costs. High interest rates (not shown) caused a 14 percent inflation rate in financing, but building costs (not shown) and rent cost were not sources of inflation, growing at below the overall inflation rate.

[281]

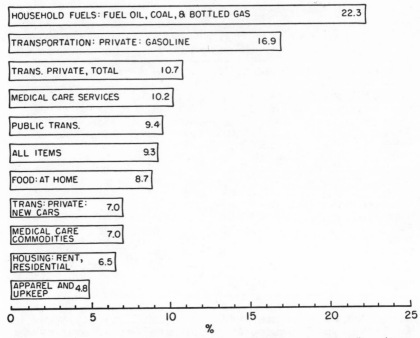

Figure 14-1. Consumer price indexes: Annual rates of inflation, various expenditure classes, 1973–Jan. 1981

One problem for national economic policy is the interaction of overall inflation with interest rates. In normal circumstances interest rates are 3 to 5 percent higher than inflation. If rising energy prices are responsible for much of the consumer price inflation, then energy price inflation has had some upward impact upon interest rates.

Possibly, high energy prices and high interest rates combine and interact to have a severe impact on automobile sales and suburban construction. For example, new housing in the early 1980s is being built at about one-half the rate in 1973. New-car sales per person are only two-thirds of the Growth Era 1973 rate, and these new cars are smaller, more fuel-efficient, and more likely to be imported. Consequently, unemployment in the early 1980s is concentrated in construction and automobiles. Approximately one in five persons in these two sectors was out of work in early 1982.[2] In sum: production and employment have fallen in the energy-dependent automobile and construction industries, and inflation has been caused or stimulated by rising energy prices.

[2]Statistics from U.S. Department of Commerce, Bureau of Economic Analysis, *Survey of Current Business*, various issues, and the *New York Times*, January 11, 1982.

The basic question, then, is whether the American pattern of high energy consumption has been transformed since the Growth Era into an obstacle to economic progress. Should the focus be on developing new fuels for automobiles? Should the focus be on developing new energy sources for electric home heating and air conditioning, or on the development of buildings which require less heat and energy?

The Marginal Cost Approach

The preceding section raised important *macroeconomic* questions: What is the relationship between GNP, national income, inflation, employment, and energy? The remainder of this chapter analyzes *microeconomic* dimensions of energy supply and use. In the next chapter, the two perspectives are integrated, at least partially, and the macroeconomic questions are examined in the context of national public policy.

The microeconomic approach relies upon an understanding of the economic meaning of marginal cost introduced in the first two chapters. Here, marginal cost is the basis for another concept, *incremental cost comparison*. It estimates the cost of using a new technology now or in the future, and compares this present or future cost with the expected rising cost of the current technology using conventional energy resources. If the new technology seems likely to be more costly than the conventional technology, then we should conclude that market economics will not permit the new technology to develop. This is efficient, in the sense that the least costly method of providing an energy-related commodity or service will be used. If it seems probable that the conventional technology will be more costly in the near future, then market economics should lead to the growth of the new technology.

Three serious obstacles can obstruct efficient market decisions. First, subsidies can lead to an artificially low price, and cause a new technology to be expanded uneconomically. Second, monopoly power in new technologies could distort market economics. It should be recalled from earlier chapters, however, that monopoly can lead to either of two contradictory results: restriction of consumption and production with excessive prices, or premature and excessive development in a growth monopoly situation. A third problem is more broadly institutional, in that market economics can be misdirected by nonmonopoly factors such as uninformed consumers, erroneous government policy, or the inability to form markets for transactions.

A preview of the conclusion: the monopoly problem is not at present serious. The other two—subsidies, and broad institutional questions—are considerably more important. If microeconomic decisions are inefficient and are made in large numbers or scale, then they have a macroeconomic impact. In the last chapter, I shall argue that inefficient microeconomics is having a severe negative effect on employment, national income, inflation, and energy conservation. But first a microeconomic survey of the major new technologies.

Gasohol

Gasohol has certain characteristics which have made it attractive as a new energy source. It is a liquid fuel, is made from a domestic source (corn), and is a renewable energy form. At its inception, it found favor to some degree with agriculture, petroleum, government, automobile, and environmental groups. It is an important example of the problem of synthetic fuel development versus conservation in transportation.

Table 3-4 shows that corn starch, sugar, and cellulose are chemical cousins to the petroleum hydrocarbons used in gasoline. This link provides the physical basis for the transformation of corn into automobile fuel. In a manner similar to whiskey distillation, ethanol can be distilled from corn. Approximately 2.5 gallons can be obtained from each bushel, and 17 pounds of leftover material can be used as a partial feed grain. Ethanol can be used directly in automobiles to substitute for gasoline, although it has certain undesirable side effects: slight increases in starting problems, vapor lock, and small mileage loss. *Gasohol* is defined as a mixture of 10 percent ethanol with 90 percent conventional gasoline. In this mixture, the undesirable burning qualities of ethanol are diluted and disappear.

The various aspects of the economics of gasohol are summarized in Table 14-2. The largest component of ethanol cost is the plant itself: a 50-million-gallon-per-year plant is estimated to cost about $115 million. A plant of this size captures most of the economies of scale, and can be located near corn-producing farm areas. Such a plant would contribute about 41¢/gallon to ethanol cost. This 41¢ is a levelized cost: it is the amount which, if charged for every gallon produced, would exactly pay off a $115 million capital investment with a 15 percent capital charge.[3]

Fuel is a significant part of the distillation process, and was estimated (with labor added) to cost 31¢/gallon. Administration and marketing add 12¢/gallon.

In early 1982, corn sold at $2.60/bushel. With one bushel yielding 2.5 gallons of ethanol, corn feedstock costs $1.04 per gallon. The leftover corn by-product can be sold as animal feed, worth perhaps 6¢ a pound. The 6.7

[3]Levelized cost, as explained in Chapter 11, is a method of allocating capital cost on an energy unit basis. First, the capital cost is amortized over the plant's expected operating life. With a 20-year life and a 15% rate of return, this amortization factor is defined by $(.15(1.15)^{20})/((1.15)^{20} - 1) = .16$. This capital recovery factor, .16, is multiplied by $115 million to give an annual capital charge of $18.4 million. As in Chapter 11, the capacity factor (here 90%) times the capacity (50 million gallons per year) gives the annual production, 45 million gallons. The levelized cost is then $18.4 million ÷ 45 million gallons, or 41¢/gallon.

Table 14-2. Gasohol economics, 1982

	¢/gallon
Market cost of ethanol	
Capital cost: $115 million plant, 50 million gallon annual capacity, 90% capacity factor	41¢
Fuel and labor	31¢
Administration and marketing	12¢
Corn purchases: $2.60/bushel, 2.5 gallons/bushel	104¢
Less feed credit (6.7 lb/gallon)	−40¢
Net market cost at ethanol distillery before tax credits	148¢
Estimated wholesale cost of conventional gasoline	100¢
Subsidies for gasohol	
Special gasohol investment tax credit	4¢
Normal investment tax credit	4¢
Value of shelter for federal gasoline tax	40¢
Value of state shelters for gasoline tax	54¢
Total subsidy	100¢

Sources: Fred Sanderson "Benefits and Costs of the U.S. Gasohol Program," *Resources,* Resources for the Future, July 1981, which makes considerable use of Ronald Meekhof et al., "Gasohol: Prospects and Implications," U.S. Department of Agriculture, Agricultural Economic Report No. 458, June 1980. Also U.S. DoE, *The Report of the Alcohol Fuels Policy Review,* June 1979; Chase Manhattan Bank, "The Petroleum Situation," October 1981; and the *New York Times.*

pounds of by-product feed reduces ethanol cost by 40¢/gallon. The net cost of these factors sum to $1.48 per gallon. In early 1982, the wholesale price of regular nonleaded gasoline from a refinery was about $1.00 per gallon. Table 14-2, then, shows ethanol cost at the distillation plant to be 50 percent higher than the refineries' average gasoline cost.

Tax subsidies for gasohol apparently are almost as much as the cost. The normal investment tax credit is added to a special energy investment tax credit, creating a 20 percent investment tax credit for gasohol. In other words, $23 million (20 percent) of the hypothetical $115 million plant is immediately written off against income tax. The two investment tax credit features will each give the producer about 4¢/gallon of ethanol.

Recall that gasohol is 90 percent conventional gasoline and 10 percent ethanol. A gallon of gasohol is exempt from the 4¢/gallon federal excise tax on gasoline. This means that one gallon of ethanol shelters nine gallons of gasoline from this tax. Similarly, several states provide state gasoline tax exemptions at an average 54¢/gallon shelter for each gallon of ethanol.

The total subsidies listed in Table 14-2 sum to $1.00 per gallon of ethanol. Other subsidies not examined here would be accelerated depreciation, loan guarantees, and direct grants. Obviously, in early 1982, gasohol without subsidies is not economically competitive with conventional gasoline.[4]

[4]One actual operating plant reported a wholesale cost of $1.65 per gallon of ethanol with the depressed corn price of $2.25 per bushel in August 1982. This was contrasted to a wholesale unleaded gasoline cost of 90¢/gallon (*Oil & Gas Journal,* September 6, 1982, p. 36). These actual figures should be compared to Table 14-2 and the text.

In addition to the economic problem, large-scale ethanol production would have serious impact on corn availability and prices. Suppose one-half of the American corn crop was directed toward ethanol. This would be about two billion bushels in the average year, and would yield five billion gallons of ethanol. In 1982, gasoline consumption in the United States exceeds 100 billion gallons. Therefore, one-half of the total corn crop could produce 5 percent of the gasoline usage.

Because gasohol itself is not economically competitive but has very large tax subsidies, it follows that gasohol producers will be corporations with large profit to shelter. This gives a major impetus to oil companies. However, the decline in demand for gasoline in the early 1980s interacts with the high cost of gasohol to create a situation in which petroleum corporations may not significantly increase their gasohol marketing.

The overall impact of gasohol use on total energy consumption has not been determined. One view is that more energy in Btu is required for production than is available in the product. Apparently, every three gallons of ethanol requires the equivalent of two gallons of fuel in the ethanol plant and more than one gallon of fuel on the farm. An alternative view perceives that automobiles may extract as much energy from a gallon of ethanol as from a gallon of gasoline, notwithstanding the considerably lesser Btu content of ethanol. This is analogous to the differences in end-use efficiencies in home heating systems described in Chapter 4. Ethanol is more efficient than gasoline in automobiles, so a gallon of each is equivalent although the Btu content of ethanol is much lower.[5]

Substituting coal for the natural gas and fuel oil presently used in ethanol distillation would not change the energy requirement. However, if coal can be used, an energy resource which is still extensive would replace declining resources.

Overall, it is improbable that automobiles can ever use corn ethanol as a significant fuel source, and since corn ethanol is the most promising of domestic biological resources for automobile fuel, it is also unlikely that any renewable source of liquid hydrocarbons will be developed as a motor vehicle or airplane fuel on a large scale.[6] In Brazil, sugar-based gasohol is widely used but still requires a high level of subsidy.[7] Brazil is well suited by agri-

[5]This discussion is from Sanderson, "Benefits and Costs of the U.S. Gasohol Programs," p. 3; he reports ethanol Btu content as 62% of gasoline's 125 Btu/gallon. Also see the U.S. Department of Energy's *The Report of the Alcohol Fuels Policy Review*, pp. 73–77, which analyzes net energy efficiencies.

[6]Ronald Meekhof, Gill Mohinder, and Wallace Tyner analyze seven crops, and consider corn the only major potential source, in "Gasohol: Prospects and Implications," U.S. Department of Agriculture, Agricultural Economic Report No. 458, June 1980.

[7]*New York Times*, October 13, 1980. The Brazilian government maintains the alcohol price at 55% of the gasoline price, exempts alcohol cars from road taxes, and prohibits conventional gasoline sales on weekends.

cultural geography to produce gasohol, and if Brazil ultimately concludes that such production is not economically feasible, one might generalize that an automobile fuel based on a renewable energy source is unlikely.

The situation is comparable to coal and nuclear power: in the absence of tax subsidies, no company or consumer would prefer gasohol on economic criteria.

Synthetic Natural Gas and Oil

Attention turns to *synthetic fuels* produced from coal which might provide the basis for new liquid or gas transportation fuel. Harry Perry, like James Griffin and Henry Steele, believes that among synthetic fuels, coal gasification is a less difficult and less costly process than coal liquefaction.[8] Indeed, this would seem to be a basic fact of chemistry and engineering; given a carbon atom in coal, it should be easier to form methane (CH_4) and simple gases than to create complex liquid hydrocarbons.

The more methane, the higher the quality of the synthetic gas. Compared to natural gas, synthetic gas has greater variation in its constituents. They include hydrogen, oxygen, carbon monoxide, and carbon dioxide in addition to methane. The composition varies according to the type of chemical process and the type of coal used. Only high-Btu, high-methane gas can be used in long-distance pipelines. The first problem for coal gasification is that the low Btu processes which are economical on a ¢/M̄Btu basis produce gas that cannot economically be transported in long pipelines.

The major economic problem is producing synthetic gas is energy efficiency. Considerable coal energy is required to produce the heat and the oxygen used in the high-temperature processes that transform coal into high-Btu synthetic gas: about 3 Btu to create 2 Btu of gas. Whatever capital cost may be, gasified coal cannot compare on an economic basis with coal itself where both are feasible for the same use. Exceptions and qualifications exist, of course. Synthetic gas may be more valuable than coal in particular industries such as glass manufacture or food processing. Air pollution control may be unimportant in a coal-producing area, but important in an urban area, where synthetic gas would be preferable. But notwithstanding these qualifications, coal gasification is not commercially feasible in the forseeable future.

Consider a hypothetical synthetic gas plant located in North Dakota, the state with the lowest cost coal in terms of $/M̄Btu. This is mine cost. Suppose the synthetic gas is shipped by pipeline to the Midwest. Table 14-3 analyzes the economics of this hypothetical plant. Its construction cost is $2 billion,

[8]See Harry Perry, "Clean Fuels from Coal," in *Advances in Energy Systems and Technology,* Vol. I, ed. Peter Auer (New York: Academic Press, 1978), p. 308. Also see James M. Griffin and Henry B. Steele, *Energy Economics and Policy* (New York: Academic Press, 1980), p. 318.

Table 14-3. Synthetic natural gas economics, 1982 dollars

Type of Cost	Cost per M̄Btu of synthetic gas
Capital cost: $2 billion direct cost, 36.5 trillion Btu/year production	$11.00
Maintenance and operating cost, $100 million yearly	$ 2.74
Coal cost, 55 trillion Btu, 75¢/M̄Btu	$ 1.13
Total production cost	$14.87
Transmission, distribution to Midwest	$ 1.50
Total cost to customers, 1982 dollars	$16.37

and it produces 36.5 trillion Btu each year. The capital cost on a levelized cost basis would be $11 per M̄Btu.[9] This assumes that the plant requires $400 million each year to pay capital and associated expenses, and this is $11 for each of the 36.5 million M̄Btu produced each year. A possible 5 percent operating and maintenance cost gives a $100 million annual estimate, or $2.74/M̄Btu.

North Dakota coal costs only $9–$10 per ton in the early 1980s. Even with its low Btu content, this is low-cost energy at 70¢–75¢ per million Btu. The plant may require 55 trillion Btu each year for steam, electricity, and heat as well as for feed stock. Surprisingly, the coal cost in Table 14-3 is the smallest part of synthetic gas cost. The total production cost in Table 14-3 is $14.87/M̄Btu. No income tax liability is assumed. As with gasohol, it appears that the investment tax credit would be 20 percent. It is certain that, as with new coal and nuclear power plants, the levelized income tax expense would be negative.

Transmission and distribution from North Dakota to Minneapolis, Chicago, and Michigan would add another $1–$3/M̄Btu to the cost paid by gas customers. The result is a customer cost of at least $16/M̄Btu. In mid-1981, wellhead natural gas cost $1.80 M̄Btu, and residential heating customers paid an average $4.50/M̄Btu.

Assuming that natural gas price decontrol continues, natural gas prices will, as outlined in Chapter 8, approach fuel oil prices on a Btu basis. At some future date, then, synthetic gas from coal may become an economic alternative to natural gas.

In general, given the actual experience of the Alaskan oil pipeline and actual new power plant construction, it may be supposed that the actual costs of synthetic gas plants will be 2–10 times the costs presently being estimated for these plants. The Rand Corporation has studied this cost escalation question in considerable detail, and the authors of the Rand study believe this to

[9]Levelized cost here is comparable to the gasohol illustration. A 15% rate of return would mean a 16% capital recovery factor, and another 4% is allowed for administrative, insurance, and property tax costs. The result: a 20% fixed charge rate, or $400 million annually. The 36.5 trillion Btu production is an 80% capacity factor for a 45.6 trillion Btu capacity plant.

be—possibly—a general rule for energy and advanced technology processes.[10] If so, actual real prices for synthetic gas would be fivefold (or more) the present level of natural gas prices.

It should be no surprise to learn that in 1982 only subsidized synthetic gas plants are being built. One such facility is the Barstow coal-gas-electricity plant.[11] Utah coal will be transported to California, gasified, and burned to generate electricity with steam turbines. The gas is also used to drive gas turbines for power generation. The financial arrangements are equally unusual. A petroleum company, Texaco, provides the conversion technology, and electric utilities provide at least $150 million investment. The instrument of this financial transfer is the Electric Power Research Institute, which is funded by customer charges throughout the country. Each American household will contribute $2 for what will be one of the most interesting efforts in synthetic technology to date. One-half of the project cost is subsidized by utility customers.

The economics of other synthetic oil and gas projects are comparable. Whether the source is coal or oil-bearing shale, these technologies cannot produce gas or petroleum at a cost comparable to the prices now charged for conventional oil and gas—even, again, given tax subsidies that place these projects in the position of having a negative tax effect on their builders.

It is evident that the economic feasibility of synthetic American oil and gas is being determined by Saudi Arabia, OPEC, and Mexico. Saudi oil, as analyzed in Chapter 5, sells at perhaps 60 times its production cost. U.S. oil prices now follow OPEC and Saudi prices. Increasingly, natural gas prices in the United States are being deregulated, and approach competing petroleum product prices on a dollar per million Btu basis. Eventually, declining oil resources in OPEC and Mexico will raise their prices to much higher real levels, and the prices (and costs) of U.S. oil and gas companies will rise. At some unknown future date, these new synthetic fuels will in fact be competitive with conventional oil and natural gas.

At this point the question of synthetic fuel economics becomes a question of the national economy. The economic dimensions of the beginning of this chapter must now be reviewed.

If automobile sales and suburban construction are depressed in the early 1980s because of the direct and indirect effects of high oil and natural gas prices, what will be the economic consequences of much higher real prices of synthetic fuels? The question becomes whether the transportation system itself can reasonably be viewed as realistic if it is expected to consist of

[10]Edward W. Merrow, Stephen W. Chapel, and Christopher Worthing, *A Review of Cost Estimation Technologies: Implications for Energy Process Plants*, Rand Corporation, July 1979.

[11]Described in the *New York Times*, December 14, 1981. Also see Dwain F. Spencer, Michael J. Gluckman, and Seymour B. Alpert, "Coal Gasification for Electric Power Generation," *Science*, 26 March 1982, pp. 1571–1576. See also footnote 37 below.

automobiles and airplanes that require liquefied coal, shale, or corn. The housing question is similar: can housing units requiring up to 100 M̄Btu each per year operate successfully on gasified lignite? The majority of knowledgeable observers in the early 1980s would probably think these new technologies are feasible on a large scale for individual customers and for the national economy. Lest my present judgment be unclear, I should note simply, I disagree.

Trains and Buses

The Boeing Aircraft Company finds that air travel is most energy efficient, and Amtrak finds that rail travel uses the least fuel per passenger-mile.[12] As noted in Table 4-3, Richard Rice, the former Greyhound Bus research director, reported that buses use the least fuel per passenger-mile. Such is the difficulty in making an independent evaluation of different kinds of transportation.

It is puzzling that in the early 1980s it is not possible to find a clear answer to the problem of fuel efficiencies of transportation modes. It is as if oil companies were planning their gasoline blends without knowing the requirements of existing automobiles.

Until the problem of independent research and evaluation is solved, the economic questions must be addressed with material little better than anecdotes. Suppose it is correct, however, that efficiently operated bus and rail provide passenger travel at less energy consumption and lower cost than do efficient auto and air travel. The United States has a higher percentage of travel by inefficient modes in the early 1980s than in the last year of the Growth Era.

There are exceptions in the United States. Los Angeles–San Diego trains carried 1.25 million passengers in 1980, a fourfold increase from 1974. In the Washington–Boston corridor, passenger travel is growing slowly, but nevertheless increasing. In general, however, rail passenger travel is not growing. Freight movement is rising slowly, but is still a smaller percentage of total freight than in 1973.

Although our information on them is superficial and economic data are absent, we know that new rail passenger lines are being developed in other countries. A congressional study reports passenger travel at average speeds of 100 mph already exists in Japan, France, Sweden, and West Germany.

[12]Boeing reports airplane travel at 18 to 28 passenger-miles per gallon and rail travel at 14-64 passenger-miles per gallon. Amtrak gives the same statistics at 55% seat occupancy as 22-33 for air and 64-157 for rail. Both estimates are discussed in Frank P. Mulvey, *Amtrak: An Experiment in Rail Service*, National Transportation Policy Study Commission, September 1978, pp. 64–73.

Switzerland, Canada, and Italy are building high-speed lines.[13] France, a country with one-fourth the population of the United States, provides rail travel for 10 times the number of intercity passengers that Amtrak carries.

The passenger rail system in the United States is not economically successful. On the average, operating expenses have been twice operating revenues. No passenger rail line has had operating revenues which equaled operating costs. Under present conditions, passenger revenue is inadequate to meet fuel and labor cost requirements for variable expenses, and provides no internal earnings for capital investment. One economic review of Amtrak concluded that, if fares were set to cover operating costs alone, they would be so high that rail passenger travel would disappear on most routes.[14]

One solution is to begin planning to expand rail travel. If rail service between major cities had a large volume, it would have lower average cost per passenger-mile. The congressional study advocates development of a passenger rail service throughout the country, and lists 20 major routes where its authors believe that population density justifies new rail development.

Urban passenger travel is, if anything, a greater problem. Population density is of particular importance in urban public transportation: the sparse suburban geography of major American cities makes rail travel less feasible and may make bus travel the only major alternative to automobile use in the near future. The problem is evident in comparing two cities at opposite ends of the density spectrum. Boston's 600,000 persons live with 12,000 persons in each square mile. Houston's 1.6 million people live 3,700 per square mile. Although gasoline has almost exactly the same cost in each city, Houston uses almost three times as many cars per family and more than 50 percent more gasoline per person. Boston's public transportation system carries 5 times as many passengers each day.[15] All large American cities in the South and West are closer to Houston's density than to Boston's.

Another problem which may be unique to the United States is the recent development of high-technology, low-use urban rail systems that are more energy-intensive than older rail systems. Charles Lave believes that the San Francisco BART system is so inefficient that emphasis should be shifted to the development of more efficient automobiles and buses for urban transportation.[16]

The Congressional Budget Office undertook an evaluation of requirements for different forms of urban transportation. (Each form of transportation is

[13]See U.S. Congress, Joint Economic Committee, "Case Studies in Private/Public Cooperation to Revitalize America: Passenger Rail," November 3, 1981, pp. 6–9, and *New York Times*, February 22, 1981.

[14]Mulvey, pp. 40, 126, 134.

[15]These data are from *New York Times*, December 9, 1979, and other sources.

[16]Charles Lave, "Negative Impact of Modern Rail Transit Systems," *Science*, 11 February 1977, pp. 595, 596.

referred to as a *mode*.) Table 14-4 summarizes part of their work. The figures for rapid transit rail, the fifth mode, are based upon the new San Francisco and Washington, D.C., systems. Observe, in column 4, that rail rapid transit requires more energy per passenger than old subways, bus travel, or van pools. Also observe (again in column 4) that typical urban auto travel uses several times the energy that the public transportation modes do.

In the last column in the table, all modes have been standardized on a 75 percent occupancy basis. The van pool falls to fourth in the ranking, and bus travel is most efficient.

The study is very thorough. For example, column 3 for manufacturing and construction estimates the energy used in manufacturing vehicles, roads, rail-beds, and bus and rail stations.

Nevertheless, the data in Table 14-4 must be qualified. It is not possible to know in the 1980s whether the present occupancy rates are guides to the future. The public transportation modes all have a present average occupancy rate of 33 percent or less. Will this be true in the future? The van pool is apparently efficient. However, van pooling requires neighborhood groups traveling to common distant work places. Is van pooling feasible on a large scale? The Congressional Budget Office study also examined the problem of access to public transportation. Bus travel is generally within walking distance, but the new San Francisco and Washington rapid transit systems frequently must be reached by auto.

Several economic factors will influence the relative efficiency of and demand for public and private transportation. Convenience and access is very significant, and automobiles and trucks will continue to be preferred to rail and bus on this criterion. Increased fuel efficiency in automobiles will be a second factor encouraging continued reliance on them. However, automobile fuel efficiency is directly proportional to the weight of the car. As cars become smaller, they also become less comfortable for long trips. This factor favors a shift from autos to air, bus, and rail.

Higher road maintenance costs are going to increase gasoline taxes (which pay for a high proportion of road upkeep), and the interaction of rising maintenance cost and declining gasoline sales will create particularly severe problems in some states. This factor will also work against highway vehicles.

Finally, the relative cost of different transportation modes may become significant. In 1979, rail freight was considerably less costly than truck and air freight per ton-mile. Rail and bus passenger travel were slightly less costly than air travel, and much less expensive than automobile costs per mile.[17] Figure 14-1 shows that price increases since the Growth Era have been particularly high for private transportation, and much less for public transportation.

[17]U.S. Department of Transportation, *National Transportation Statistics*, September 1981, pp. 42, 44, 104.

Table 14-4. Urban passenger travel energy requirements, Btu per passenger-mile

(1) Travel mode	(2) Fuel	(3) Mfg. & constr.	(4) Total pres. use	(5) Typical no. of seats per vehicle	(6) Present occupancy	(7) Energy at 75% occup.
Average car	7,860	2,300	10,160	5	28%	3,790
One-person car	11,000	3,220	14,220	5	20%	3,790
Van pool	1,560	460	2,020	12	75%	2,020
Bus	2,610	210	2,820	50	23%	865
Rapid transit rail	3,570	980	4,550	72	29%	1,770
Old rail, subways	2,540	560	3,100	72	33%	1,380

Source: U.S. Congress, Congressional Budget Office, Urban Transportation and Energy: The Potential Savings of Different Modes, December 1977, pp. 8, 20, 33, 35.

The density question introduced earlier is linked to several of these factors: access, occupancy rates, and cost. This may be the most important economic factor influencing the feasibility of rail and public transit.

In the Growth Era, America moved to a low-density suburban geography. Typically, this meant single-level houses with considerable automobile travel required to reach employment, schools, and commerce. Will the 1980s see movement to higher-density urban areas, residences closer to work, two- or three-story houses or apartments, and parks instead of lawns? It is precisely this speculative question that must be answered before transportation policies can be made.

The data on inflation and unemployment given earlier in this chapter indicate that future transportation decisions will have an extensive effect on the national economy. The question reappears in the next, concluding chapter.

Consider Western Europe to be France, West Germany, Sweden, Denmark, Norway, Belgium, the Netherlands, and Switzerland. Since the end of the Growth Era, Western Europe has continued to develop its public transportation, and GNP per capita moved ahead of the American level. In the United States, automobile and air transportation continued to dominate the other modes; the United States continued to surpass Western Europe in per-capita energy consumption, and has fallen from first place to ninth in per-capita GNP. Whether this is coincidence or causal relationship requires serious consideration.

Coal Potential in Rail Systems

Automobile, truck, and bus transportation all rely to a significant degree upon imported petroleum. Is it possible that rail transport could be reorganized on a different technological basis so that it could utilize American coal resources? There are two possibilities by which coal can be used for locomotive power: in electric engines, and in coal-fired steam engines. The latter technology was used throughout the 1800s and into the early 1900s. The former—electrified railways—are widely used throughout Europe and in some American cities.

Each technology has serious economic problems which place diesel locomotives in an economically preferred position. The relative efficiency of the engines has great significance: a diesel locomotive is about twice as efficient in extracting energy from its fuel source as is the coal steam locomotive.[18] The latter also has a higher capital cost and is more expensive in terms of

[18]This discussion is based upon data in Gibbs and Hill, Inc.. *Electrification Feasibility Study*, 3 vols., prepared for Consolidated Rail Corporation (Conrail), November 1979, and M. Dayne Aldridge, Thomas Campbell, and Alfred Galli, "Coal as a Rail Fuel: An Assessment of Direct Combustion," West Virginia University Research Center, August 15, 1980.

maintenance and fuel handling. The single advantage of the coal steam locomotive is the cost of the fuel. Coal energy on a Btu basis would in the 1980s
cost about one-fourth of the cost of diesel oil energy. In total, however, the
diesel locomotive that can now be purchased has a lower total cost per ton-
mile than the estimated cost of a possible coal steam locomotive. Very roughly, the Btu cost of diesel oil would have to be about nine times the present Btu
cost of coal before the coal steam locomotive would be less expensive on a
cents/ton-mile basis.[19]

Electric rail has greater energy end-use efficiency than either diesel or coal
locomotives. The major expense in expanding electric rail would be the
electrification of any route where electric locomotives are to be used. In part
because the tonnage and passenger travel carried by American railways are
less than those in Europe, original fixed investment is less affordable. Expansion of electric rail in the United States is unlikely until (1) higher diesel oil
and gasoline prices increase the cost of motor vehicle transportation, thereby
increasing the demand for rail transportation, and (2) higher diesel oil prices
make electric systems more competitive for the future heavier traffic rail
lines. Neither seems likely in the near future.

Solar, Wood, Hydro, and Renewable Energy

Solar energy means the adaptation of solar radiation for uses which have in
the United States been the ordinary province of conventional fossil and nuclear fuels. In theory, the annual potential solar energy reaching earth is 3
million Q, about ten thousand times the level of present total world consumption. Some of this energy is, of course, already used. It warms the earth and
provides the basis for the photosynthetic process in plants. In general, every
kind of renewable energy is probably derived from solar radiation.

The potential for efficient utilization of solar energy is one of the most
difficult economic problems in energy policy. As with nuclear power, advocates and critics of particular processes are defining the pathways of public
debate.

Table 14-5 reports the most optimistic estimates of solar energy potential,
published by the Solar Energy Research Institute (SERI).

Current use of renewable energy is estimated in Table 14-6. The contrast
between present and SERI-projected future renewable energy is great: a factor

---

[19]This conclusion is based upon my interpretation of the two studies cited above. Although
delivered low-sulfur coal at $1.79/MBtu ($44.34/ton) is much less costly than diesel oil at $7.97/
MBtu ($1.10/gallon), the difference in efficiencies is so great as to give the diesel locomotive the
lower fuel cost in dollars per thousand ton-miles. The electric locomotive uses less direct energy
than the diesel, but it loses efficiency in the electricity generation stage. In addition, the track
electrification investment is very sizable per thousand ton-miles. The net result is a clear advantage in lesser cost for the current diesel locomotive system.

Table 14-5. Optimistic estimates of renewable energy, 2000, in Q

Sector	Solar	Biomass	Wind	Photo-voltaic	Hydro-power	Total
Residential buildings	1.9	1.0	1.1	0.5	—	4.5
Commercial buildings	0.4	—	—	0.3	—	0.7
Industry	2.0	5.5	—	—	—	7.5
Agriculture	—	0.7	—	—	—	0.7
Transportation	—	5.5	—	—	—	5.5
Utilities	—	—	3.4	—	3.7	7.1
Total	4.3	12.7	4.5	0.8	3.7	26.0

Source: Solar Energy Research Institute (SERI), *Building a Sustainable Future,* 2 vols., Committee Print, U.S. House Committee on Energy and Commerce, April 1981, Vol. 1, Table 2.

of five. In residential heating, however, SERI projects no increase in wood burning. This, as one of the major current uses of renewable energy, is a subject for careful economic analysis later in this chapter.

Industry *biomass* in Table 14-5 refers to increased use of woodfuel as a substitute for fossil fuel in the forest industry's manufacture of pulp, paper, and wood products. Transportation biomass is gasohol. If the economic evaluation earlier in this chapter is correct, significant utilization of gasohol is unlikely. Wind energy in Table 14-5 is expected to be in the form of wind-generated electricity. This, too, is improbable on an economic basis, according to Sam Schurr.[20] The photovoltaic transformation of solar energy into electricity is also not economically feasible under presently forseeable circumstances. The process would apparently require a tenfold reduction in the cost of photovoltaic cell manufacture to become economically competitive.[21]

The level of hydropower generation in the early 1980s is unchanged from the early 1970s, an annual 275 to 300 billion kWh, depending upon precipitation patterns. In assessing aggregate consumption, it is common to measure hydropower in terms of the fossil fuel it displaces at about 10,400 Btu/kWh. So this level of hydropower generation is equal to 3 Q per year. Since higher conventional and nuclear generating costs make expensive small hydro generation more competitive, the SERI increase of 75 billion kWh by the end of the century is plausible.[22]

[20]See Sam H. Schurr et al., Resources for the Future, *Energy in America's Future: The Choices before Us* (Baltimore: The Johns Hopkins University Press, 1979), pp. 311–313.

[21]Ibid., pp. 308, 309. Also Robert Stobaugh and Daniel Yergin, eds., *Energy Future: Report of the Energy Project at the Harvard Business School* (New York: Random House, 1979), pp. 258–262. Atlantic Richfield's Solarwest Electric (Santa Barbara) advertises a photovoltaic system costing $15,000 which can provide 3.8 kWh per day in the Southwest or 2.4 kWh per day in the Northeast. For the Southwest, the levelized capital cost without installation or maintenance is $1.73/kWh. Smaller systems have a higher cost per kWh.

[22]Small hydropower projects are expensive. A 3-megawatt facility being built in Ithaca, New York, will cost $3.8 million and generate 8 million kWh annually. Computed at the same 15.5% fixed charge rate used in Chapter 11, the capital cost alone is 7.4¢/kWh.

Table 14-6. U.S. use of renewable energy, 1982, in Q

Woodfuel in residences	1
Forest by-product use	1
Hydropower	3
Solar space and water heating	0.05
Total	5

Note: Hydropower generation is expressed in terms of conventional fuel displaced. This is about 10,435 Btu/kWh.

Sources: U.S. DoE, *Monthly Energy Review,* February 1980 and November 1981, and *Estimates of Wood Energy Consumption,* 1982; National Academy of Sciences, *Energy in Transition 1985–2010,* Final Report of the Committee on Nuclear and Alternative Energy Systems, 1980, p. 351.

Having ruled out much of the SERI optimistic projection on the basis of microeconomic analysis, I must note the remaining applications of renewable energy. These are woodfuel for residences, and solar heating for residential and commercial buildings. The problem here is to delineate the social and market economics of these heating sources, and to provide some boundary for public policy choices.

Solar space-heating techniques are often divided into two broad categories: passive and active. *Passive solar heating* uses a building itself for solar heat collection and storage. Many passive methods of solar heating are also a form of conservation because these methods simultaneously reduce heat loss and increase the utilization of solar energy. For example: locating windows on southern rather than on wind-chilled northern and western sides of buildings reduces the amount of interior heat lost through windows. In addition, solar energy use is increased by having windows that face south. Creating an area of a few hundred square feet of south-facing windows is termed *direct gain.* This design technique may be the simplest kind of solar heating, and actual new houses with this design in the northern United States are reported to have a third of their heat energy supplied by this method.[23]

Thermal storage walls are large structures using concrete, sand, or water to store heat. One design uses seventeen-foot tubes, each tube holding nearly one ton of water. This design is apparently efficient in the sense that it captures, stores, and releases solar energy. However, some persons may not be wholly comfortable sharing a room with several tons of water.

Greenhouses or sunspaces are a third form of passive solar heating that capture and store solar heat. They do not appear to be economical if judged solely on a $/MBtu basis.

One result of the design analysis research that has been undertaken on passive solar heating is to emphasize the interaction of conservation and solar energy. If the heating requirement is reduced from 100 MBtu annually to

[23]SERI, 2:155–163.

20–30 M̄Btu, passive solar systems can supply a large fraction of total energy.

Active solar heating is the term used to describe the collector systems which heat and circulate air or water throughout a building. Table 3-8 compared the economics of solar space heating with other heating methods in Long Island, New York. Recall that the solar system cost received two considerable tax subsidies: a 40 percent federal income tax credit, and an exemption from local property taxation. The solar tax subsidy shown there is much more than the personal tax subsidies received by any other home heating system. Nevertheless, solar space heating is more costly to the owner than are natural gas or heating oil systems, although it is less so than either electric resistance or electric heat pump systems.

If social cost is to be taken into account, then the corporate tax subsidies described in Chapters 11 and 12 as applicable to coal and nuclear power plants must be considered as well. When Kathleen Cole included these corporate subsidies in her assessment, she found that the solar space-heating system had the highest total social cost. For example, she found the levelized cost of a solar heating system with an electric resistance back-up to be about $3,400, and the unsubsidized cost to be about $5,300. The least expensive system was natural gas: $1,300 levelized cost per year with subsidies, and $1,700 per year without subsidies.[24] This illustration is instructive. Active solar space heating in and around New York City will probably not be economical on a customer cost basis or a social cost basis in the forseeable future.

In other areas, however, active solar heating is economically feasible. It is, indeed, feasible on Long Island, for water heating. The important difference between space and water solar heating is that a solar system can provide 67 percent of the annual 20 M̄Btu requirement for heating water, but only 50 percent of the annual 84 M̄Btu required for space heating. The cause is climatic: hot water is used throughout the year, and solar energy can provide almost all the necessary heating energy in the summer. In contrast, the need for space heating is greatest in the winter when received solar energy is lowest.

Cole found the levelized cost of solar hot water systems to be less than electric hot water for both customer costs and social costs.[25] However, natural gas continued to be less expensive. Apparently, natural gas prices will have to double in real dollars before a solar hot water system with a natural gas back-up becomes less costly than ordinary natural gas hot water systems. As we saw in Chapter 8, however, this doubling in real natural gas prices will eventually occur.

[24]Kathleen Lynn Cole, "Tax Subsidies and Comparative Costs for Utilities and Residential Heating in New York," M.S. Thesis, Cornell University Department of Agricultural Economics, 1981, p. 148.
[25]Ibid., p. 152.

This summary of the complicated economics of active solar space and water heating leads to an important conclusion: Active solar heating may be geographically limited because of variations in solar insolation, heating energy requirements, and the costs of competing energy sources. Figure 14-2 maps the amounts of solar radiation received throughout the country. It uses a metric unit, the langley, which is one calorie per square centimeter. This is equal to 3.7 Btu per square foot. I expect that active solar space heating is economically competitive in the 500-plus region: Arizona, New Mexico, and part of California, Nevada, and Texas. Solar hot water systems may be economically efficient in the 450-plus regions, a much larger area, covering the southwestern United States and southern Florida.

Generally speaking, where natural gas is available, it will be less costly further south of the map lines. Electric resistance space and water heating will be more costly further north of these solar energy lines. As natural gas, oil, and electricity become more costly in the 1980s and 1990s, the boundaries of comparative advantage will move northward.

These generalizations are not wholly speculative: Roger Bezdek and colleagues investigated the comparative economics of regional solar space and water heating in 1979.[26] Their in-depth study evaluated apartments and houses in four locations: Colorado, Los Angeles, Boston, and Washington, D.C. They found that solar space heating was not competitive with any conventional heating system in single-family homes in any location. Neither solar space or water heating was competitive with natural gas systems in 1979 in any location in either houses or apartments.

The positive comparisons found that solar water heating was less costly than electric water heating in both homes and apartments in Los Angeles and in the Colorado location, Grand Junction. In these positive comparisons, however, the edge to solar hot water disappeared if the solar tax credit was eliminated. For Los Angeles, on the other hand, solar space heating has the economic advantage against electric space heating in apartments, even without a solar tax credit.

Cole analyzed water-heating costs in Tucson and Miami Beach. She found that, if the solar system can provide all of the hot water, then it is less costly in these cities in a new house than any other form of water heating. But there are only about 300,000 solar hot water systems in the United States, probably concentrated in California, Florida, New Jersey, and New York.[27]

The puzzling question is the absence of new solar installations in those regions where they are competitive on a customer market cost basis. New

[26]R. H. Bezdek, A. S. Hirshberg, and W. H. Babcock, "Economic Feasibility of Solar Water and Space Heating," *Science*, 23 March 1979, pp. 1214–1220.

[27]Cole's analysis of solar water heating in Tucson and Miami Beach was provided for this chapter. The estimate of installations is via personal communication, Theresa Flaim, Solar Energy Research Institute, January 1982.

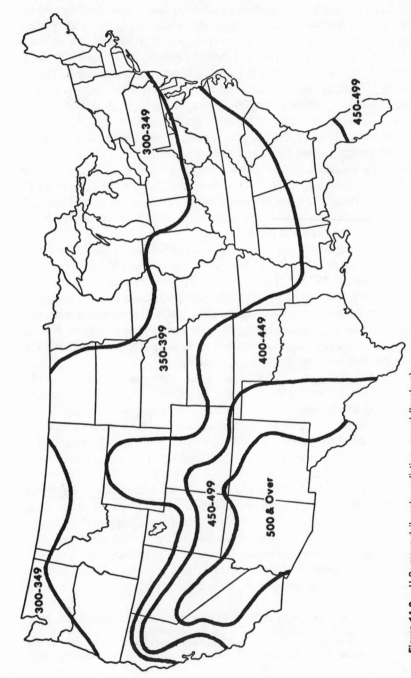

Figure 14-2. U.S. mean daily solar radiation, annual (langleys). One langley equals one calorie per square centimeter, or 3.7 Btu per square foot. *Source:* U.S. Environmental Data Service, *Weather Atlas of the World,* 1975.

electric space and water systems are being installed in two-thirds of the apartments and one-half of the houses being built.[28] The primary reason is evident in Table 3-8. The purchase costs of $10,000–$15,000 for a solar space system and $3,000 for a solar water system make the solar processes many times more expensive in original capital investment. As Chapter 3 concluded, builders will weigh public preference and initial cost. In the absence of buyer preference for solar systems, builders will install the system which costs them least. This will always be electric resistance space and water heating.

A second obstacle to the development of solar energy in feasible locations is public perception. In much of the region where solar installations are competitive, air conditioning is more important than either space or water heating. It may be that the natural attention given to air conditioning has partially obscured interest in solar water heating.

The 4.3 Q of solar heating in Table 14-5 may be economically justified by the year 2000, but a major change in public and builder attitudes must precede this growth from less than .05 Q in 1982 to the 4.3 Q level by the end of the century.

Wood Heating

The economic factors that determine the comparative cost of wood heating are the impact of air pollution, the availability and cost of wood, and the cost of conventional heating systems. Wood heating receives no tax credit, making it the only energy source not so promoted. Nevertheless, wood heating has grown rapidly since 1973 and has surpassed the one Q level.

The air pollution problem has not yet become significant. Although sulfur is absent in wood, particulate and unburned gas emissions are very high. Whether home wood burning will come to have air pollution regulation will depend upon its future growth.

Michael Slott has studied the economics of wood burning in a manner comparable to the Cole analysis discussed earlier.[29] His assumptions and results are given in Table 14-7. Observe that regulated natural gas would be the least costly over the 20-year period. In Slott's analysis, regulated natural gas means the price increases at 12½–13 percent annually, parallel to price increases in other fuels. Unregulated gas means that the Btu price of gas rises to the oil price in the base year, $7.18/MBtu. Unregulated natural gas in an

---

[28]U.S. DoE, Energy Information Administration, *Monthly Energy Review*, March 1980, pp. i–v.

[29]Michael Slott, "Economic Aspects of Woodfuel Use in New York State," M.S. Thesis, Cornell University Department of Agricultural Economics, 1982. Printed as Cornell University Department of Agricultural Economics A.E. Res. 82-11, February 1982.

Table 14-7. Wood heating economics, Central New York, 1981–2001

Type of system	System purchase cost	System efficiency	Fuel cost $/M̄Btu 1980	Cost total annual uninsulated	Cost total annual insulated
N.G. (regulated)	$2500	.8	$4.02	$1930	$1452
N.G. (unregulated)	2500	.8	7.18	3153	1764
Coal	3500	.7	3.70	2160	1642
Wood, purchased	3000	.5	4.40	3198	2124
Wood, free	3000	.5	0	405	405
Heat pump	5500	1.8	15.38	3468	2446
Electric resistance	1800	1.0	15.38	5030	2950
Oil	3200	.7	7.18	3654	2367
Insulation	2000		4.92		

Note: Insulation reduces annual energy requirements from 116 M̄Btu annually to 58 M̄Btu.
Source: Slott, 1982. Slott's energy requirement estimates differ slightly from Cole's estimates, shown in Table 14-8.

uninsulated house has an annual equivalent cost of $3,153 per year. As would be expected from Chapter 3, oil and electric heating costs are also high, matching unregulated natural gas in the $3,000-plus range.

The surprise in Table 14-7 is the lowest-cost ranking gained by coal burning. If Slott is correct on this point, it should be expected that coal may become widely used as a residential fuel in areas where air pollution regulations permit.

The focus in Slott's work is on wood burning. He finds that if well-seasoned wood is purchased at $100 per cord, and yields 22.5 M̄Btu/cord, the cost is in the $3,000 annual range in the future.

The growing use of woodfuel is probably explained by the free wood entry. Here, the annual cost is only the after-tax payment on the stove installation, or $405 per year regardless of insulation. "Free wood" is unrealistic. Costs are involved in chain saws, gasoline, hauling, and hospital bills. But skilled woodcutters, for whom these would be only a few dollars per cord, can obtain home heating at a very low money cost. In rural areas where wood is accessible at little cost to the woodcutter, it is certain that wood burning for space heating will continue to increase. The present national estimate of one Q is certain to rise. The SERI projection cited in Table 14-5 of nonincreasing woodfuel usage seems unlikely. Woodfuel use will grow.

Wood has also been studied as a fuel source for electric utilities. Experience to date in Vermont and Wisconsin indicates that it costs utilities about the same as coal on a $/M̄Btu basis. The sulfur content of woodfuel is insignificant although particulate control may be more difficult. It is premature to decide whether woodfuel will be competitive with coal for rural utilities. It is feasible in some areas now, but sufficient experience with it has not yet accumulated to permit an evaluation of its potential in wide use.[30]

Before we conclude our consideration of woodfuel economics, it should be emphasized that Slott (as did Cole) reports that insulation makes a considerable difference. A $2,000 investment reduces annual space-heating energy requirements by one-half, to 58 M̄Btu. Over the 20-year period, insulation saves more than $1,000 annually for the oil and electrically heated homes, for the homes using purchased wood, and for a home using unregulated natural gas. Given the present economic circumstances for renewable and conventional energy resources, it seems clear that insulation is economically efficient.

In this section, a review of economic data shows that wood burning is economically competitive with conventional fuels in much of the rural temperate regions of the country. The preceding section reviewed the economic data on active solar heating, and speculated that the geographic margin of competitive economics is in the Southwest and Florida, and that this margin is moving northward.

Taken together (because wood belongs in the solar group), the data make clear that renewable solar energy has a growing role among American energy resources.

Conservation and Space Heating

The solar section reported the interaction between passive solar techniques and conservation; the last section observed that insulation is economically efficient. The chapter concludes with a review of the most important new residential energy technology, conservation.

A basic component of engineering economics for residences is the annual energy requirement. This has been referred to in earlier chapters, especially Chapter 3. The basic relationship is defined by Equation (1):

$$(1) \qquad\qquad E = 24 * DD * HL$$

E is the annual space-heating energy requirement, measured in million Btu per year. DD is *degree days,* a measure of coldness for every locality. A degree day represents the difference in temperature between a typical residence at 65°F and the outside temperature. Take a typical winter average temperature over 24 hours in San Francisco: 40°F. This gives 25 degree days,

[30]The cost per M̄Btu for wood-burning power plants is regularly reported in the monthly U.S. DoE, Energy Information Administration's *Cost and Quality of Fuels for Electric Utility Plants.* Also analyzed by Slott in his chap. 6.

calculated by 65°–40°. These degree days summed over a year measure the annual coldness. Degree days may be 3,000 in San Francisco, or 8,900 in North Dakota.

The *heat loss* HL in Equation (1) is measured in Btu per hour which escape from the residence with a one-degree differential between inside and outside temperature. The number 24 (from hours per day) in Equation (1) is necessary to complete the numerical conversion to Btu per year.

As an illustration, assume an old San Francisco apartment with a heat loss of 1,000 Btu per hour for each degree Fahrenheit below 65°F. Using the equation, and 3,000 degree days per year for San Francisco, the result is $E = 72$ MBtu space-heating requirement.

In contrast, consider an old, poorly insulated house in central New York with the same heat loss of 1,000 Btu per hour per degree below 65°F. With 7,000 degree days in central New York the result is E at a 168 MBtu annual space heating requirement.

The heat loss is itself a sum of two types of loss: conduction through surfaces and *air loss* through cracks in or around doors, windows, ceilings, and walls.

Conduction loss is measured as the amount of heat energy passing through a square foot of surface if the outside temperature is one degree Fahrenheit below the inside temperature. An insulated wall, for example, may have a conduction loss of 0.045 Btu per square foot per degree. This is its *U-value*. An uninsulated wall may have a U-value of 0.23.

A final concept used in measuring insulation is the *R-value*. This measures the resistance to heat loss. Its numerical value is simply 1/U. So an insulated wall with a U-value of 0.045 has an R resistance rating of 22.2. Obviously, more insulation translates into higher R, lower U conduction loss, lower *HL* heating loss, and lower *E* space-heating energy.

One consequence of the growing popular interest in solar energy and conservation is the increasing effort to explain these basic concepts—and one of the resulting publications is the basis for this summary here.[31]

With this introduction, we can analyze the microeconomics of conservation for space heating for an individual house. Table 14-8 is taken from Kathleen Cole's work.[32] It calculates space-heating requirements for different levels of insulation. For the uninsulated house, heat losses per hour sum to 1,449 Btu/h. Equation (1) is used to estimate the annual space-heating requirement. Central New York has 7,000 degree days, and the result is an annual space heat requirement of 243 MBtu.

---

[31]Bruce Anderson, *The Solar Home Book* (Harrisville, N.H.: Brick House Publishing, 1976). The same subject is covered, although with less detailed explanation, in Helga Olkowski, Bill Olkowski, Tom Javits, et al., *The Integral Urban House* (San Francisco: Sierra Club Books, 1979).

[32]Cole, Appendix C.

Table 14-8. Heat loss, Btu/hour, Btu/year, annual levelized cost

Heat loss from—	Uninsulated house	Medium insulation	Strong insulation
Doors and windows		With storm windows, doors	Same: storm windows, doors
air loss	254 Btu/h	71 Btu/h	71 Btu/h
conduction loss	138 Btu/h	81 Btu/h	81 Btu/h
Conduction losses			
walls	262 Btu/h	3.5 inches insulation 79 Btu/h	6 inches insulation 51 Btu/h
ceilings	375 Btu/h	6 inches insulation 75 Btu/h	9 inches insulation 48 Btu/h
floor with carpet	420 Btu/h	388 Btu/h	3.5 inches insulation 101 Btu/h
Total heat loss per hour per °F (HL)	1449 Btu/h	694 Btu/h	352 Btu/h
Total heat, annual, 7000 DD ($E = 24 * DD * HL$)	243 M̄Btu	117 M̄Btu	59 M̄Btu
New home, insulation purchase cost	$0	$2,000	$4,000
Levelized cost	$0	$2.25/M̄Btu	$3.09/M̄Btu
Retrofit: assume 50% higher annual cost		$3.38/M̄Btu	$4.64/M̄Btu

Note: This material is from Cole, Appendix C. The mortgage factor is 14.2% with 13% interest. As before, levelized cost equals the insulation purchase cost times the mortgage factor, this product being divided by the reduction in space heating. Example: for strong insulation in the house, $4000 * .142/184 = $1.89/M̄Btu. This does not include the tax effects. Table 14-7 shows insulation cost in going from medium insulation to strong insulation as $4.92/M̄Btu. The comparable figure from this table would be $4.90. That is calculated by: ($4,000 − $2,000) * .142/58 = $4.90.

Suppose natural gas costs $5/M̄Btu and the furnace is 80 percent efficient. The furnace will require 304 M̄Btu of natural gas, and the annual heating bill will be $1,520. Electric resistance heating at the same M̄Btu energy level would be 100 percent efficient, but at a cost of $15.40/M̄Btu the annual bill is $3,742. Very few houses in central New York are completely heated by electricity and wholly uninsulated.

The basic changes made to achieve medium insulation are the addition of storm windows and doors, 3½ inches of insulation in the walls, and 6 inches in the ceilings. Without insulation, the walls have an R rating of R-4. The wall insulation itself is R-11 but fills an R-1 air space. The total wall value is R-14 (i.e., R-4 plus R-11 minus R-1). Similar calculations are made for ceilings and floors. Uninsulated ceilings are R-4, and uninsulated floors are R-4.5.

Inverting the R-values gives the U ratings for walls, ceilings, and, so on. With the U-values of energy loss per square foot per °F temperature differential, the hourly heat losses are calculated. Take, for example, the 375 Btu/h heat loss through ceilings in the uninsulated house. This is found as follows: The ceiling is R-4, so U is 1/R, or .25 Btu/h/ft. The house has 1,500 square feet of ceiling (and floor) space, so the hourly heat loss value is 375 = 1,500 * .25. The other values in the table are constructed with the same relationships.

Observe that moderate insulation reduces the annual requirement to 117 M̄Btu. Using the levelized cost relationship for the 13 percent interest rate, we find that moderate insulation costs the homeowner $2.25/M̄Btu. This may be compared to the fuel costs per M̄Btu in the preceding table.

Strong insulation further reduces heat loss. Wall insulation is R-22, ceilings are R-31, and the floor is R-15. The result is that 59 M̄Btu of space heat is needed. Insulation here has a levelized cost of $3.09/M̄Btu. This increased insulation is also clearly economically justified. It should be remembered that insulation once installed is immune to inflation. Fuel prices do not have the same stability.

The same analysis can be applied to apartments and condominiums and to old and new residences. In a broad generalization, one could say that insulation costs in apartments are probably about 25 percent lower per M̄Btu. Insulation costs in old houses are probably 30–50 percent higher than the new home costs shown in Table 14-8.

The data in that table can also be generalized to other locations by use of Equation (1). Washington, D.C., for example, has 4,200 degree days. The estimate would be 35 M̄Btu annually for strong insulation.

The same method can be applied to the degree days for any locality given in Table 14-9. Degree days for nearby locations can be estimated from the values in this table. With this information, a general guideline of the value of insulation can be calculated for any residence in the United States. The qualitative conclusion will always be the same: conservation of space heat by insulation is always economically efficient for every part of the United States where space heating is important.

As natural gas and other energy prices increase, levels of insulation beyond the strong insulation level will become economically efficient. Recall that strong insulation meant R-22 walls, R-31 ceilings, and R-15 floors. Take this level of insulation and add: electronic furnace ignition in place of gas pilot, insulated furnace ducts, triple pane windows, south-facing windows, and a heat exchanger. The SERI study discussed above concluded that such a house with somewhat less insulation and in a typical location would require just 13 M̄Btu per year for space heating.[33] Other studies of space-heating conservation conclude that potential reductions in space heating are of this magnitude. The General Accounting Office finds that an electrically heated house built in

[33]SERI, 2:111. The assumed location is in Washington, D.C., which has a winter climate typical of the country.

Table 14-9. Degree days in U.S. cities

Alabama	Birmingham	2,600	Nevada	Las Vegas	2,700
	Mobile	1,600		Reno	6,300
Alaska	Anchorage	10,900	New Hamp.	Concord	7,400
Arizona	Phoenix	1,800	New Jersey	Atlantic City	4,800
	Prescott	4,400		Trenton	5,000
Arkansas	Little Rock	3,200	New Mexico	Albuquerque	4,300
Calif.	Los Angeles	2,100	New York	Albany	6,900
	San Diego	1,400		Buffalo	7,100
	San Francisco	3,000		New York City	
Colorado	Denver	6,300		(Central Park)	4,900
Conn.	Hartford	6,200	N. Carolina	Charlotte	3,200
	New Haven	5,900		Raleigh	3,400
Delaware	Wilmington	5,000	N. Dakota	Bismarck	8,900
Florida	Jacksonville	1,200	Ohio	Cincinnati	4,800
	Miami Beach	100		Cleveland	6,400
	Orlando	800		Columbus	5,700
Georgia	Atlanta	3,000	Oklahoma	Oklahoma City	3,700
Idaho	Boise	5,800		Tulsa	3,900
	Pocatello	7,000	Oregon	Portland	4,600
Illinois	Chicago	6,200	Penn.	Philadelphia	5,100
	Springfield	5,400		Pittsburgh	6,000
Indiana	Indianapolis	5,700	Rhode Is.	Providence	6,000
	South Bend	6,400	S. Carolina	Columbia	2,500
Iowa	Burlington	6,100	S. Dakota	Sioux Falls	7,900
	Des Moines	6,800	Tennessee	Memphis	3,200
Kansas	Dodge City	5,000		Nashville	3,600
	Wichita	4,600	Texas	Dallas	2,400
Kentucky	Louisville	4,700		El Paso	2,700
Louisiana	New Orleans	1,400		Houston	1,400
Maine	Portland	7,500	Utah	Salt Lake City	6,100
Maryland	Baltimore	4,700	Vermont	Burlington	8,300
Mass.	Boston	5,600	Virginia	Norfolk	3,400
	Worcester	7,000		Richmond	3,900
Michigan	Detroit	6,200	Washington	Seattle	4,400
Minnesota	Duluth	10,000		Spokane	6,700
	Minneapolis	8,400	Washington	D.C.	4,200
Miss.	Jackson	2,200	W. Virginia	Charleston	4,500
Missouri	St. Louis	5,000		Elkins	5,700
Montana	Great Falls	7,800	Wisconsin	Green Bay	8,000
	Missoula	8,100		Madison	7,900
Nebraska	Norfolk	7,000		Milwaukee	7,600
	Omaha	6,700	Wyoming	Cheyenne	7,300

Source: U.S. Environmental Data Service, *Weather Atlas of the World,* 1975, p. 196.

the future in the South should be so insulated as to require only 6 M̄Btu or less per year. This is about one-tenth present usage. A 1981 Oak Ridge National Laboratory study gives the energy requirement for a well-insulated house as 29 M̄Btu annually.[34]

[34]From U.S. General Accounting Office, *Analysis of Trends in Residential Energy Consumption,* July 9, 1981, p. 55, and Dennis L. O'Neal and Judy L. Jones, "An Energy and Economic Evaluation of the Single-Family Residential Building Energy Performance Standards," Oak Ridge National Laboratory, November 1981, p. 13. Their value of 216 MJ/m^2 is—I hope— correctly translated as 29 M̄Btu annually for a 1,500 ft^2 house also located in Washington, D.C.

The evidence is clear. With existing methods, space-heating requirements can be reduced to one-eighth those of pre-1973 houses in the northern United States. Although the level of technological knowledge in conservation air conditioning is just being developed, it is probable that air-conditioning requirements can be reduced by a similar magnitude.[35]

Our apartments and houses now use 17 Q total energy each year. Ten Q are used for heating and cooling, 5 Q for appliances, and 2 Q for hot water heating. Several studies find that conservation and increased efficiencies are also economically justified in appliances and hot water heating.[36]

The general implication is that conservation technologies are economically efficient, and that microeconomics favors expansion of energy-conserving technologies.

Industry Organization

There is no single industry that embraces the resources, technologies, and corporations studied in this chapter. Table 14-10 summarizes the activities of the major oil companies in four areas.

Four companies have been active in gasohol sales. Texaco has been the nation's leading retailer. As indicated in this chapter, the gasohol tax subsidies are particularly attractive to petroleum companies in years of high profitability. This, of course, is in addition to the established distribution and marketing systems which oil companies possess for gasoline.

Coal gasification and liquefaction are of interest to oil companies because these fuels would compete with conventional natural gas and petroleum. Table 14-10 (adapted from Chapter 6) shows only Tenneco and Amerada Hess without coal resources. While the major oil companies are obviously in a position to develop synthetic fuels, Chapter 10 has shown that petroleum companies are only one group of participants in the coal industry. It is unlikely that, in present circumstances, major oil companies would have the potential or motivation to establish a monopoly position in synthetic fuels.

The role of public subsidies is crucial here also. Each of the three major projects now being developed involves oil and gas companies, and each receives considerable public subsidy.[37]

[35]O'Neal and Jones, p. 25; SERI, 2:125.

[36]SERI, 2: 137–150; U.S. GAO, pp. 58–60.

[37]The Barstow, California, coal-gas-electricity project discussed in this chapter is based upon a Texaco coal gasification system. The developers are Texaco, Bechtel, General Electric, and Southern California Edison. The Electric Power Research Institute provides $150 million, which is collected from utility customers under the approval of regulatory authorities. A coal gasification plant similar to the illustration in Table 14-3 is under construction by a consortium of the American Natural Resources Company, Tenneco, and two other companies. This project has guaranteed federal financial support if it is unprofitable. This form of federal assumption of losses, if any, is also provided to the major shale oil effort involving Union Oil of California. All three projects, then, include a major oil company and public financial subsidy. Sources: American Natural Resources Company, Securities and Exchange Commission Form 10-K Report; *New York Times*, December 14, 1981, and November 6, 1981.

Table 14-10. Major petroleum company participation in new energy resources, 1980–82

Company	Gasohol	Coal resources (basis for coal liquids and gas)	Copper resources (used in active solar heating)	Photovoltaic cell research
Exxon		X		X
Mobil		X		X
Texaco	X	X		
Standard/California	X	X	X	X
Gulf		X		
Standard/Indiana	X	X	X	
Atlantic Richfield		X	X	X
Shell		X		X
Conoco/Du Pont		X		X
Phillips	X	X		X
Tenneco				
Sun		X		
Occidental		X		
Standard/Ohio		X	X	X
Getty		X		
Union/California		X		
Marathon/U.S. Steel		X		
Ashland		X		
Amerada Hess				
Cities Service		X	X	

Sources: New York Times, December 14, 1981, on gasohol; Table 6-4 on coal resources; Table 14-11, below, on copper resources; and Annual Reports on solar activities.

Copper is an important raw material for active solar systems, and is a potential material for use in photovoltaic cells. America's largest copper producers are owned by major oil companies. Table 14-11 lists the copper companies now owned by oil companies, and their 1980 production. (Both production and consumption have declined since 1980.) Evidently, five major oil companies were responsible for 73 percent of American copper production. If active solar space and water heating develop in the Southwest and Florida, these companies are in a potential position to influence that development.

Photovoltaic cells are not commercially feasible on a large scale in the forseeable future. The research activities noted in Table 14-10 pose no problem for potential monopoly influence at present. There is no significant activity on the part of major petroleum companies in other solar energy applications in the early 1980s.

Overall, the only important problem here is one of potential rather than present reality. The dominant position of major oil companies in U.S. copper production should be considered in studying public policy toward solar energy development. At present, there is no evidence to suggest that solar heating is

Table 14-11. Major petroleum companies and U.S. copper production, 1980

Petroleum company	Copper company	1980 mine production (tons)
Standard Oil of Ohio	Kennecott Minerals	335,914
Atlantic Richfield	Anaconda Copper	212,318
Standard Oil of Indiana	Cyprus Mines	194,153
Cities Service	Tennessee and Miami Copper	105,600
Standard Oil of California (20% ownership of AMAX)	AMAX	55,582*
Total copper production by 5 companies		903,567
Total U.S. mine production was 1,293 thousand American tons; the 5-company proportion was 73%.		

*This figure does not include AMAX's production of 127,942 tons of recycled copper.
Sources: Annual Reports; U.S. Department of Commerce, *Survey of Current Business; New York Times,* March 18, 1982.

being either restricted or prematurely developed because of the position of oil companies in copper production. The depressed status of copper mining and the absence of growth in solar heating interact to place the problem, if it ever exists, in the future.

Summary and Conclusions

On economic criteria, the development of new liquid and gaseous fuels is not feasible in the near future. Gasohol and coal gasification each receive very great public subsidies, and each of these processes involves major oil companies as participants. Table 14-12 summarizes the data examined in the microeconomic comparison. Ethanol production cost is 50¢/gallon more than the refinery cost of conventional gasoline. However, the subsidy in the form of investment tax credits and tax exemptions is approximately $1 per gallon. This makes gasohol marginally competitive for major oil companies.

Corn-based ethanol is the most feasible source of renewable liquid transportation fuel. It seems unlikely that automobiles, trucks, and airplanes would be able to operate at present levels if gasohol were to be a major source of fuel. One-half of the domestic U.S. corn crop could provide 5 percent of the national gasoline consumption.

Coal gasification is the most feasible form of producing synthetic fuel from coal. The cost of synthetic gas would be, without subsidies, about five times the present cost of natural gas. Increasing natural gas prices will narrow this margin, but synthetic gas will not be economical on a commercial basis in the near future.

Although gasohol and synthetic gas were analyzed in this chapter because they are the most feasible forms of renewable and synthetic new energy forms, this should, perhaps, be restated as a conclusion that gasohol and synthetic gas are less infeasible. By inference, and on the basis of other information, we can say that such other new energy forms as liquefied coal, shale oil, and renewable natural gas are also not likely to be economic on a large scale in the forseeable future. A similar conclusion arises from a review of existing data on power generation from solar energy and windmills. Both are much more costly than power generation from new coal or hydro plants.

Three new technologies using renewable energy have an economic basis for expansion. These are wood burning in the rural forest regions of the country, hydropower generation, and active solar space and water heating in the Southwest and in Florida. Limited increase in direct gain and other passive solar designs is economically justified throughout the country. Present renewable energy use is about 1 Q each for wood fuel providing space heat in residences and energy for the wood products industry. Hydropower generation is 300 billion kWh annually, which displaces 3 Q of conventional energy. The direct solar energy contribution is currently less than .05 Q. Significant increase in each category should be expected in those specific regions where these renewable technologies are less costly than conventional technologies.

I conclude that it is unlikely that new supply technologies will be adequate

Table 14-12. Illustrative microeconomic comparisons of current and alternative technologies, early 1980s

Economic activity	Current technology	Alternative technology
Automobile fuel, cost at refinery	Gasoline $1.00/gallon	Ethanol $1.48/gallon
Gas, cost to customers	Natural gas $4/M̄Btu Unregulated $7/M̄Btu	Synthetic gas from coal $16.37/MBtu
Energy requirement for urban travel, 75% occupancy	Auto 3,800 Btu/pm Bus 900 Btu/pm	Old trolley, subway 1,400 Btu/pm New rapid transit 1,800 Btu/pm
Freight cost, intercity, 1979	Air 41¢/tm Truck 13¢/tm	Rail 3¢/tm
Future heating cost, 1981–2000, annual cost including installation, 58 M̄Btu per year	Electric resistance $3,000 Electric heat pump $2,400 Natural gas, deregulated $1,800 Oil $2,400	Insulation $283 Wood, self-cut $400 Wood, purchased $2,100 Solar, 85% eff. $1,900 Solar, no tax credit $2,400

Note: The solar entries are adapted from Cole, and assume 85% efficiency which may be appropriate in the Southwest and South Florida. The system is assumed to cost $14,000, and use electric resistance back-up heat for 9 M̄Btu. Values are shown with and without the solar tax credit. The estimate for wood, self-cut, equals the free wood cost of $203 in Table 14-7, plus $197 as an allowance for the saw, truck, and gasoline expenses in cutting five cords per year. Freight cost data are from U.S. Department of Transportation, *National Transportation Statistics.* Other data are from tables in this chapter.

in terms of economic efficiency. It will probably be necessary to reorganize transportation and space heating in order to maintain satisfactory living standards. Rail freight and passenger transportation will need to be developed, and greater use made of urban, suburban, and intercity buses, because bus and rail systems have less energy requirements per passenger-mile and per ton-mile than do air, automobile, or truck modes.

Although electric rail systems are widely used in Europe and in U.S. cities, major expansion of electrified rail for intercity transportation is probably not feasible in the near future. At some period, higher petroleum prices will increase the cost of motor vehicle and air transportation sufficiently to cause increased demand for rail transportation. The joint effect of higher rail traffic and higher diesel oil cost will at some point make increased rail electrification feasible. Since electric rail can operate on coal-generated electricity, this technology will make use of domestic U.S. resources.

Coal steam engines have considerably higher overall cost than diesel locomotives. This is primarily because the coal steam engine is about one-half as efficient as the diesel engine in obtaining work from the energy in its fuel. If petroleum should double in price from 1982 values while coal remains at present prices, then coal steam locomotives may merit more serious economic analysis.

Overall, it appears that the transportation policy of the near future should focus on expanded bus travel within and between cities, and increased rail transportation on those 20 or so routes where passenger and/or freight usage would make this feasible.

As for space heating, new residential units should be built which require less than one-third of the typical 100 M̄Btu per year now used. Significant investments in insulation and in design and construction techniques will probably add about $5,000–$10,000 to the cost of new residences. The cost per M̄Btu will be above $5 for the last $1,000 invested, and average $3 to $4 per M̄Btu for the first several thousand dollars. It is clear that, economically, conservation in space heating is preferable to heating 100 M̄Btu residences with conventional or synthetic fuels.

Industrial organization in new technologies gives little evidence of monopolistic structure or behavior. Concern may be justified in the future because five major oil companies hold about three-fourths of domestic copper production. This position gives these companies a potential dominant interest in solar energy technologies. In 1982, however, there is no evidence to suggest that copper prices are being raised and production restricted in the pursuit of monopoly profit. The experience of the uranium cartel described in Chapter 11 indicates the potential of future problems.

There are several economic obstacles that will retard the development of efficient conservation, transportation, and renewable technologies. The major ones are (1) cost: probably, in total, measured in trillions of dollars, (2) suburban geography, a serious barrier to rail and bus transportation, (3) the continued growth in electric resistance heating, and (4) the existence of considerable excess electric generating capacity in the eastern United States, which leads utilities to accept continued growth in electric space and water heating.

Overcoming these obstacles is essential. Declining domestic oil and gas resources, international petroleum monopoly, and doubts about nuclear power's reliability have an interactive effect with the national economy. Inflation in its high growth period from 1973 to 1981 centered on energy prices, and unemployment has been closely linked to recession in automobile manufacture and suburban construction. American national income as GNP per capita has lost the first position it held in 1973; the United States now ranks behind all the continental countries of Northern and Western Europe. The higher level of energy consumption in the United States appears to be a major cause of the deterioration of the American economy since 1973.

The objective, then, is to define a public policy which is efficient in terms of microeconomic markets, and has a potentially positive impact upon employment and income, and a negative impact upon inflation. This is the task of the final chapter.

15

Public Policy: One View

In approaching this concluding chapter, I have been perplexed by the choice between organizing it on the basis of current policy discussion (deregulation, divestiture, taxation), or on the basis of the conclusions I have reached on a different and larger set of public policy questions. I have chosen the second approach. Increasingly, the conventional policy debate is too circumscribed. In general, I think the logical inferences of conservative economists are usually correct on specific points, and that liberal critics too frequently argue from an ideological perspective rather than from a factual understanding. Having alienated readers that identify with the liberal perspective, I must now note that the area of policy debate has itself been defined by the conservative perspective, and I am considerably more concerned about the consequences of a falsely narrow definition of problems than I am about inaccurate generalization on rather minor points.

The path taken in this chapter is to mark the conservative prescriptions that appear to have a sound empirical context, and then change the focus to delineate the problems that are of pressing urgency, yet remain—currently— beyond the domain of policy discussions.

Divestiture and Decontrol

Given the descriptive economics in preceding chapters, it should be unsurprising that divestiture has little economic justification. Divestiture generally means a policy of not allowing any large corporation to hold significant operations at two or more stages of an industry. In petroleum, advocates of divestiture envisioned separating each of the major oil companies into separate companies for crude oil and natural gas sales, refining, transportation, and marketing. This, of course, is vertical divestiture: the separation of an existing corporation along technological stages in the processing of a common

[314]

group of products. Horizontal divestiture means corporate separation of different energy sectors such as solar, nuclear fuel, oil, and coal, or separation of energy activities from non-energy enterprises. Here, for example, Mobil would separate itself from Montgomery Ward.

The logic of the divestiture position can be summarized in three steps. First, assume, as in Chapters 1 and 2, that a competitive industry is socially optimal or Pareto optimal, that it creates the framework in which consumers and producers achieve maximum efficiency and maximum net social value. Second, assume that the petroleum industry is a profit monopoly, maximizing excess profit while restricting production and consumption. It follows as a third step in this logical development that divestiture can create a competitive industry. This causes an improvement in economic efficiency and social welfare.

Such reasoning appears, for example, in the Senate Judiciary Committee Report which recommended Congressional approval of oil company divestiture: (1) Divestiture will create competition in the petroleum industry; (2) Competition "spurs innovation, promotes productivity, prevents undue concentration of economic, social and political power, and helps preserve a democratic society"; (3) Divestiture would reduce monopoly, which would reduce prices, unemployment, and inflation.[1]

This chain of logic presupposes that economic competition is feasible as the basic framework of the petroleum industry. But the analysis in earlier chapters has described the economic incentives which create vertical integration, joint ventures, unitization, trading in refined products, and management and ownership affiliations. These incentives operate with equal strength for public or private oil companies, and for American or European companies. The pattern of cooperative activity and vertical integration would emerge again among the divested corporations. Simply stated, economic competition between atomistic, noncooperating oil companies would be so inefficient as to approach impossibility.

As for horizontal divestiture, the technological and economic linkage between oil and natural gas is unavoidable. Coal production and marketing is similar to oil and natural gas production and marketing. There is no evidence that oil companies with coal property have withheld coal, and there is some evidence that these oil companies have expanded production, maintained relatively good safety records, and contributed to the relative degree of competition in the coal industry.

For nuclear fuel, the policy problem, as I see it, is that in the late 1970s the oil industry joined the nuclear industry in promoting nuclear power before it was logical to do so. Problems in reactor safety, spent fuel disposal, and

---

[1]U.S. Senate, Committee on the Judiciary, *Petroleum Industry Competition Act of 1976*, Report to accompany S.2387, June 28, 1976, p. 161.

reactor decommissioning should have been resolved before nuclear power expanded to its present level. The problem is not that the oil company position in nuclear fuel may have retarded nuclear power growth below desirable levels because of profit monopoly activities. In fact, oil companies may have contributed to the premature growth of nuclear power by subsidizing nuclear fuel activities from profit in oil and natural gas.

Divestiture, horizontal or vertical, seems to have little to recommend it on an economic basis.

Deregulation means a reduction in the domain of economic activity which is subject to administrative and legislative determination. In the late 1970s and early 1980s, the primary focus of the deregulation process has been price controls. A secondary one has been environmental regulation. As applied to energy resources, this has meant the deregulation of crude oil and petroleum product prices in the United States, and the gradual deregulation of the producer wellhead price of natural gas.

Pipeline transportation of oil and natural gas continues to be subject to federal regulation by the Interstate Commerce Commission, and interstate and wholesale electricity sales prices are regulated by the Federal Energy Regulatory Commission. Prices charged to final customers for electricity and natural gas are regulated by state commissions. The major exception here is the publicly owned electric utilities, which are generally immune to federal or state price regulation. Coal prices are not regulated, nor is the price of nuclear fuel, although exceptions to this generalization might be perceived with respect to rail rate regulation which influences delivered coal cost, and the operations of the government-owned enrichment plants for nuclear fuel.

All of this, of course, is summation of information in earlier chapters. In analyzing deregulation as an economic concept, it is necessary to distinguish the type of deregulation (price or environmental protection) and the sector (petroleum, electricity, and so on).

My conclusion is that price control of oil and natural gas in the United States is neither feasible nor desirable. OPEC and other government oil-exporting agencies determine the supply-price relationship for crude oil. On the demand side, the most effective American response to date has been to reduce petroleum product consumption as a result of recession and reduced economic output. If OPEC oil sells at $30 per barrel, it is undesirable for American crude oil to sell at $15 per barrel. This leads, inevitably, to a "drain America first" policy and causes an acceleration in the continuing depletion of U.S. oil resources.

If it is not logical to control American crude oil prices at a level below international prices, then it is not logical to control natural gas prices. If home heating oil sells at $7/M̄Btu, natural gas should be allowed to sell at the same price. Further, the common corporate ownership of oil and gas production means that, in practical terms, regulated price differentials between gas and petroleum are not a politically realistic policy.

In terms of conservation, it makes economic sense to allow higher oil and gas prices to discourage further growth in consumption. Consider the reverse: effective price control which lowered oil and gas prices would lead to demand above the 1978 levels. It is impossible to suppose that, at lower prices, the American oil and gas industry could reach even 1978 production levels. Further price regulation would increase demand, and create a certain increase in oil and gas imports.

The problem free market pricing creates, of course, is that American gasoline and natural gas prices are strongly influenced and in some years determined by OPEC pricing. With Saudi Arabian oil costing 50¢/barrel to produce, the monopoly profit at a $30/barrel price is significant.[2] In the Growth Era, American oil companies successfully obtained low-cost oil imports. Such is not the case in the 1980s.

In examining divestiture and price decontrol, I have adopted the industry perspective on these points. Economists use the term "workable competition" to describe a concept with this meaning: there is no clear public policy that would alter the organization of an industry in a manner certain to create greater social benefits than costs. The concept seems to assume implicitly that normal profitability, declining prices, and maximum increasing consumption are desirable and attainable goals of U.S. public policy. Given these widely held assumptions, I would conclude that the petroleum and coal industries are workably competitive, and reserve judgment on natural gas and nuclear power.

However, if allowed to reject the assumptions, I conclude that global monopoly and declining resources negate the attainability of the goals of increasing consumption and declining real prices. Much of the energy industry (now including electric and gas utilities) was organized during the Growth Era to achieve accelerating consumption and production rates with declining real average costs and prices. It is now evident that these goals are no longer feasible for much of the energy industry.

Economic Growth after the Growth Era

The year 1973 marked a watershed between two qualitatively different periods in economic growth. From 1965 to 1973, real per capita GNP had an average annual growth of 2.8 percent. From 1973 to 1982, the average growth was 1.2 percent. The pre-1973 growth rate was lowest in 1970, the year in

[2]Table 5-5, above.

which defense expenditures began to decline during the conclusion of the Vietnam War. The post-1973 rate was highest in 1976–78, when real energy prices were stable and energy consumption was increasing.[3] The implication is that reduced economic growth is caused by or associated with reduced growth in energy consumption. Yet the comparison of Sweden and the United States in Chapter 4 had exactly the opposite implication. It was suggested there that lesser energy consumption was a cause of Sweden's greater economic growth in the 1970s. The Nordhaus theory, discussed in Chapter 5, postulates a global linkage between OPEC pricing, petroleum consumption, and economic growth. In the preceding chapter, on the other hand, the growth in GNP per capita in the 1970s in Europe appeared to be linked to Europe's lesser energy consumption, and the fall of the United States from first to eleventh place seemed associated with higher energy consumption.

The paradox is clear: in the United States, growth in national income and in energy consumption have moved together, rising and falling in a loose relationship. In Europe, economic growth has been more rapid and reached higher levels with lower energy consumption. The key to the paradox may be that Europe is less dependent on energy consumption because of its greater emphasis on public and rail transportation, urban geography, better-insulated housing, and greater reliance on renewable energy.[4]

The Growth Era concept must now be considered again. It is now clear that expansion of coal and nuclear power will increase both the market cost and the social cost of electricity. Electricity shares with oil and gas the phenomenon of experiencing its historical minimum cost in the early 1970s.

The available data on the economics of new technologies show that the unsubsidized costs of synthetic liquid or gas fuels exceed the cost of conventional energy by considerable amounts. It now seems clear that there is no source of energy that can supply automotive fuel, natural or synthetic gas, space heating, or electricity at costs or prices comparable to those which existed at the end of the Growth Era.

Globally, it is impossible that any large region in the world will ever again attain the high level of per capita energy consumption that formerly existed in the United States. It is probable that world oil production reached its historical maximum in 1979, the same year that saw maximum consumption in the United States.

Therefore, public policy must simply abandon the impossible goal of returning to an era of accelerating production and consumption of automobiles, suburban housing, and related energy-intensive activities.

[3]Data from U.S. Council of Economic Advisors, *Economic Report of the President*, February 1982.

[4]According to U.S. Department of Energy, Energy Information Administration, *1980 International Energy Annual*, September 1981, 17% of the energy produced in Europe results from renewable hydropower.

The maintenance or expansion of present U.S. living standards depends upon the degree to which macroeconomic policy adapts to the micro-economics of resource feasibility. The Growth Era is not simply a century-long correlation of declining energy prices and increasing income and energy consumption. It is a historic concept; it marks a period which terminated. The economic goals and policies of that Era cannot be a useful compass for the present or future.

My intention is to sketch out a pattern of energy consumption and production which reflects the microeconomic knowledge gained about energy conservation potential, declining American resources, and the economics of synthetic and renewable resources. The emphasis is upon proposals for positive solutions.

Energy Conservation and Economic Growth

The righthand column of Table 15-1 lays out a pattern of energy consumption and production typical of the 1980s. The column to its left indicates one feasible approach to reduced energy consumption. I have labeled it simply the "A" plan to avoid a pejorative or superlative adjective.

Table 15-1. Reorganizing the economy with lesser energy use, annual energy consumption, quadrillion Btu(Q)

	"A" Plan	Current (1980s)
Residential houses & apartments		
Conventional space heating	3.8	7.7
Wood space heating	2.0	1.0
Solar space heating	0.5	0
Solar water heating	0.5	0
Water heating, conventional	2.3	2.3
Air conditioning	0.8	1.1
Appliances	4.0	4.0
Electric lights	1.0	1.0
Total	16.1	17.1
Business & commercial bldgs.		
Heating	3.5	4.6
Air conditioning	2.2	2.2
Lighting	2.2	2.2
Water heating & equipment	1.3	1.3
Total	9.2	10.3
Transportation		
Automobiles	8.0	12.2
Airplanes, passenger	1.1	1.5
Buses	2.3	0.1
Trains, passenger	2.3	0.1

(continued)

Table 15-1. Continued

	"A" Plan	Current (1980s)
Trains, freight	1.2	0.6
Truck, freight	1.5	1.9
Barge, ship, pipeline	1.7	1.7
Military air	0.5	0.5
Total	18.5	18.6
Industrial		
Construction	1.2	1.2
Mining	2.1	2.1
Agriculture	1.1	1.1
Food	1.3	1.3
Paper, conventional	1.5	1.5
Paper, woodfuel	2.0	1.1
Chemical	6.3	6.6
Petroleum	3.4	3.7
Stone and glass	1.7	1.7
Steel and metals	5.3	5.6
Other manufacturing	4.4	4.4
Total	30.3	30.3
Total consumption	?74.2?	76.3
Energy production and imports		
Nuclear power	0	2.9
*Woodfuel, residential	2.0	1.0
*Woodfuel, paper and industrial	2.0	1.1
*Active solar space & water heating	1.0	0
*Hydropower	3.3	2.9
Natural gas production	15.0	19.8
Crude oil, gas liquids	15.4	20.4
Coal production	24.5	18.9
Coal exports (−)	−6.0	−2.9
Natural gas imports	2.0	0.8
Petroleum imports	15.0	11.4
Total production and imports	?74.2?	76.3
Energy consumption by energy type		
*Active solar space & water heating	1.0	0
*Woodfuel, residential	2.0	1.0
*Woodfuel, paper and industrial	2.0	1.1
*Hydropower	3.3	2.9
Nuclear power	0	2.9
Natural gas	17.0	20.6
Petroleum products	30.4	31.8
Coal	18.5	16.0
Total	74.2	76.3
(*Note: Renewable energy total	8.3	5.0)
(Possible electricity in above totals	25.0	25.0)

Sources: Current statistics are adapted from SERI, *Building a Sustainable Future,* and U.S. DoE, EIA, *Monthly Energy Review.*

Remember that it is an illustration. For example: the 80 million residential houses and apartments now in use are consuming an annual average 96 M̄Btu for conventional space heating. This includes the electric utility fuel required to generate the electricity to heat electrically heated residences. Also, currently, about one Q or a little more is provided by wood space heating. So the national average space heating now is about 109 M̄Btu per residence.

Let us increase the number of residential units by 20 million, so that the total number of houses and apartments will be 100 million, and also suppose that an additional 20 million new units are built to replace old units. Now, insulate the remaining 60 million old units to reduce their average annual space heating by one-fourth, to 82 M̄Btu. Make the 40 million houses and apartments which are built sufficiently well insulated and designed so that, on the average, a new unit requires 35 M̄Btu annually for space heating.

The result of these two calculations gives a total residential space heating requirement of 6.3 quadrillion Btu (6.3 Q). Now, double woodfuel use in the rural temperate regions of the country to 2 Q. Also install solar space heating in the Southwest and in Florida so that one-half Q is provided by active solar space heating. This leaves 3.8 Q needed for residential space heating by conventional electricity, natural gas, or oil.

Notice, in these calculations, that it is not required that electric heat pumps replace electric resistance heating, which is, as Chapter 3 shows, economically sensible and reduces energy use. Nor is it required that solar space heating be used in the North (which is, as Chapters 3 and 14 indicate, not clearly economical).

In transportation, suppose that (a) car passenger-miles are reduced by one-seventh by shifting to bus and rail, (b) car occupancy increases by one-seventh, and (c) miles per gallon increases by one-seventh. The result would be a 34 percent reduction in fuel use. Granting this logic throughout urban and intercity automobile use, the 8 Q in automobile transportation in Table 15-1 is consistent with 1,083 billion intercity passenger miles in the "A" plan in Table 15-2. The remainder of Table 15-2 has the same consistency with Table 15-1.

Basically, Table 15-2 says that intercity travel can continue to increase while energy consumption drops, if several policies are adopted simultaneously:

(1) All average vehicle and aircraft fuel efficiencies increase by a modest amount, here one-seventh.
(2) Occupancy rates increase for all travel, from a small one-seventh for automobiles to a twofold increase for rail travel.
(3) Passenger-miles decrease slightly for autos (one-seventh), and increase slightly for air travel (one-ninth).

Table 15-2. Intercity passenger travel, billions of passenger-miles per year

	"A" Plan	Current
Automobile	1,083	1,263
Bus	946	27
Rail	385	11
Air	239	215
Total	2,653	1,517

Source: see Appendix 15-A.

(4) Bus and rail travel increase by prodigious increments, handling 50 percent of intercity travel in the future.

(5) Total travel increases 75 percent.

Basically, this means, in spite of the many real obstacles, developing rail and public transportation in the United States in a manner comparable to European transportation.

The Energy Production section of Table 15-1 illustrates how the 74.2 Q consumption could be supplied. In line with the conclusions reached in Chapters 5, 6, and 8, domestic crude oil and natural gas production may decline considerably. In Table 15-1, the assumption is a 25 percent decline. Observe that petroleum imports increase even with the shift to more fuel-efficient transportation systems.

Note that increasing coal production by 5.6 Q allows exports to grow to 6 Q, while domestic consumption also increases. The last section of the table gives the hypothetical energy consumption by energy type.

Renewable energy constitutes 8.3 Q of the "A" plan. No recognition is made of such potential technologies as photovoltaic cells, windpower, gasohol, and others of current interest. While this correctly delineates the microeconomic conclusions of Chapter 14, it should be noted that many persons studying or advocating solar energy expect a much larger contribution from renewable energy. For example, the Solar Energy Research Institute evaluated the economic potential of solar and renewable energy, and found that potential to be between 18 and 31 Q per year. This optimistic view (Table 14-5) is not shared here.

Also, since the analysis in Chapters 10–13 has led me to conclude that no new nuclear plants will or should be built, future energy supply in the "A" plan requires no nuclear power supply.

More existing studies expect or recommend continued growth in energy consumption and production. For example: the Exxon, DRI, and federal studies summarized in Chapter 4 all recommend 100 Q as a planning basis. The exception is the National Academy of Sciences Demand Panel studies.

Having noted this weight of professional opinion in opposition to my own, I shall make no further reference to it until the last section of this chapter.

The basic elements of the policy I am proposing are (1) reduced space-heating requirements through new and retrofit insulation and design, (2) development of transportation so that increases in passenger and freight go almost entirely to public transportation and rail, and (3) increased use of renewable resources that are appropriate to each region of the country. The emphasis on these three elements assumes that industry continues to make logical decisions about energy use. The engineering economics of efficient energy use allows the industrial sector to make individual decisions that accumulate to sensible national policy. Industrial energy consumption has declined since 1973, and, although real GNP has grown somewhat, industrial energy consumption per dollar of real GNP has declined. It is the problem areas discussed above which require the greatest attention.

It is interesting to speculate about the investment cost associated with this policy. Before proceeding to do so, however, I should note the employment impact—which I judge on the whole to be clearly positive. Table 15-3 draws upon the studies made by Bruce Hannon and associates at the University of Illinois, and the work at the Brookhaven National Laboratory reported by M. D. Rowe and P. J. Groncki.[5] Both studies use input-output analysis to examine the full direct and indirect effects of economic impact. For example: the substitution of rail travel for air travel is supposed to reflect increases in rail crews, reduction in flight crews, increased steel employment, changing mine labor employment, and so on. The Hannon studies examine the consequences of reducing total energy consumption by 1 Q, and then estimate the employment effect of this energy change. In the abstract, this makes each Hannon entry separable and additive: making all of the six changes would reduce energy consumption by 6 Q and increase employment by 2.56 million persons.

In Part B of Table 15-3, Rowe and Groncki study the employment requirements of producing 1 Q of residential heating energy. In Part B, change is estimated by subtraction. For example, replacing 1 Q of nuclear electricity with 1 Q of active solar space heating would have a net positive employment effect of 260,000 persons (i.e., 370,000 − 110,000).

Undoubtedly there is much further research to be undertaken on the employment question. I note, simply, that both research groups represented in Table 15-3 have been sponsored at or by America's nuclear energy research

[5]See Bruce Hannon, "Energy, Labor, and the Conserver Society," *Technology Review*, 79:5 (March/April 1977). See also M. D. Rowe and P. J. Groncki, *Occupational Health and Safety Impacts of Renewable Energy Sources*. Brookhaven National Laboratory, 1980; their findings are also reported in Chester Richmond, Phillip Walsh, and Emily Copenhauer, eds., *Health Risk Analysis* (Philadelphia: Franklin Institute Press, 1981), p. 66.

[323]

Table 15-3. Estimates of employment effects for change of 1 Q in consumption

	Additional employment
A. Hannon study: Substitution effects	
Rail travel for plane travel	930,000
Rail travel for intercity car use	700,000
Freight train for truck	675,000
Urban bus for car	210,000
Rail construction for highway construction	30,000
Insulation for home heating oil	15,000
B. Rowe & Groncki study: Heating systems	
Active solar space heating	370,000
Active solar hot water heating	220,000
Coal-fired power generation	190,000
Nuclear power generation	110,000
Passive solar heating	140,000

Sources: Hannon, and Rowe & Groncki.

centers, the National Laboratories. There is no institutional bias to color the assessments of rail and public transportation, insulation, and renewable energy. Each has, on the whole, positive employment effects. No evidence is available to weight these changes qualitatively: bus driver versus car mechanic, nuclear engineer versus solar plumber. The data now available simply indicate the positive direction of the quantitative impact.

Cost, Finance, and Taxation

Thus far I have argued that the "A" plan requires less conventional energy, no nuclear power supply, more renewable energy, and greater employment. Since this kind of plan is in the early 1980s not yet seriously studied, the cost can only be illustrated. A hypothetical estimate would be at least one trillion dollars: $1.4 trillion in Table 15-4.

For public and rail transportation, there is no present guideline for routes which should be developed and expanded in the 1980s. The magnitude of potential investment can only be indicated by analogy. Suppose, for each household, the public/rail investment is the equivalent of the cost of a new automobile, perhaps $7,000. For 80 million households the result is $560 billion. This, surely, is not correct. Yet it may give a correct indication of the magnitude of new investment that would be required for expanding the rail and bus system to provide the indicated level of transportation.

Reasonable data exist on the cost of conservation in space heating, and Kathleen Cole's work (see Chapter 14) provides a hypothetical example of $4,000 cost per new residential unit in order to achieve a 57 MBtu annual heating requirement. Assume that a higher $5,000 per unit achieves a 75

Table 15-4. A representation of the magnitude of investment in Plan "A"

Indication of investment	Illustrative cost
Transportation, development of public and rail system, e.g., Table 15-2 $7,000 per household	$560 billion
Space heating and air conditioning, attain 25% reduction in old buildings and 75% reduction in new buildings, e.g., Table 15-1	
Retrofit 60 million existing residences, $2,500 each	$150 billion
Build 40 million new residences, $5,000 each	$200 billion
Increase of renewable energy resources from 5 Q to 8.3 Q	
Residential woodfuel, $4,000 per unit, 20 M̄ units	$80 billion
Solar space, water heating, avg. $15,000 per unit, 20 M̄ units	$300 billion
Paper, lumber, woodfuel, investment for .9 Q increase	$70 billion
Hydropower, 21,000 new MW @ $1,000/kW	$21 billion
Total magnitude of illustration	$1,381 billion

percent reduction to 35 M̄Btu annually. Observe in Table 15-4 that retrofitting provides much less conservation per dollar than conservation in new buildings. The total investment required is $350 billion.

The investment illustration for renewable resources is similarly broad. The solar and wood heating costs are also based on Cole's and Michael Slott's work. For increased energy production from wood use in the paper, lumber, and other wood product industries, simply assume the investment cost is comparable to residential woodfuel investment on a Btu basis. For hydropower, take 21,000 MW as the basis for .5 Q new sources, at a cost of $1,000 per kW. Result: $21 billion for new hydropower.

Regardless of the consequences of repetition, I shall stress again that this outline is based upon the best information available, and this information is inadequate. Inadequate, in this context, seems rather too positive. The only comparable kind of estimate is that of the Solar Energy Research Institute which, in its detailed analysis, reported that up to 26 Q could be provided by conservation and renewable technology at an investment cost of $800 billion.[6] Although the SERI work has provided considerable insight into technology and economics, I find the economics to be quite in error in terms of optimism and cost. Hence the illustration for Plan "A" provides, in comparison, substantially less renewable and conserved energy at a significantly greater cost.

Generally, conservationists can be expected to criticize this hypothetical future because of the high cost estimated for it. Corporation energy specialists are less likely to criticize the high cost, but will probably dispute the conclusion that future energy consumption at a level of 70–80 Q is feasible and preferable. I arrive at these conclusions, however, from the information available.

[6]Solar Energy Research Institute (SERI), *Building a Sustainable Future*, 2 vols., Committee Print, U.S. House Committee on Energy and Commerce, April 1981, 1: xi–xxx. Compare to Table 14-5, above.

Let us suppose $1.4 trillion is a reasonable figure. Accurate, perhaps, within one or two trillion dollars. How can this be financed? With a proper recognition of the microeconomics of comparative incremental cost, space-heating conservation and renewable energy technologies may develop without further special incentives. The key here is to solve the kind of institutional problem defined in Chapter 3. The decision which is least costly for a builder or landlord is not the least costly choice for an owner-occupant, renter, or society. The problems in transportation are institutional and financial. Where is $560 billion which could be invested in transportation? There is, in the early 1980s, only one source: the petroleum industry. In Chapter 6, revenues from the 20 major companies were seen to be $566 billion in 1981, considerably above the federal government's non-Social Security budget receipts of $424 billion. The profit rate (net income to shareholder equity) exceeds that of any other industry.

And in the late 1970s and early 1980s, the largest single public or private beneficiary of petroleum profitability is probably the British government. Through its BP/Sohio ownership, it controls the one billion dollars estimated to have been paid by Sohio to BP for 1981.[7] Sohio also continues to have the highest return, as shown in Table 6-1.

Simply stated, my suggestion is to nationalize the monopoly profit now arising from OPEC monopoly pricing as it causes monopoly pricing of American oil and natural gas resources.

The corporate income tax as now structured is simply inadequate as a source of new revenue. Figure 15-1 shows that the corporate income tax provided a high of 35 percent of federal receipts in 1952, but has declined to its present 11 percent. I expect the tax revisions of 1981 to cause this corporate contribution to approach 5 percent in the mid-1980s. Social insurance and personal income taxes continue to increase their proportion of federal taxes, now exceeding 80 percent.

Several proposals follow, which will be discussed in turn.

(1) Create government participation in share ownership of the major oil companies.

(2) Organize a U.S. National Energy Corporation to negotiate with oil-exporting countries and OPEC, and buy and sell oil, natural gas, and hydropower which is imported into the United States.

(3) Eliminate all tax exclusions, deductions, and credits now held by government energy corporations.

(4) Eliminate the corporate income tax in its present form as derived from net income with exclusions, deductions, and credits.

(5) Introduce a uniform value-added tax.

[7]New York Times, August 19, 1981. Also Standard Oil of Ohio, 1981 Annual Report, p. 41.

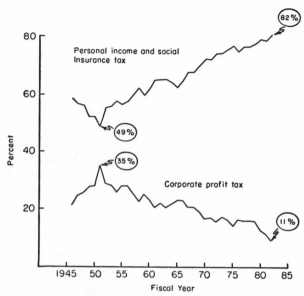

Figure 15-1. Percentage contributions to federal government receipts.
Source: Economic Report of the President, 1968 and 1982.

(6) Eliminate special tax provisions relating to oil companies and utilities: the windfall profit tax, the small producer depletion allowance, and required normalization.

(7) Eliminate the special solar tax credit for individuals.

(8) Institute a significant federal severance tax on domestic energy production and imports, equally applicable to public and private corporations.

(9) Encourage the direct election of utility company directors on a service area basis.

(10) Analyze the institutional and legal problems preventing the economic development of efficient housing, rail and public transportation, and renewable technologies.

(11) Define tariffs on imported goods according to differentials in pollution control and wages between the United States and country of origin of the imported good.

(12) Scheduling: subject these and other proposals to careful study, and introduce a unified program in the middle and late 1980s, in the 1985–1990 period.

(1) Partially nationalize the major oil companies. The intention is to nationalize the profits, not the problems. BP/Sohio, CFP/Total, and Kuwait/Santa Fe are better precursors of government participation than is the Chrysler loan program. Significant ownership permits the pursuit of three related

[327]

goals. First, the government would acquire considerable financial resources. Assume that national stock ownership in the 20 major oil companies will, in the average year, yield $20 billion.

A second goal concerns the political dimension of political economy. Government share ownership reduces or eliminates the ability of segments of the petroleum industry to employ economic power to define public policy on energy. The 1980 elections indicate the ability of major oil companies to influence the congressional committees that are responsible for economic policies affecting those companies. In Oregon, energy political action committees contributed $35,000 in amounts exceeding $1,000 each to the opponent of then-Representative Al Ullman.[8] Ullman's district in eastern Oregon has no significant oil, gas, or coal. However, Ullman was chairman of the House Ways and Means Committee, the most important committee in the House of Representatives in terms of tax legislation affecting energy companies. Ullman was defeated. In Texas, $220,000 was contributed by the energy industry to the opponent of then-Representative Robert Eckhardt.[9] Eckhardt represented the industrial and refinery area of Houston, Texas. Nonetheless, he had supported petroleum price regulation and windfall profit tax legislation. Eckhardt was defeated; he would have become chairman of the Energy and Environment Subcommittee of the House Interior Committee.

These two contributions are apparently the largest made by any economic group in the 1980 elections,[10] and were given to candidates who defeated incumbents who had been active in supporting legislation opposed by the energy industry.

Historically, energy companies have actively opposed their congressional critics and supported candidates who supported the industry. Kermit Roosevelt went from Iran (see Chapter 5) to the Gulf Oil Corporation, where he was employed as the Gulf vice-president for Government Relations. In that capacity, he was the supervisor of Claude Wild, later a key figure in providing Gulf Oil cash for the various Watergate operations.[11] Like then-President Nixon, the executive leadership of Gulf resigned, and Gulf was investigated by a special internal committee for the Securities and Exchange Commission.

Three major oil companies (Ashland, Gulf, and Phillips) and the chairman of Occidental were convicted of illegal cash contributions of $100,000 each to the campaign committee which financed the Watergate activities.[12] In the

---

[8]Gannett News Service, *Ithaca Journal*, March 1, 1982.

[9]Robert Sherrill, "A Texan vs. Big Oil," *New York Times Magazine*, October 12, 1980.

[10]*Ithaca Journal*, March 1, 1982.

[11]John J. McCloy, Nathan W. Pearson, and Beverly Matthews, *Report of the Special Review Committee of the Board of Directors of Gulf Oil Corporation*, in the U.S. District Court, District of Columbia, Securities and Exchange Commission (Plaintiff) v. Gulf Oil Corporation and Claude C. Wild, Jr. (Defendants), December 30, 1975.

[12]Summarized in Robert Engler, *The Brotherhood of Oil: Energy Policy and the Public Interest* (Chicago: University of Chicago Press, 1977), chap. 3.

1970s, at least eight major oil companies were convicted of, had executives resign because of, or signed consent judgments arising from improper political activities.

An abundance of analysts have enthusiastically investigated the political strength of energy corporations.[13] The most detailed study is that undertaken by the McCloy Committee for Gulf's board of directors. John J. McCloy, an attorney, had been board chairman of the Chase Manhattan Bank and had represented each of the seven largest major oil companies. His committee report is entitled "Report of the Special Review Committee of the Board of Directors of Gulf Oil Corporation." It is unsensational, formal, specific, and persuasive.[14]

Public participation in share ownership is intended to reduce this sort of activity.

A third objective of public ownership in petroleum is to give institutional recognition to the concept that national resources are, in the abstract, owned by the nation.

The most commonly voiced criticism of this proposal is that of the inherent problems in government operation. In refutation, it is only necessary to consider, again, Standard Oil of Ohio. This corporation is the fastest growing and most profitable American oil company and is the second largest producer and largest owner of proven crude oil reserves in the United States. It is controlled by the British government. Government oil corporations are more common than private oil corporations in Europe. Public ownership of petroleum companies exists in the United Kingdom, West Germany, France, Italy, and Norway[15] and also in Canada and Mexico.

The best method for the government to acquire shares cannot be firmly stated without further study. Three obvious methods exist: (1) open market purchase of the common stock of the major companies; (2) confiscation of a fixed proportion of existing stock and compensation of present owners; (3) issuance of new national stock to a fixed amount for each company.

The first method has several virtues. It is well established. Du Pont, U.S.

---

[13]In addition to the Engler and Sherrill publications cited above, see Ralph Nader and Mark J. Green, eds., *Corporate Power in America* (New York: Grossman, 1973) (especially the chapter by former Senator Fred Harris), and Morton Mintz and Jerry S. Cohen, *America, Inc.: Who Owns and Operates the United States* (New York: Dial, 1971). Much material about political contributions and legislation as influenced by the Standard Trust at the turn of the century can be found in W. A. Swanberg, *Citizen Hearst* (New York: Scribner, 1961). An academic article which explores the relationship between company size and effective taxation in the petroleum industry is Lester M. Salomon and John J. Siegfried's "Economic Power and Political Influence: The Impact of Industry Structure on Public Policy," *American Political Science Review*, 71 (September 1977), 1026–1043.

[14]See footnote 11.

[15]See "State-Owned Oil Companies," chap. 9 in Raymond Vernon and Yair Aharoni, *State-Owned Enterprise in the Western Economies* (New York: St. Martin's, 1981); and Leslie E. Grayson, *National Oil Companies* (New York: Wiley, 1981).

Steel, and British Petroleum each acquired a major American oil company through stock purchase, although the British Petroleum/Sohio arrangement involves a special class of stock. Stock purchase also assures that present owners would receive a compensation equal to the stock market valuation of their shares. Finally, open market purchase does not impose legal coercion upon the shareholders or management.

The present shareholders are investment, insurance, and banking corporations as delineated in Chapter 7. Open market purchase would, in essence, mean the replacement of private capital by public capital.

The cost of this method is uncertain. Stock market prices fluctuate over wide ranges. In addition, market purchases by the government would place an upward force on all stock prices. The range in stock market prices is evident if one compares prices at the end of 1980 (a high-profit year, as were 1979 and 1981), with prices in early 1982 (a period of reduced petroleum sales and normal profit). At the end of 1980 (when there were 3.6 billion shares of common stock of major companies), the average price of a share was $61. In April 1982, the price was $32. The difference in stock market valuation is significant. Total stock value at the end of 1980 was $223 billion. The April 1982 value was $117 billion. Shareholders' total equity was $137 billion at the end of 1980.[16]

The largest obstacle to this method of acquisition might be a political action. It cannot be known whether existing financial company shareholders would move to cease stock sales to block acquisition.

The second method of acquisition is confiscation with compensation. Each shareholder holding 100 or more shares could be required to sell one-half of the shareholdings. Compensation could be based upon shareholders' equity, or upon stock market price. U.S. Steel adopted this method in acquiring a major segment of its Marathon stock. U.S. Steel had purchased 51 percent of Marathon stock at $125 per share. This was considerably above the $50–$75 per share range in which Marathon had sold prior to U.S. Steel's move toward acquisition. After acquisition, U.S. Steel issued $100 face-value notes to the remaining 29 million shares. These notes are due in 1994, and pay 12½ percent. Since 12½ percent was below then-current interest rates, these U.S. Steel–Marathon bonds were traded in bond markets at $75 to $78 each. While a bond with a value of $75–$78 would seem fair compensation for a share of stock which had sold earlier at $50–$75, the owners of these 29 million shares have been seeking a higher compensation.

Since Marathon's dividend had been $2.00 per share, the new note-holders receive six times their previous annual dividend payment. This ought to soften the perceived loss.

[16]The data used in this discussion are from the Annual Reports, and the *New York Times*, April 6, 1982. The data on the Marathon acquisition are from the United States Steel Corporation *1981 Annual Report*, p. 32, and the *New York Times*, March 9 and 12, 1982.

In the 1980s, shareholders' equity includes the cumulative effect of OPEC monopoly pricing. Recall from Table 7-4 that Exxon equity increased from $12 billion in 1972 to $25 billion in 1980. This is entirely attributable to the growth in retained earnings. Retained earnings, in turn, have been much increased by the high revenues collected from oil and natural gas customers, and the unusually high profit earned since 1973. Viewed from a critical perspective, much of the growth in shareholders' equity since 1973 derives from monopoly pricing and profit. If compensation is less than shareholders' equity, it means that monopoly profit from previous years is being confiscated.

As with the first method of acquisition, the primary obstacle here is also political. Each large shareholder of each major oil company would be in a position to choose between cooperation and noncooperation.

The third method of acquisition follows the process used by British Petroleum in acquiring control of Standard of Ohio. British Petroleum owns 1,000 shares of special stock, which are equivalent to 126 million shares of ordinary common stock. In this approach to nationalization, each major oil company would be required to issue special shares of new national stock in amounts that would create a 50 percent interest in combined private and national stock.

The obstacle here is the certain legal objection by the major oil companies. Litigation would endure for years.

One purpose of this partial nationalization is (as indicated) access to the financial resources which flow from monopoly pricing of declining natural resources. One objective of the new national shareholders should be to require a significant increase in the dividend payout of funds earned from operations. In 1980, only $8.5 billion of the $54 billion earned was paid out in dividends. Suppose 75 percent was an average dividend payment: this, in 1980, would have been $41 billion. Of this, one-half would be received by private shareholders and one-half by national shareholders.

Note that the present private shareholders would receive higher annual dividends under this form of nationalization than if the present ownership structure continues.

What percentage of the stock of the majors should be owned by government? In Western Europe, there is considerable variation. The British National Oil Corporation and the Italian and Norwegian oil companies are wholly government-owned. British Petroleum and the national oil companies in France and West Germany combine private and public ownership.[17] I favor continuation of significant private holdings.

Is 10 percent government ownership sufficient? Chapter 7 explained that this figure is generally viewed as constituting an ownership interest sufficient

[17]See Grayson, p. 234.

to have considerable influence on or control over management decisions. However, the objective proposed here is, basically, financial: the attainment through dividend payments of the monopoly profit that major oil companies earn through the OPEC monopoly prices charged to their customers. On this basis, I advocate 40–60 percent share ownership.

National shareholdings would have to be managed by some new organization. Several types are possible. One might be patterned after a holding or investment corporation or corporations in which one-to-five public corporations would manage the national stock in the major oil companies. This, obviously, is a public capital plan patterned after the private capital structure of present ownership. National dividends received by the public corporation(s) would be remitted to the federal government. This latter arrangement is similar to the manner in which the British government receives payment for its ownership of British Petroleum.

One irony of these modified nationalization proposals is that the British government's interest in America's most profitable oil company would be affected. In addition, the recent acquisitions of major oil companies by U.S. Steel and Du Pont would themselves be partially acquired.

Considerable detail has been offered about these proposals, but they remain in need of much more detailed planning. In my judgment, the remainder of the proposals given here are not feasible unless some form of nationalization of the profit of the oil and gas industry occurs. The major reasons are twofold. In political terms, continued private ownership of American oil, gas, and coal reserves will, logically and inexorably, lead these corporations to oppose conservation, public transportation, and renewable energy sources. In financial terms, the profit being earned from the sale of petroleum and natural gas should be a major source of funding for conservation, rail and public transportation, and renewable energy.

(2)   Organize a U.S. National Energy Corporation to negotiate with OPEC and oil-exporting countries. As the global market is now structured, the United States economy is seriously affected by political and economic decisions in which the national welfare has no representation.

Consider the incentives applicable to Sohio/BP/British government in Alaska. Sohio sells all the oil it produces there to other oil companies. Its economic interest lies entirely with maximum feasible oil prices in the United States. In this circumstance, the British government will find its revenue from Sohio maximized to the extent that Sohio sells at the maximum price. The United States finds itself in the 1980s in a curiously mirrored reflection of its earlier controlling economic position in the Middle East. The largest domestic oil interest is motivated to obtain the highest possible price, and this company is controlled by a foreign government.

[332]

A second international problem arises from the economic incentive that the private companies have to accept high prices in exchange for secure access to foreign crude oil. In early 1982, Aramco continued to accept high volumes of Saudi Arabian crude at a price $5 per barrel above world prices.[18]

The rapid fluctuation of oil and gasoline prices about a rising trend creates considerable difficulty for the American economy. As was discussed in the context of the Nordhaus theory in Chapter 5, and again at the beginning of Chapter 14, recession in the automobile industry, construction, and the national economy follows each unexpected major price increase.

The solution is a national oil company to represent the American interest in stabilizing price increases in some predictable, incremental system. The economic reality is declining availability of resources, rising cost of extraction, and monopoly control in OPEC. Crude oil price increases could have been negotiated from 1972 to 1982 to attain the same 1982 value without the boom-and-bust pattern which actual prices followed.

(3) Eliminate all tax exemptions now held by publicly owned utilities. The present tax benefits are not economically justified; public utilities are engaged in the same economic activities as private utilities. The major tax benefits are tax-exempt low-interest bonds, low- or no-interest government investment, and exemption from both property and corporate income taxes. Because of the significant capital investment in hydropower and generation plants, property taxes collected from generating facilities should probably be items of state rather than local revenue. If public utilities provide the same actual tax per dollar of revenue that the private utilities did in 1980, this would increase total tax receipts by about $1 billion.

That the increase would be so low is in part due to the low tax rate now being paid by privately owned utilities. With the tax change introduced in points four and five, the tax on publicly owned utilities would be higher.

One defense offered for the special tax status of public utilities is that they merit or need special protection. This argument can cut several ways. If public utilities need special protection, does this imply they are less efficient? The answer is unequivocal and based upon empirical evidence: public utilities are at least as efficient as private utilities. The nature of the difference between the two forms is explored more fully under point 8.

(4) Eliminate the corporate income tax. As Figure 15-1 indicated, the present federal corporate income tax is becoming a nuisance rather than a major revenue source. This is particularly relevant for capital-intensive energy corporations since—as explained in Chapter 12—the investment incentives create tax benefits for capital-intensive industries.

[18]*New York Times*, March 8, 1982.

Table 15-5 indicates actual federal income tax payments by major oil companies and by all private electric utilities for 1978–80. The after-tax return for the utilities was a low 10 percent each year. For the petroleum companies, the profit rate rose to 22 percent in 1980. However, actual tax payments were only 18–20 percent of pre-federal income tax net income for the petroleum companies. Recall that the statutory rate on taxable income was 48 percent in 1978–79 and 46 percent in 1980. For the utilities, the actual federal corporate income tax rate was 7–9 percent of pre-tax net income. The 1981 tax act further reduces income tax liability for petroleum and utility corporations.

Perhaps the bottom line in each part of Table 15-5 is of the greatest interest. Each year's actual federal corporate income tax payment is only 1–2 percent of operating revenues. This is less than a typical local sales tax. Considering

Table 15-5. The federal income tax and major energy industries, $ billion

	1980	1979	1978
Petroleum			
Number of companies in group; includes gas, coal, nuclear operations	20	26	26
Federal income taxes			
Reported expense	$11.7	$8.9	$5.9
Actual year's payment	$8.2	$6.0	$3.6
Profitability			
Net income after reported tax expense	$29.7	$23.5	$13.9
Stockholders' equity	$136.7	$124.9	$108.7
Rate of return	22%	19%	13%
Actual tax payments as percent of			
Net income plus reported tax expense	20%	19%	18%
Operating revenues	2%	2%	1%
Utilities, investor-owned, electric			
Federal income taxes			
Reported expense	$4.4	$3.8	$4.0
Actual year's payment	$1.3	$0.8	$1.1
Profitability			
Net income after reported tax expense	$10.5	$9.3	$8.6
Stockholders' equity	$103.7	$94.6	$85.9
Rate of return	10%	10%	10%
Actual tax payment as percent of			
Net income plus reported tax expense	9%	7%	8%
Operating revenues	1%	1%	1%

Note: for petroleum companies in 1980, the data are for the 20 major companies described in Chapter 6. For 1979 and 1978, the data are from U.S. DoE, EIA, *Performance Profiles of Major Energy Producers 1979,* July 1981. The EIA group of 26 includes all of the major oil companies, and adds Coastal States, Kerr-McGee, Superior, and American Petrofina. It also includes Burlington Northern, a major coal producer, and Union Pacific, another railroad company which has extensive coal and oil interests. The data for electric utilities are from Edison Electric Institute, *Statistical Yearbook of the Electric Utility Industry* and probably summarize the financial circumstances of 99–100% of the 200 privately owned electric utilities. Net income for utilities was not adjusted for AFUDC or for deferred taxes and credits.

the complexity of the Internal Revenue Code as it applies to corporate income, a tax that collects 1–2 percent of revenues is basically a nuisance, in that it creates much administrative work for corporations and government while producing almost no financial benefit.

(5) Introduce a value-added tax. In its simplest form, value-added is that portion of the cost of a product which has been added by the company selling the product. It is the difference between selling price and raw materials. A 5 percent value-added tax is a 5 percent tax on wages, interest charges, normal profit, and monopoly profit. It is a form of corporate income tax that places a uniform burden upon every factor of production.

For illustration, consider electric utilities in 1980. They spent $36.3 billion on fossil fuel, and $900 million on nuclear fuel. These fuel purchases total $37.2 billion, and are added to the normal depreciation of $6.6 billion.[19] Taking actual operating revenues of $95.3 billion, value-added in this illustration is revenues less fuel purchases and normal depreciation, or $51.5 billion. A 5 percent value-added tax would equal $2.6 billion. Actual payment in 1980 was $1.3 billion. A 5 percent value-added tax would double actual tax payment. But, because of the normalization accounting described in Chapter 12, customer costs would decline. Normalization, it will be recalled, bases revenue determination upon reported expense rather than actual expense.

Such a tax's application to petroleum can be illustrated with Exxon. In 1980 revenue was $110.4 billion. Crude oil and product purchases were $60.9 billion and normal depreciation and depletion was $2.3 billion.[20] Value-added would be $47.2 billion, and a 5 percent value-added tax would be $2.4 billion. Exxon's actual 1980 federal corporate income tax payment was $1.2 billion. Extrapolation to the major companies would mean that a value-added tax would have collected $13.3 billion in 1980.

Implicit in this value-added proposal is that tax accounting be standardized with company accounting. The same definitions are to be used in each, eliminating the current wasteful and frustrating practice of having several legitimate and required sets of accounts for the same assets or expenses.

One interesting effect of the value-added tax is that it removes the incentive to employ capital instead of labor. It is indifferent and impartial as to size: If Company A buys all its raw material from Company B, or if A and B are both vertically integrated, or if Company B absorbs A and is the sole company with the same total revenue, the value-added tax collected from A and B will be identical in each instance. The current corporate profit tax has considerable

---

[19]Data from Edison Electric Institute, *Statistical Yearbook of the Electric Utility Industry*, annual.

[20]Data from Exxon *1980 Annual Report*. In actual practice, purchased materials will be in other categories in addition to crude oil and product purchases.

incentive for energy companies to organize their corporations according to tax benefits. Value-added taxation eliminates this.

The major problem in Europe with the value-added tax is enforcement.[21] Some small enterprises or individuals do not report their full economic activity. Because of the simplified record-keeping and the potential for full congruence with company accounting, it seems unlikely that compliance with value-added taxation will present as many problems as does the corporate income tax.

This value-added tax should be applied to publicly owned utilities, to the major petroleum companies which should be partly government-owned, and to private oil, utility, and other companies.

(6) Eliminate special petroleum and utility tax provisions. Given the preceding points, there is no need to continue the depletion allowance for small producers, or the windfall profit tax, or required normalization for utilities. The windfall profit tax in 1981 was only $16 billion for the major oil companies (Table 6-4). It must be the single most complicated provision in the present tax code, making a per-barrel tax on oil vary according to perhaps 50 different factors linked to amount of production, weight, year of initial production, year of current production, type of well, type of owner, geographic location, and inflation rate.[22] Since the preceding points have (a) given 40–60 percent of dividends to government through share ownership, (b) eliminated the corporate income tax, and (c) introduced a value-added tax that rises with profit, there is no logical need to continue this unfortunate provision.

Normalization for utilities, as noted above, will also be irrelevant with a value-added tax that is treated in full conformity with normal company accounting practices.

(7) Eliminate the special solar tax credit for individuals. The solar tax credit is intended to offset the tax benefits accruing to conventional finite energy sources, thereby making solar space and water heating more competitive. The preceding points would effectively dismantle the present tax incentives for conventional and nuclear energy production and use. Therefore, justification for the solar credit is removed.

As I noted in Chapter 14, I am puzzled by the lack of growth of solar space and water heating in the Southwest and Florida. However, one reason for the

---

[21]U.S. General Accounting Office, *The Value-Added Tax in the European Economic Community*, December 5, 1980, surveys enforcement problems in Europe. Also see Henry J. Aaron, ed., *The Value-Added Tax: Lessons from Europe* (Washington: Brookings, 1981).

[22]U.S. Congress, Public Law 96-223, 96th Congress, "Crude Oil Windfall Profit Tax Act of 1980," April 2, 1980, 94 Stat. 229. Also see Ben W. Bolch and William W. Damon, "The Windfall Profit Tax and Vertical Integration in the Petroleum Industry," *Southern Economic Journal*, 47:3 (January 1981), 788–791. The original legislation phases out the tax sometime in the 1990s.

absence of solar heating development may be the solar research budget. As
Table 15-6 indicates, no government-supported research in the future will be
directed to the economically competitive forms of solar energy: passive design techniques and active space and water heating. The Carter administration, as the table indicates, had also focused on noncompetitive activities. In
both the Carter and Reagan administrations, photovoltaic research, an area
discussed in Chapter 14 in the context of the solar activities of petroleum
companies, received the greatest support. Solar research at the federal level
has, I think, been mistakenly directed. It should focus on material, design,
and process improvements for the simple methods of active and passive solar
space and water heating. Energy companies will develop photovoltaic cells
and other future technologies when and if they become economic.

(8) Institute a significant federal severance tax on domestic production and
imports.

With the level of production costs for OPEC oil being 50¢ to $1.50 per
barrel, a crude oil price of $35 per barrel provides a monopoly "tax" of
$33.50 to $34.50. There is no economic or ethical reason why the United
States should accept this pattern. If the 2 billion barrels of imported oil pays,
on the average, a monopoly tax of $25 per barrel, the monopoly tax collected
by oil exporters to the United States is $50 billion annually.

Imported crude oil and products should be taxed at a level in the range of
$10–$15 per barrel. On an approximate basis, the monopoly profit can be
shared between producing and consuming countries. This action would, of
course, not be popular with OPEC. It would be the responsibility of the
National Energy Corporation (point 2 above) to manage the problems. Future

Table 15-6. Federal government research budgets for solar energy, $ million

	Actual fiscal year 1980	Requested fiscal year 1982
Active space and water heating	37	0
Passive residential heating	12	0
Active solar cooling	12	8
Passive solar cooling	8	9
Photovoltaics	150	63
Thermal towers	143	44
Alcohol fuels	22	10
Biomass	33	21
Wind	61	19
Ocean energy	43	0
Other and administration	29	19
Total solar research	550	193

Source: U.S. House of Representatives, Committee on Appropriations, Subcommittee on Energy
and Water Development, Hearings, *Energy and Water Development Appropriations*, 1981, Part 5.

Table 15-7. Energy production on federal and Indian land, 1979 and 1950

Energy resource	1979 amount	Percent of national total	
		1979	1950
Crude oil and condensate	462 million barrels	14.9%	5.2%
Natural gas	5.9 trillion cubic feet	32.8%	2.2%
Coal, total	84.9 million tons	10.9%	1.5%
Low-sulfur western coal resources		60%	—

Sources: U.S. DoE, EIA, Energy Policy Study, Vol. 8, *The Use of Federal Lands for Energy Development,* March 1980, pp. 10, 12, and Vol. 14, *Energy Taxation: An Analysis of Selected Taxes,* September 1980, pp. 41, 42.

increases in OPEC monopoly profit should be matched by equivalent increases in U.S. import taxes.

A significant fraction of domestic energy production already occurs on federal land or Indian land under federal supervision, and this proportion is increasing. Table 15-7 shows 1979 amounts and change since 1950. Energy production from public and Indian land should be subject to the same tax system as imports and production from private land. There is no economic argument arising from efficiency or equity why corporations may obtain resources from publicly owned land at low cost with little taxation, and pay foreign producer governments a monopoly tax which is 50–99 percent of the price.

Assume an import tax of $15 per barrel, or about 36¢ per gallon of crude oil or product. This is equivalent to a tax of $2.60/MBtu. With this as a basis, a national severance tax on finite energy resources can be defined which varies, roughly, according to type of ownership, type of energy, and whether the source is domestic or imported. An example is given in Table 15-8. The table is not intended to be complete. It indicates one approach to a national severance tax which recognizes that national resources are national wealth. Imported resources are taxed as are resources from public and Indian land. Resources from privately owned lands are one-half the public land tax to allow for royalties and state severance taxes.

State severance taxes on energy are extensive. In 11 states, taxes on oil, gas, and coal provided $2.6 billion for those states in 1979.[23] For example: the average crude oil tax in Texas was 46¢ per barrel; the natural gas tax in Louisiana was 7¢ per kcf. On coal, the Ohio tax is 4¢ per ton, but the North Dakota tax is 65¢ per ton.

This severance tax on finite resources should be viewed in the context of the preceding proposals. For example: there have been many problems with

[23]U.S. DoE, EIA, Energy Policy Study, Vol. 14, pp. 39, 40.

Table 15-8. An example of a national severance tax on finite energy resources

| Energy resource | U.S. origin | | Imported |
	Private land	Federal or Indian land	
Oil, tax per barrel	$7.50	$15	$15
equivalent tax per M̄Btu	$1.30	$2.60	$2.60
% 1982 price ($30/bl)	25%	50%	50%
Natural gas, tax per kcf	49¢	98¢	98¢
equivalent tax per M̄Btu	50¢	$1	$1
% 1982 price ($2.30/kcf)	21%	43%	25% ($4 price?)
Coal, tax per ton	$6	$12	n.a.
equivalent tax per M̄Btu	27¢	54¢	
% 1982 price ($38/ton)	16%	32%	

Note: The coal price is delivered to electric utilities.

accurate reporting of production on federal land, and this with a royalty rate which is usually only 12.5 percent.[24] Point 1 creates a federal participation in major oil company ownership, and point 5 institutes a value-added tax that would apply to profit gained by improper severance tax reporting. National shareholding and value-added taxation reduce the motivation for false severance tax reporting.

The severance tax would apply to foreign governments producing American energy resources: Sohio, to be specific, would be included in this and the other proposals. In the early 1980s the Table 15-8 severance tax on national finite energy resources would collect $85 billion. Much of this would displace existing monopoly profit.

While Table 15-8 shows that the natural gas and crude oil severance taxes are comparable on a percentage basis, natural gas price decontrol will alter this. As natural gas prices rise to comparable oil prices on a Btu basis, a 50¢/M̄Btu severance tax on natural gas should increase accordingly.

A tonnage tax on coal would appear to penalize the low-Btu western coal. However, western coal seams are much thicker (100 feet plus, recall, compared to 3 feet in the East) and have lower production costs. The tonnage severance tax therefore has less regional impact than might be supposed.

In theory, severance taxes could be framed as Btu taxes. In practice, however, physical units are easier to count—and harder to hide.

(9) Encourage direct election of members of boards of directors of utility corporations. The present problems facing electric utilities have a considerable impact on the nation's economic welfare. How can conservation be

[24]U.S. GAO, *Oil and Gas Royalty Collections—Longstanding Problems Costing Millions,* October 29, 1981.

promoted when many companies have excess capacity? Even if no additional nuclear plants are built, what shall be done with the spent fuel and the contaminated reactors from existing plants? If the beneficiaries of air pollution control are, as is the case with acid rain, in some distant state, who pays? How can the present tax incentives for new capacity be reconciled with the economic imperative of reduced requirements? How should rate schedules be framed?

These economic issues are inherently political, and demand political mechanisms for their resolution. If one-third of the membership of boards of directors are directly elected, these issues become part of an open process.

Increasingly, the private utilities have become dependent upon public capital. Table 15-9 portrays existing data in a new format for the 1975–80 period. Public sources, under these definitions, provided 34 percent of the new capital for privately owned utilities.

New debt and equity are conventionally defined. Tax incentives are estimated as the difference between (a) the corporate income tax rate applied to net income before federal income tax expense, and (b) reported current actual tax payment. AFUDC was excluded from taxable income. This $14.3 billion estimate of public tax incentives is necessarily understated because there are no consistent public statistics indicating where reported tax expense is reduced by investment tax credits. As explained in Chapter 12, AFUDC is not actual company investment, but a rate base allowance granted by regulatory commissions. It totaled $15.5 billion for this period.

This is the basis for concluding that public sources funded 34 percent of the new capital for privately owned electric utilities. The implication is that the interaction of public funding with the existence of major policy questions provides a basis for public participation in these decisions.

Table 15-9. New capital for private utilities, 1975–80

	$ billion
Private sources	
New debt	$35.6
Common stock	13.1
Preferred stock	8.9
Total	$57.6
Public sources	
Taxpayers: Tax incentives	14.3
Regulation: Allowance for construction	15.5
Total	$29.8
Total new capital from private and public sources	$87.4
Total new capital from public sources, percent	34%

Source: Edison Electric Institute, *Statistical Yearbook,* 1980, pp. 70, 72. Other paid-in capital is not included here, nor are any investment tax credits reported as income rather than as an expense adjustment.

A stronger policy would be public ownership. However, the publicly owned electric utilities have not proven themselves superior in any important way. For example, their growth in nuclear power is identical to that of the private utilities.[25] The country's largest source of air pollution in the middle 1970s was apparently the Tennessee Valley Authority. The Authority also owned the individual plants that were the largest single sources of sulfur oxides, nitrogen oxides, and particulates. Not surprisingly, TVA ranked fifteenth out of 15 companies studied for air pollution emission control.[26]

Independent studies over the last 35 years have generally found that the public utilities operate with lower costs per kilowatt hour.[27] Many of these studies do not properly distinguish the subsidies accruing to public power, or the access to low-cost hydropower. A full 41 percent of public electricity is generated from hydropower, while the private utilities produce only 4 percent of their power from this source. Donn Pescatrice and John Trapani, who studied public and private utilities which had only conventional steam plants, found that the public utilities had one-fourth to one-third lower costs per kWh. They did not, however, analyze the tax subsidy contribution to this lower cost.

But it seems clear that there is no empirical basis to support the inference that publicly owned utilities are less efficient. Appendix 15-B summarizes selected comparative data on public and private utilities.

One aspect of public ownership has not been widely discussed. Certain public utilities have been able to involve the public to a considerable extent in their decisions. Seattle City Light was part of a major public debate on nuclear power and conservation in the 1970s. Apparently, formal discussion groups were common, and these issues were important in a municipal election. As a result, Seattle City Light ended its nuclear planning, and initiated serious conservation efforts. As of 1980, average consumption per residential customer remained below the 1972 peak.[28] The Electric Department in Burlington, Vermont, figured (via a bond issue) in a municipal election which

[25]In 1980, public utilities had 13% of their steam generation in nuclear plants. The private utilities had 12%. See Appendix 15-B.

[26]From Ronald H. White, *The Price of Power Update* (New York: Council on Economic Priorities, 1977).

[27]See: American Public Power Association, "Public Power Costs Less," *Public Power*, 39:3 (May-June 1981), 14–16; Phillip Fanara, Jr., James E. Suelflow, and Roman A. Draba, "Energy and Competition: The Saga of Electric Power," *Antitrust Bulletin*, 25:1 (Spring 1980), 125–142; Robert A. Meyer, "Publicly Owned versus Privately Owned Utilities: A Policy Choice," *Review of Economics and Statistics*, 57:4 (November 1975), 391–399; Robert A. Meyer and Hayne E. Leland, "The Effectiveness of Price Regulation," *Review of Economics and Statistics*, 62:4 (November 1980), 555–566; Donn R. Pescatrice and John M. Trapani III, "The Performance and Objectives of Public and Private Utilities Operating in the United States," *Journal of Public Economics*, 13:2 (April 1980), 259–276.

[28]Seattle City Light, *1980 Annual Report;* also see the discussion by Peter Henault from Seattle City Light in Public Interest Economics West, *The Economics of Alternative Energy Technologies in California* (San Francisco, November 1979), pp. 119–125.

[341]

authorized renewable resource development in wood, waste, and hydro-power, and also ended the prospect of nuclear construction.[29]

Anecdotes, however, are unsatisfactory. It is equally possible to point to public utilities which are committed to nuclear power and have little interest in conservation. The interesting point here is that Burlington and Seattle made their basic decisions with public participation, and avoided many of the problems now common to the industry. The implication is that participation matters; and that public election of some board members of utilities by service area may bring these decisions into a better public focus.

(10) Expand legal and economic research in the institutional problem areas. The major areas where economic reorganization appears efficient are embarrassingly deficient in serious analysis. Building design, conservation, practical solar technologies, rail and other public transportation, and environmental protection are simply not part of the main effort in energy research.

In general, corporate research logically focuses on technological innovation which the corporation can efficiently and profitably develop. Hence fuel economy in automobiles is well studied, and improving. Work on advanced methods of exploration for and recovery of resources is well funded, and is proving successful.

Federal energy research has an embarrassing propensity to prove itself unproductive and inefficient. Notwithstanding the major problems discussed in Chapter 13, federal research on air pollution and nuclear waste problems is being reduced. On the other hand, gasohol, synthetic coal gasoline, and solar satellites are all areas that have been popular with federal research administrators. Corporate research modestly declines to fund this kind of work alone, and Chapter 14 indicates the eminent good sense of corporate research administrators.

Generally, I am willing to advocate public enterprise in the public interest, but the current federal research policy gives one pause. The serious problem, of course, is that economic progress in transportation, space heating, and renewable technologies cannot proceed without a technological basis and institutional framework that makes efficient decisions into feasible decisions. This, stated abstractly, is the kind of research needed.

(11) Establish a social tariff on imported commerce which varies according to wage and pollution control standards in the United States and in the country of origin.

Table 13-6 indicates that environmental protection for air, water, and land may, on the average, now constitute about two-thirds of the cost of a new coal generating station. Environmental protection costs for coal burning in steel

[29]Burlington Electric Department, *Annual Reports*, 1978–81.

and other industries must be comparable. I suggest that import tariffs should penalize goods manufactured in areas of the world with lower air pollution standards.

There are three reasons. First, foreign steel made without any or with less particulate or sulfur oxide control will be less costly, and will undersell American steel manufactured in plants that meet U.S. air pollution and environmental standards. A social tariff would reduce this advantage. Presently, the United States may be simply shifting pollution to other areas, reducing it within the country but leaving the world amount essentially unchanged.

Second, a U.S. tariff system based on pollution standards will motivate foreign governments and their citizens to look more favorably at pollution control in their own countries.

Third, a U.S. tariff on imported goods which is based upon environmental protection will be a strong incentive for the development of international air pollution standards. And this, of course, is the only logical step. As climate change and acid rain emerge as global air pollution problems, global standards will be developed. A social tariff in the United States, or the serious consideration of it, will hasten this international process of developing world pollution control.

A similar set of arguments applies to wage differentials. It is illogical to develop a minimum wage in the United States, and then encourage importation of labor-intensive consumer goods that may be manufactured at wages one-fifth of the American minimum. Here also, a U.S. social tariff based upon differentials will stimulate a global approach to the problem.

In the early 1980s, the United States is importing about $125 billion worth of manufactured commodities annually,[30] largely automobiles, metals, and machines. To understand the magnitude of the social tariff, suppose that, under U.S. air pollution and wage standards, these goods would have cost 40 percent more to manufacture. Assume the social tariff is 50 percent of the environmental protection and wage differential. The tariff revenue would be $25 billion if the import total stayed at $125 billion.

It should not be assumed that this proposal is directed primarily at Japan. While detailed comparisons are presently unavailable, my present conclusion is that air pollution control standards in Japan are comparable to U.S. standards. However, it should be noted that a Japanese automobile imported into the United States may have been assembled in Japan from iron which was sintered in the Philippines without significant air pollution control. (Sintering is a process of ore treatment which removes major impurities before smelting. It may be the stage of the steel-making process that produces the greatest emissions of sulfur oxide and other air pollutants.) That sintered ore may have

[30]U.S. Department of Commerce, Bureau of Economic Analysis, *Survey of Current Business*, January 1982, p. S-20.

been made from iron ore from Brazil and coal from South Africa.[31] The Japanese automobile may be achieving part of its advantage because of wage and pollution control differentials in the Philippines, Brazil, and South Africa.

Japan has entered the ranks of world leaders in GNP per capita (Table 14-1). Three Arab countries have as well. The average global GNP per capita was $2,510. There is no evidence at this date to support an argument that world pollution and wage standards would reduce world living standards, or that a U.S. social tariff would cause such a result in Japan or globally. Overall, global standards mean less world air pollution, less uneven income distribution, and a reduction in the rate of depletion of exhaustible world energy resources.

(12) Schedule. The "A" plan for major new investment and the 11 points to reorganize ownership and finance are linked together in Table 15-10. The first seven items summarize the estimated magnitude that could be associated with each change. The sum is $166 billion—an illustration, recall, of the scale of financial reorganization I think is necessary.

Table 15-4 showed the total investment associated with the "A" plan as $1.38 trillion. Suppose 25 percent of the items in Table 15-4 are in the private sector, particularly home construction cost related to reduced heating and cooling requirements. Seventy-five percent ($1.04 trillion) is a federal responsibility. Suppose this is distributed and financed over the period 1985–2000 in such a manner that the cost is 15 percent of the total each year. For example: 20-year treasury mortgage bonds with a 13.5 percent interest rate might yield 14.7 percent annually. The "A" plan annual cost, then, is 15 percent of $1.04 trillion, or $156 billion per year.

The illustrative revenue in Table 15-10 is a comparable $166 billion.

The twelve proposals for economic and financial reorganization just discussed should be viewed in the context of their objectives. The "A" plan works to achieve macroeconomic goals for employment, income, and price

[31]Japanese critics of Japanese steel make these points with severity, but with a frustrating absence of aggregate data; for example, Endo Nao, "The Report on a Fact-Finding Tour to Mindanao," and Yoshiwara Toshiyuki, "The Kawasaki Steel Corporation: A Case Study of Japanese Pollution Export," both in *KOGAI*, 5 (Spring-Autumn 1977), 4–24. Jumpei Ando compares U.S. and Japanese air pollution standards in "SO_2 Abatement for Coal-fired Boilers in Japan," March 1982, and "SOx and NOx Removal for Coal-fired Boilers in Japan," May 1982, both prepared for the U.S. Environmental Protection Agency. He finds Japanese standards to be more stringent. In other publications, Charles Geisler discusses the general problem in "Exporting Pollution: The Case of Japan," *Western Sociological Review*, 8:1 (Spring 1977), 1–15. A U.S. Department of Commerce study analyzed the impact of pollution control costs on U.S. copper production, concluding that environmental protection adds one-third to the cost of U.S. production. An associated displacement of one-third of U.S. copper production by imported copper is directly attributed to these environmental protection costs. See U.S. Department of Commerce, "U.S. Pollution Control Costs and International Trade Effects: A 1979 Status Report," September 1979.

Table 15-10. Potential magnitude of annual national revenue from reorganization

	$ billion
Government participation in share ownership	$ 20
End exemptions for publicly owned utilities	$ 1
Value-added tax	
At 5% on value added by corporations	$100
Less elimination of corporate profit tax	−$ 65
Net increase (about 10–15% from energy companies)	$ 35
Eliminate special tax provisions for depletion allowance, windfall profit tax, solar tax credit, and utility normalization: included in value-added tax	
National Severance Tax	
Imported oil	$ 36
Domestic oil	$ 32
Natural gas	$ 13
Coal	$ 4
Total	$ 85
Social tariff: Air pollution & wage differential	$ 25
Total illustrated new revenue	$166
Comparison: Annual federal cost of "A" plan	$156

stability by developing new policies for public and rail transportation, reduction in energy use for space heating and air conditioning, and efficient, geographically appropriate growth in renewable energy resources. The macroeconomic policies have a microeconomic grounding in the analysis in Chapter 14. Table 14-12 summarizes comparative costs. The guideline developed in that chapter is that the incremental cost of a new technology should be less, in the near future, on an unsubsidized basis, than the comparable conventional technology. Chapters 14 and 15 together emphasize the multiple relationships between microeconomic feasibility, resource use, macroeconomic objectives, and financial reorganization.

This last chapter began by observing that the policy conclusions here differ markedly from those held by most economists. The general view is that energy consumption will or should increase to 100 quadrillion Btu by the end of the century, that synthetic fuels can replace conventional energy, that transportation can continue to focus on automobile and air travel, and that market economics has and should define the proper development of energy conservation and renewable energy.

The process of preparing this book and participating in applied economic research on these subjects has led me to a different conclusion. It is inevitable that transportation will be reorganized and residential housing made more efficient, and that finite resources will continue to decline. The Growth Era in energy production and consumption spanned more than a century, but it is part of history. Energy policies in the 1980s must address technological realities, the proper role of private ownership, and the global nature of the American economy.

Appendix 15-A. Transportation: Assumed Ratios, "A" Plan to Current Values

Mode	Resulting fuel consumption ratios	Ratio of total passenger-miles traveled	Ratio of increased occupancy	Ratio of improved vehicle-miles per gallon
Automobiles	0.66	$\frac{8}{7}$	$1\frac{1}{2}$	$1\frac{1}{2}$
Bus	23	35	$1\frac{1}{3}$	$1\frac{1}{2}$
Rail	15.3	35	2.0	$1\frac{1}{2}$
Air	0.73	$1\frac{1}{8}$	$1\frac{1}{3}$	$1\frac{1}{2}$

$$\frac{F_t}{F_o} = \frac{PM_t/vmpg_t}{PM_o/vmpg_o} \div \frac{\text{occupancy rate}_t}{\text{occupancy rate}_o}$$

Example, Tables 15-1 and 15-2, automobiles:

$$\frac{F_t}{F_o} = \frac{8.0\ Q}{12.1\ Q} = \frac{1{,}083\ \text{billion}\ PM/1\frac{1}{2}}{1{,}263\ \text{billion}\ PM/1} \div \frac{1\frac{1}{2}}{1}$$

Appendix 15-B. Selected Economic Data on Public and Private Utilities

	Public utilities	Private utilities	Total utility industry
Average prices, ¢/kWh			
residential	3.7	5.4	4.9
commercial	3.6	5.5	5.1
industrial	2.4	3.7	3.4
all customers	3.2	4.7	4.4
Sales, billion kWh	366	1,732	2,098
Generation, billion kWh	503	1,783	2,286
Percentage of generation from hydropower	41%	4%	12%
Percentage of steam generation from nuclear power	13%	12%	13%
Growth rates, annual			
1960–73	6.7%	7.3%	7.2%
1973–80	3.1%	3.0%	3.0%
Average residential consumption, kWh/year	10,239	8,458	8,862

Data are from the Edison Electric Institute, *Statistical Yearbook*, 1980.

Glossary

Active solar heating: a collector system used to heat and circulate air or water through a building to provide space and/or water heating.

Air loss: the amount of heat energy escaping from a building in air flow through doors or air leaks in or around windows, ceilings, or walls. It is measured in Btu/hour/°F temperature difference between inside and outside.

Allowance for funds used during construction (AFUDC): an allowance which is added to actual construction cost in determining the value of utility plant when it is allowed in the rate base. The allowance is in concept an interest charge, and repays the utility for use of debt and equity while new plant is being built.

Anthracite: hard black coal with a carbon content above 92 percent. Clean burning.

Anthropogenic: of human origin; distinguishes man-made air pollution and radiation from natural sources of these substances.

Average cost: the cost of producing one unit, arrived at by dividing total cost by total production. Marginal cost will be above or below average cost according to whether average cost is rising or falling.

Back-end: a colloquial expression referring to actual or proposed stages of the nuclear fuel cycle after spent fuel is removed from an operating reactor.

Back-up: in a solar space or water heating system, the conventional system that provides heat when solar energy is not sufficient.

Benchmark crude: a specific crude oil which is available in sufficient quantity to be used as a reference in pricing other crude oil. In the mid-1980s, Saudi Arabian light crude oil is such a "benchmark" in international pricing.

Biomass: an agricultural or animal resource that can be utilized to produce solid, gas, or liquid fuels. Examples are woodfuel, methane from manure, and ethanol from corn.

Bituminous: the most important class of coal now being widely used. It is dark brown or black in color, high in carbon and energy content, low in moisture, and usable in continuously operating boilers.

Glossary

Board of directors: the group of persons elected by shareholders to make general company policy and review its implementation.

Breeder reactor: a reactor that creates more fissionable material than it consumes. Reprocessing of spent fuel, as from all reactors, is necessary. Breeder reactors may increase the available energy from existing uranium supplies 60 times.

Capacity factor: the average utilization of the capacity rating of a power plant. Most commonly it is expressed as a percentage or a decimal fraction (e.g., 70%, or .70), and often describes operation over a year's time.

Carbon monoxide: the carbon-oxygen molecule CO, an air pollutant. Primary sources are automobiles and heavy trucks.

Cash flow: the actual amount of money income and outgo. It can be seen as revenue and other income less operating expenses, interest charges, and actual current tax payments. In other words, it is equivalent to net income plus depreciation, depletion, and deferred taxes. It gives a better picture of current available funds than does net income.

Chief executive officer: the chairman of a company's board of directors, generally responsible for policy.

Chief operating officer: the highest ranking executive person responsible for ongoing management of a company.

Coal worker's pneumoconiosis: a disease common to underground coal miners before the implementation of the Coal Mine Health and Safety Act of 1969. It affects lung capacity, and is commonly referred to as black lung disease.

Coefficient of performance: the ratio of heat energy utilized to energy supplied. It is similar to an efficiency ratio, but is applicable to heat pumps as well as systems which convert one energy form into another.

Combination utility: a utility that sells both electricity and gas.

Commercial rivalry: competition between firms which does not include price competition, but does include advertising and product differentiation.

Competitive industry: an industry in which no single producer or customer can influence the prices at which the product is bought or sold.

Concentration ratio: for a defined geographic area and product, the percentage of reserves, production, or sales owned or controlled by a given number of corporations.

Condensate: liquid hydrocarbons condensed from petroleum or gas vapors.

Conduction loss: the amount of heat energy which passes through a square foot of surface with a one °F temperature differential between inside and outside. It is measured as the U-value.

Construction work in progress (CWIP): term denoting the sum of money spent on plant and equipment under construction but not yet in use.

[348]

Control, energy resources: the exercise of management decisions over the production of a resource. Applicable both to royalty oil or gas for the producing company, and to net ownership interest.

Conversion: the third step of the nuclear fuel cycle: the solid uranium oxide (U_3O_8) in the yellowcake from milling is converted to a gaseous UF_6 at high temperature.

Converter reactor: a commercial nuclear power reactor that consumes more fissionable material than it creates. In the United States, only light water converter reactors are presently in wide commercial use.

Debtholder: an individual or corporation lending money to a company, most commonly, the purchaser and owner of bonds which guarantee that the company will pay the debtholder a fixed sum at a specific future date.

Decommissioning: the shutting down of a nuclear power plant with assurance of public safety from radiation exposure to contaminated components of a reactor. Three general approaches are dismantlement, storage, and entombment.

Deferred taxes: tax liabilities on current income for which payment is deferred to later years.

Degree days: a measure of coldness which is used in estimating the amount of space heating required for a specific building and locality. For one day, the degree days equal the difference between 65°F and the actual average (lower) temperature for the day. The annual degree days equal the sum of the degree days for each of the 365 days in a typical year.

Demand curve: graphic representation of the demand function, which is a relationship showing the amount of a good that buyers will wish to purchase at different prices. It is affected by population, prices of competing goods, income, and advertising.

Demand function: in terms of energy, the quantitative relationship between the amount of energy consumed and the economic factors which affect that consumption: personal and national income, population, conservation, and the availability of alternatives.

Depletion: the exhaustion of a natural resource through production.

Depreciation: the accounting charge which reflects the using up of a fixed asset.

Direct gain: the increase in usable solar radiation resulting from the location of windows on south-facing walls.

Diseconomies of scale: average cost of production increases as the level of production increases.

Dismantlement: making the site of a nuclear power plant safe for unrestricted use by shipping all hazardous radioactive material to a depository.

Dose-response relationship: a quantitative expression of the relationship between exposure to given concentration levels of a toxic substance and the proportion of exposed population which experiences a specific health effect.

Glossary

Drawback: in the early years of the petroleum industry, a charge levied on rail shipments of independent oil producers. The railroad then paid the money to the Standard Trust, in order to gain the trust's oil shipment business.

Economies of scale: average cost of production declines as the level of production increases.

Efficiency: the proportion of energy supplied which is utilized in accomplishing work. If a power plant has an efficiency ratio of 32 percent, then 32 percent of the fossil fuel energy is transformed into electricity, and 68 percent is lost. If a gas furnace has an 80 percent efficiency, then .8 of every gas Btu goes into the residence as heat. Every energy process can also be viewed as a system with several stages. A *stage efficiency* gives information about energy use at a single stage. A *system efficiency* represents the interaction of the several stages.

Elasticity: approximately the percentage change in demand associated with a 1 percent change in an economic factor. (E.g., if the income elasticity for electricity demand is +0.6, a 1 percent increase in income will cause electricity demand to increase by .6 of 1 percent).

Energy conservation: a general term related to changes in energy consumption patterns. Several of the more common definitions are: (1) a reduction in use associated with a price increase or a decline in income; (2) less-than-expected use; (3) a substitution of one energy form for another; for example, the replacement of oil heating by direct solar heating or natural gas heating; and (4) a significant shift in energy requirements caused by such actions as increased insulation, improved fuel efficiency in automobiles, or a shift to public transportation.

Enhanced recovery: techniques that augment natural oil or gas flow, thereby increasing the recovery rate.

Enrichment: the fourth step in the nuclear fuel cycle: the gaseous UF_6 is passed through cylindrical membranes to increase ("enrich") the proportion of fissionable U-235 in the product. In natural uranium, U-235 constitutes about .711 of 1 percent of the total. Enriched uranium contains about 3 percent U-235.

Entombment: burying a nuclear power reactor in concrete until harmful radioactivity has decayed.

Exchange agreement: provides for the trading of one company's crude oil or products for another company's crude oil or products. A form of barter.

External social cost: an unintended by-product of economic activity with negative impact on human welfare, such as health damage to the public and employees, property loss, and environmental degradation.

Fabrication: the fifth stage of the nuclear fuel cycle: enriched uranium is transformed into a uranium oxide metal. This is fabricated into fuel pellets, which are in turn placed in fuel rods.

Finite resources: energy resources whose periods of formation are measured in geologic time and theoretically may be completely exhausted by human use: oil, natural gas, coal, and uranium.

Flow-through: the immediate passing on of tax benefits received by a utility to its customers in the form of lower rates, required by regulation. A utility must base its revenue requirements upon actual current expected corporate income tax payments.

Fossil fuel: energy resources which are the fossil remains of early life forms: coal, natural gas, and oil.

Front-end: another colloquial expression referring to the first four stages of the nuclear fuel cycle, before fuel is placed in a reactor core.

Fuel adjustment clause: a provision in rate schedules which governs the automatic increase or decrease in customer rates according to changes in costs for coal, oil, natural gas, or nuclear fuel to the utility.

Funds earned from operations: current revenues from sales of a company's products. Excluded from this concept are deductions for depreciation or depletion as well as deferred taxes which are not actually paid in a current period.

Gasohol: a blend of 10 percent ethanol and 90 percent conventional gasoline. It may be used in present automobiles without modification of the engine.

Gathering system: a network of pipelines which collects oil or gas from individual wells and transports it to a main pipeline or to another transportation system or processing facility.

Growth monopoly: an industry with monopoly power such that price is considerably different from marginal cost. This power is used to pursue maximum sales, which are limited only by the requirement that there be no long-run loss, or by the appearance of negative marginal revenue.

Half-life: the length of time necessary for the decay of one-half of the atoms of a given radionuclide.

Heat loss: the Btu which escape from a building in one hour with a one °F differential between inside and outside temperatures. It is the sum of conduction and air losses.

Horizontal integration: the coordination of multi-energy activities within a single corporate management. Generally, in the context of oil companies, it means the production of natural gas, coal, nuclear fuel, copper, and minerals, or solar and alternative energy forms.

Hydrocarbon: a chemical term indicating the organic basis of oil and natural gas in hydrogen and carbon.

Incremental cost comparison: a comparison of the present or future costs of services or commodities as provided by different technologies or energy sources.

Inside director: a person working for a company in a management capacity who sits on the board of directors of the company.

Interstate gas: generally, gas shipped by pipelines which are used to transport gas from a producing state to a consuming state.

Glossary

Intracompany purchases: purchases by one operating division or subsidiary of a parent company from another.

Intrastate gas: generally, gas sold and consumed within the state in which it is produced, and not transported by an interstate pipeline.

Investor-owned utility: an electric or gas utility organized as a private corporation with institutional and individual shareholders.

Levelized cost: a method of allocating costs in which capital cost is spread over the expected operating life of a plant and charged to each kilowatt hour. The result gives the amount which, if collected as a capital charge for each kilowatt hour, would exactly pay a loan and interest payments for the assumed capital investment, interest rate, years of operation, and generation. It frequently includes a small allowance for expenses linked to investment such as property tax type payments or a decommissioning allowance.

Light water reactor: a reactor that uses ordinary water in its steam generation and cooling systems, and requires enriched uranium fuel.

Lignite: the lowest-quality commercial coal, high in moisture and low in carbon and heat content.

Liquefied natural gas (LNG): natural gas which has been liquefied under high pressure at very cold temperatures to reduce its volume to 1/600 of normal gas.

Liquefied petroleum gas (LPG): a product of the natural gas cleaning process. LPG, widely known as "bottled gas," is primarily propane and butane which is liquefied under moderate pressure at normal temperatures.

Load factor: a measure of variability in customer's use of electricity. For a given time period, it is the ratio of the average kilowatts used over that time period divided by the maximum kilowatts used at any one point in the time period.

Macroeconomic: concerning dimensions of the national economy such as income levels, inflation, and employment.

Marginal cost: the additional cost of producing another unit.

Marginal revenue: the additional revenue to the firm from selling another unit. In a monopolistic industry where the firm or group of firms controls production and price levels, marginal revenue will always be less than price. In a competitive industry, marginal revenue for a firm is always identical to price.

Marketed production: the amount of gas available for consumption after producers have used portions of their production in repressuring, venting, and flaring. This is sometimes termed "wet" gas; "dry" gas results from the removal of natural gas liquids from wet gas.

Microeconomic: concerning economic aspects of the decisions of individual consumers, employees, and corporations. The focus is upon product prices, production cost, and demand.

Microgram: one millionth of a gram, represented by μg.

[352]

Milling: the second stage of the nuclear fuel cycle: uranium oxide U_3O_8 is separated from ore by chemical processing and physical crushing. The result is called "yellowcake."

Mine-mouth power plant: a coal-burning power plant located at the site of a surface or deep mine.

Mining: the first stage in the preparation of nuclear fuel. Typically, the mined ore is .1 of 1 percent of U_3O_8.

Mode: in transportation economics, the system or type of transportation: e.g., rail, air, truck, bus, automobile, barge, or ship.

Monopolistic competition: a situation in which an industry has characteristics of both monopoly and competition. There are few firms, and each sets its marginal revenue equal to marginal cost. However, price and average cost are equal, and entry by new firms is unrestricted.

Monopoly profit: profit in excess of a normal 13–15 percent rate of return on invested capital, brought about by one company (or group of companies) increasing its profit by withholding production and raising prices.

Natural gas liquids: liquid hydrocarbons that have been removed from natural gas. They include ethane, propane, butane, and small amounts of gasoline, kerosene, jet fuel, and fuel oil.

Natural monopoly: an industry characterized by significant economies of scale over broad levels of production and large geographic areas. Competition in such a context is believed to be inefficient, and the single company monopoly produces at minimum average cost.

Net income: the general definition of profit. It usually equals revenue less operating costs, interest expenses, depreciation and depletion charges, and allocated tax expenses.

Net ownership: the actual equity interest of an individual company in an oil or gas field, well, or property. The net ownership interests (in percents) sum to 100% for all the partners for any property.

Net social value: the total gain to society from production and consumption of a commodity. It includes all the positive value of consumption from all consumers, and is reduced by external social cost, subsidies, and the company's market cost. Maximum net social value is the social optimum.

Nitrogen oxide: a gaseous air pollutant, primarily nitrogen dioxide, NO_2. Not necessarily harmful by itself, it interacts with organic compounds to produce ozone, and is a secondary factor in acid rain. Major sources are automobiles, coal-burning electric utilities, and natural gas burned by industry.

Normalization: a method of allocating tax expense whereby a utility collects revenue as if its tax liability were essentially unaffected by tax benefits in the year in which the benefits are received. The company may invest this revenue in new plant and equipment, and return the tax benefit to customers by amortizing it on a straight-line basis over the life of the facility which creates the tax benefits.

Nuclear fuel cycle: the stages through which nuclear fuel passes in its preparation, its utilization in the reactor, and disposition of the spent fuel.

Order of magnitude: an expression indicating a 10-fold difference. For example: a million tons is "an order of magnitude" more than 100,000 tons.

Outages: reductions in generating capacity caused by unplanned events or emergencies.

Outside director: a person belonging to a company's board of directors who is not an employee of the company.

Pareto optimality: the ideal of market system efficiency in this very limited sense: no individual or group of individuals can be made better off without someone else being made worse off. (Named after Vilfredo Pareto, an Italian economist and sociologist.)

Particulates: very small solids which are an air pollutant; the primary source of them is industrial processes.

Passive solar heating: the collecton, storage, and release of solar energy for space and/or water heating by means of building design and construction.

Pathway: the sequence of events through time and space in which a toxic substance is created, passes through the environment, has physical contact with humans, and causes a health effect.

Person-rem: a standardized unit measuring the biological impact of radiation damage.

Physical-quality-of-life index: a measure of international living standards based upon life expectancy, infant mortality, and literacy.

Potential resources: estimates not specific to individual areas, and based upon general geological knowledge about the possible distribution of the resource.

Power pool: a group of interconnected utility companies that plan and/or operate their generation stations and supply power in the most economic and reliable manner for members of the pool. They may have jointly owned plants and equipment, and exchange, buy, and sell electricity among themselves.

Processing receipts: a form of exchange that comes into being when two companies agree to refine each other's oil in their own refineries. The oil which A owns but B refines and delivers to A is a processing receipt for A, and is a processing delivery for B.

Product differentiation: the process of individualizing essentially identical products of different firms irrespective of price differences, to create brand loyalty. Buyer preference for one product over another may result from advertising, durability, service, tradition, or customer idealism. (In a class discussion it was suggested that the degree or type of external social cost associated with a particular company may be a legitimate type of product differentiation for some customers—e.g., electricity from nuclear power compared to electricity from a windmill. While this is not part of the traditional definition, it does illustrate the same concept.)

Profitability: the rate of return on invested capital. This most commonly means the ratio of net income to stockholders' equity.

Proven reserves: amounts of resources which are known to be recoverable under existing economic conditions and technology. They are located in specific known reservoirs that have been tested and have demonstrated the ability to produce.

Public utility: an electric or gas system owned by a municipality, a state, or a federal agency or authority. Cooperatives are usually not defined as public utilities.

Public Utility Regulatory Policies Act (PURPA): an act passed by Congress in 1978, intended to promote conservation by electric utilities and their customers. Provisions of the act apply to rate schedules, fuel adjustment clauses, advertising, consumer intervention, and purchases from small hydropower and cogeneration facilities.

Radionuclide: a specific isotope of an element which can decay spontaneously into another isotope; it is radioactive.

Rank: a means of classifying different physical types of coal according to carbon content, heat value, moisture, and other properties.

Rate base: the value of a utility's investment, which a regulatory agency determines should be the basis for the company's allowed rate of return. It is usually actual cost plus AFUDC, less accumulated depreciation. It may be adjusted for inflation. Property allowed in the rate base is usually supposed to be in actual use.

Rate of return: the profit rate that a regulatory agency permits a utility to earn on its investment in property in the rate base.

Rate schedule: the charges made to customers for given levels of consumption of electricity as measured in kilowatt hours. A large commercial or industrial customer will also be charged according to the levels of electric capacity (measured in kilowatts) which that customer requires.

Real price: a price divided by an inflation index to make prices in different years comparable. For example, the average price of electricity for residences was 26.6 mills/kWh in 1955 and 34.5 mills/kWh in 1976. The consumer price index with 1976 = 1.00 was .47 in 1955, so the real price of electricity to residences was 56.4 mills/kWh in 1955, expressed in 1976 dollars.

Recovery rate: the proportion of the original resource in place that will be extracted over the producing life of an area. Applicable to oil, natural gas, coal, and uranium.

Renewable resources: energy resources that can be used without depletion if carefully managed. The major forms are solar energy, windmills, hydro facilities, and wood and biomass fuel.

Reprocessing: a stage in the nuclear fuel cycle in which spent fuel is processed to remove usable uranium and plutonium. Reprocessing in conjunction with breeder reactors might potentially increase the amount of energy released by a given amount of uranium by 60 times. Reprocessing has been halted in the United States because of contamination problems and the potential for theft and terrorist use of nuclear material and explosives.

Retained earnings: the portion of net income retained by a company and not paid out in dividends.

Royalty gas: natural gas produced and controlled by a company but owned by the owner of the land (or its mineral rights) under which the gas is found. Similar to royalty oil.

Royalty oil: oil produced and controlled by a company but owned by the owner of the land under which the oil is found.

R-value: for insulation, the measure of resistance to conduction heat loss. It is the opposite or inverse of conduction, so $R = 1/U$.

Shareholder: the legal owner of one or more shares of capital stock of a corporation. Also frequently termed "stockholder." The shareholders, as formal owners, have the legal authority to define company policies, elect directors, and approve or disapprove management's actions.

Social cost: the cost incurred by society in using a product. It includes market cost of production by the company, external social cost, and subsidies.

Social optimum: the level of production and consumption for an industry with maximum gain in social value less social cost. This will be the production level where marginal social cost and marginal social value—measured by the price—are equal.

Social value: the value to society of using a product. In economic terms it is frequently supposed that social value is measured by the area under a demand curve, which shows the value to all consumers of every unit of consumption. Social value is always greater than revenue, and the price of a product measures marginal social value.

Solar energy: the utilization of solar radiation for a purpose which is frequently provided by a conventional energy source. In this sense it becomes a measurable component of aggregate energy consumption.

Spent fuel: fuel removed from operating reactors. Also called waste fuel.

Spot market: transactions in oil, coal, or uranium sales which are not included in long-term contracts.

Storage and dismantlement (also called "mothballing"): decommissioning procedure under which a closed nuclear plant will be guarded for some period (perhaps 100 years) while much of the radioactivity decays, and then dismantled.

Straight line depreciation: a way of figuring depreciation that makes each year's amount always equal to original cost (less salvage value, if any) divided by the expected operating life of the plant or equipment.

Subbituminous: a rank of coal intermediate between lignite and bituminous.

Subsidy: a government expenditure or reduction in tax liability that lowers the costs of producing or purchasing particular commodities. Both cash payments and tax reductions are subsidies.

Substitutability: the characteristic of interchangeability in end-use between most energy forms.

Sulfur oxide: a gaseous air pollutant which is primarily sulfur dioxide, SO_2. It is the major factor in acid rain, and its most important source is coal burning for power generation.

Surface mining: the form of coal mining in which access to coal seams is obtained by removal of the overburden of topsoil, earth, rock and other material. Coal is removed by power shovels and other heavy equipment. The term "strip mining" is still frequently used.

Synthetic fuels: liquid or gas fuels which may be substituted for conventional petroleum products or natural gas, and are made from some other resource such as coal, oil shale, or biomass.

Synthetic gas: a gas manufactured from coal or oil that may be substituted for or mixed with natural gas.

Tax benefits: investment incentives in the federal income tax laws; investment in new plant and equipment reduces tax liability. The two major forms of incentives are the investment tax credit and accelerated depreciation.

Trade purchases: sales of oil or products between oil companies.

Underground mining: the form of coal mining that leaves the land surface generally undisturbed. Coal seams are reached by underground shafts or tunnels. The term "deep mining" is synonomous.

Unitized operation (unitization): a process or agreement by whicn multiple owernship interests in an oil field are managed as a single unit in order to achieve more efficient production of the total resource.

Utility holding company: a parent corporation which owns all or a majority of voting shares in a utility. Utility holding companies may be entirely focused on utilities (e.g., American Electric Power), or may be diversified into real estate, agriculture, etc. in addition to their utility interests.

U-value: the measure of conduction loss, in Btu per square foot for a one °F difference between inside and outside temperatures. Compare *R-value*.

Vertical integration: a petroleum company is vertically integrated if it is active in production, refining, and marketing. Exploration and transportation are also commonly included within the meaning of the concept. A simple rule of thumb for defining a vertically integrated company might be that each stage of its physical operations is within approximately 50 percent of the adjacent stage's size in terms of average number of barrels per day.

Glossary

Volatile organic compounds: air pollutants from the combustion of gasoline and oil and vapors of gasoline and solvents. They interact with nitric oxide to form ozone, O_3. Primary sources are automobiles and industrial chemical processes.

Wheeling: one electric utility system using its transmission facilities to transport power to a second utility. The wheeled power is owned by the second utility.

Windfall profit tax: an excise tax on barrels of crude oil produced in the United States. The tax varies according to type of well, owner, age of well, selling price, and other factors.

Index

This index was prepared by Lucrezia Herman.

[359]

Index

Index

Library of Congress Cataloging in Publication Data

Chapman, Duane.
  Energy resources and energy corporations.

  Includes index.
  1. Energy industries—United States. 2. Power resources—United States. 3. Energy
policy—United States. 4. Energy industries. 5. Power resources. 6. Energy policy. I.
Title.
HD9502.U52C49 1983   333.79   82-74022
ISBN 0-8014-1305-2